Instructor's Manual
to accompany

INTRODUCTORY ASTRONOMY & ASTROPHYSICS
fourth edition
ZEILIK • GREGORY

Paul A. Heckert
Western Carolina University

Thomas J. Balonek
Colgate University

Saunders College Publishing
Harcourt Brace College Publishers

Fort Worth   Philadelphia   San Diego   New York   Orlando   Austin
San Antonio   Toronto   Montreal   London   Sydney   Tokyo

Copyright © 1998 by Harcourt Brace & Company

All rights reserved. No part of this publication may be reproduced or transmitted in any form or by any means, electronic or mechanical, including photocopy, recording, or any information storage and retrieval system, without permission in writing from the publisher.

Requests for permission to make copies of any part of the work should be mailed to: Permissions Department, Harcourt Brace & Company, 6277 Sea Harbor Drive, Orlando, Florida 32887-6777.

Portions of this work were published in previous editions.

Printed in the United States of America

Balonek & Heckert: Instructor's Manual to accompany *Instroductory Astronomy & Astrophysics, 4e*. Zeilik & Gregory.

ISBN 0-03-024557-5

789 023 7654321

PREFACE

Our aim in writing this Instructor's Manual for Zeilik and Gregory's **Introductory Astronomy & Astrophysics**, fourth edition, was to provide a set of detailed solutions to the end of chapter problems which would be useful to instructors and students. For many students this may be a first course in astronomy; instructors may not have taught introductory astronomy at this level. A detailed solutions manual provides an important complement to the textbook, providing insights into the methods of astronomical research. Instructors may find it useful to make selected solutions available to the students.

We intended this manual to be of use also to advanced undergraduate students or graduate students who wish to review fundamental astronomy and astrophysics. It is our hope that instructors will share this manual with interested students.

To this end, we decided to include here the original questions from the text and all steps in the solutions to the problems so that the manual will be a learning tool rather than just an answer check. By providing step-by-step numerical and descriptive solutions, the reader will be able to follow the reasoning which we used in solving the problem. Also, any errors which we have made in the solutions will be more readily detected and corrected. We have included cross-references to the equations and sections of the text that are needed to solve each problem. In some multiple-step problems we have carried an extra significant digit during the calculations to avoid round off errors. Where we have found errors in the first printing of the fourth edition of the text, we have made comments in the problem solutions and summarized these corrections at the end of the Manual. These errors may be corrected in later printings of the text, but we include them here in case instructors or students use the first printing.

In any work of this length and complexity there are undoubtedly mistakes which will have escaped our notice. We would appreciate being informed of any errors which occur in the manual, as well as suggestions on alternate solutions. We thank those who sent in comments and corrections to the previous editions, including K. Kwitter, J. Napolitano, L. Likkel, and R. Gelderman. Please send corrections, suggestions, or inquiries to us at the addresses given below.

P.A.H. would like to acknowledge the support and understanding that he received from his family -- Sue, Jessica, Carina and Paul Stephen -- during the many hours that were spent working on this manual. T.J.B. would like to thank Jim Lloyd for his encouragement and support during the past thirteen years -- and wishes him best in his retirement. We dedicate this work to our families and our students, who have shared in our study of the Universe.

Paul A. Heckert
Dept. of Chemistry & Physics
Western Carolina University
Cullowhee, NC 28723
email: heckert@wcuvax1.wcu.edu

Thomas J. Balonek
Dept. of Physics and Astronomy
Colgate University, 13 Oak Drive
Hamilton, NY 13346
email: tbalonek@colgate.edu

Contents

Chapter 1	Celestial Mechanics and the Solar System	1
Chapter 2	The Solar System in Perspective	11
Chapter 3	The Dynamics of the Earth	29
Chapter 4	The Earth-Moon System	40
Chapter 5	The Terrestrial Planets: Mercury, Venus and Mars	52
Chapter 6	The Jovian Planets and Pluto	63
Chapter 7	Small Bodies and the Origin of the Solar System	74
Chapter 8	Electromagnetic Radiation and Matter	88
Chapter 9	Telescopes and Detectors	103
Chapter 10	The Sun: A Model Star	114
Chapter 11	Stars: Distances and Magnitudes	123
Chapter 12	Stars: Binary Systems	133
Chapter 13	Stars: The Hertzsprung-Russell Diagram	143
Chapter 14	Our Galaxy: A Preview	155
Chapter 15	The Interstellar Medium and Star Birth	165
Chapter 16	The Evolution of Stars	179
Chapter 17	Star Deaths	192
Chapter 18	Variable and Violent Stars	207
Chapter 19	Galactic Rotation: Stellar Motions	219
Chapter 20	The Evolution of Our Galaxy	232
Chapter 21	Galaxies Beyond the Milky Way	246
Chapter 22	Hubble's Law and the Distance Scale	258
Chapter 23	Large-Scale Structure in the Universe	268
Chapter 24	Active Galaxies and Quasars	277
Chapter 25	Cosmology: The Big Bang and Beyond	293
Chapter 26	The New Cosmology	303
---	Comments and Corrections to Problems and Text	309

Chapter 1: Celestial Mechanics and the Solar System

1-1. Assume that the orbital plane of a superior planet is inclined 10° to the ecliptic and that the planet crosses the ecliptic moving northward at opposition. Make a diagram similar to Figure 1-1B, showing the retrograde path of this superior planet.

The planet will be south of the ecliptic prior to opposition and north after opposition. The planet will begin retrograde motion prior to opposition (south of the ecliptic), will be mid-way in its retrograde motion when it crosses the ecliptic (at an angle ≈10°), and will be north of the ecliptic when it resumes its normal (west to east) motion. Because of the planet's orbital inclination and since it crosses the ecliptic near opposition, the apparent path will not be looped as shown in Figure 1-1B, but rather will be "zig-zag" in shape.

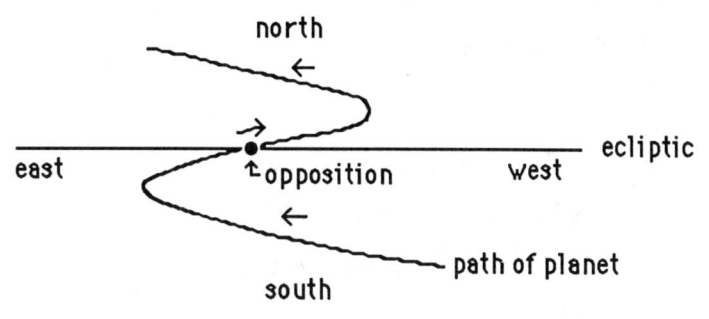

1-2. Imagine that you are observing the Earth from Jupiter. What would you observe the Earth's synodic period to be? What would it be from Venus? (*Hint:* See Appendix 3.)

The synodic period for a planet as seen from the Earth is the same as the synodic period of the Earth as seen from the planet. This can be understood, for example, by noting that the times of opposition of Jupiter as seen from the Earth are identical to the times of inferior conjunction of the Earth as seen from Jupiter. (See Appendix Table A3-1 for the synodic periods of the planets.) The Earth's synodic period as seen from Jupiter would be 398.9 Earth days; the Earth's synodic period as seen from Venus would be 583.9 days.

These values can also be calculated from the sidereal periods (Appendix Table A3-1) using the equation (for Earth being an inferior planet as viewed from Jupiter):

$$1/S_{Earth} = 1/P_{Earth} - 1/P_{Jupiter}$$

where S is the synodic period, and P is the sidereal period (see Section 1-1A).

$$1/S_{Earth} = 1/(365.26 \text{ days}) - 1/(4333 \text{ days})$$
$$= 2.51 \times 10^{-3} \text{ days}^{-1}$$
$$S_{Earth} = 398.9 \text{ days}$$

As seen from Venus the Earth is a superior planet, so

$$1/S_{Earth} = 1/P_{Venus} - 1/P_{Earth}$$
$$1/S_{Earth} = 1/(224.70 \text{ days}) - 1/(365.26 \text{ days})$$
$$= 1.71 \times 10^{-3} \text{ days}^{-1}$$
$$S_{Earth} = 583.9 \text{ days}$$

1-3. (a) Explicitly carry out the derivation of Equation 1-4, showing all the appropriate steps. Refer to Figure 1-7 for the definition of the variables.

Subtracting the two equations prior to Equation 1-4 in the text:
$$r'^2 = (x + ae)^2 + y^2 = x^2 + 2aex + a^2e^2 + y^2$$
$$r^2 = (x - ae)^2 + y^2 = x^2 - 2aex + a^2e^2 + y^2$$
$$r'^2 - r^2 = \qquad\qquad 4aex$$

Using Equation 1-1:
$$r = 2a - r'$$
we get by substitution:
$$r'^2 - (2a - r')^2 = 4aex$$
$$r'^2 - 4a^2 + 4ar' - r'^2 = 4aex$$
$$-4a^2 + 4ar' = 4aex$$
$$-a + r' = ex$$
$$r' = a + ex$$

Substituting into our first equation:
$$(a + ex)^2 = x^2 + 2aex + a^2e^2 + y^2$$
$$a^2 + 2aex + e^2x^2 = x^2 + 2aex + a^2e^2 + y^2$$
$$a^2 - a^2e^2 = x^2 - e^2x^2 + y^2$$
$$a^2(1 - e^2) = (1 - e^2)x^2 + y^2$$

From Equation 1-2:
$$a^2(1 - e^2) = b^2$$
and $(1 - e^2) = b^2/a^2$
so substituting,
$$b^2 = (b^2/a^2)x^2 + y^2$$
$$1 = x^2/a^2 + y^2/b^2 = (x/a)^2 + (y/b)^2 \quad \text{--- which is Equation 1-4}$$

(b) On graph paper, plot the following polar equations: Equation 1-3 for an ellipse, Equation 1-6 for a parabola, and Equation 1-7 for a hyperbola.

Equation 1-3: ellipse $\quad r = a(1 - e^2)/(1 + e\cos(\theta))$

Equation 1-6: parabola $\quad r = 2p/(1 + \cos(\theta))$

Equation 1-7: hyperbola $\quad r = a(e^2 - 1)/(1 + e\cos(\theta))$

The curves will vary, depending on the values of the eccentricity used. Let's take $e = 0.5$ for the ellipse, and $e = 2.0$ for the hyperbola.

	$\theta = 0°$	30°	60°	90°	120°	150°	180°
ellipse, $e = 0.5$	0.50 a	0.52 a	0.60 a	0.75 a	1.00 a	1.32 a	1.50 a
parabola, $e = 1$	1.00 p	1.07 p	1.33 p	2.00 p	4.00 p	14.9 p	---
hyperbola, $e = 2$	1.00 a	1.10 a	1.50 a	3.00 a	---	---	---

```
         / hyperbola e = 2.0, a = 1.0
        /
       / parabola p = 1.0
      /
                    ellipse e = 0.5, a = 1.0
```

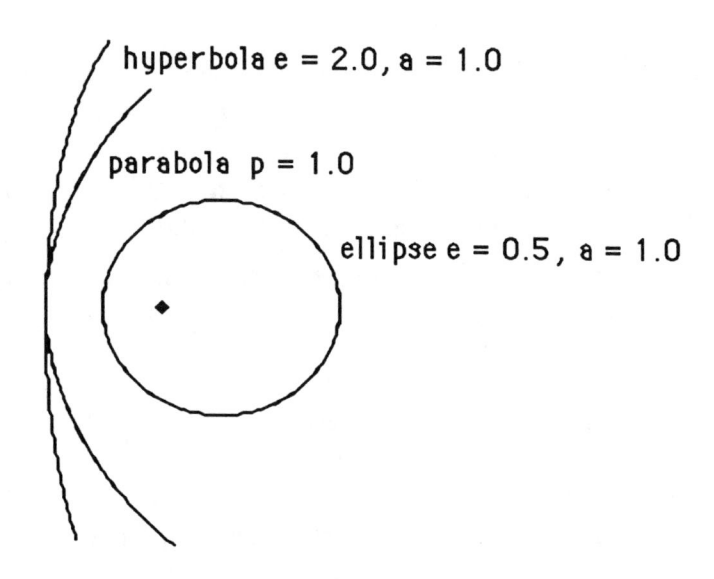

1-4. In terms of the gravitational acceleration g at the surface of the Earth, find the surface gravitational acceleration of

(a) the Moon (M_m = 0.0123 $M_\oplus$, R_m = 1738 km)

The gravitational acceleration is given (from Equation 1-15) by:

$$a = F_{grav} / m = G M / r^2$$

Compared with the acceleration at the surface of the Earth,

$$a_{planet} / a_{Earth} = [M_{planet} / (r_{planet})^2] / [M_{Earth} / (r_{Earth})^2]$$
$$= (M_{planet} / M_{Earth}) (r_{Earth} / r_{planet})^2$$

For the Moon, using the Earth's equatorial radius,

$$a_{Moon} / a_{Earth} = (M_{Moon} / M_{Earth}) (r_{Earth} / r_{Moon})^2$$
$$= (0.0123 \, M_\oplus / 1.0 \, M_\oplus) (6{,}378 \text{ km} / 1{,}738 \text{ km})^2$$
$$= (0.0123) (3.67)^2$$
$$= 0.17$$

(b) the Sun ($M_\odot$ = 2 x 10^{30} kg, $R_\odot$ = 7 x 10^8 m)

For the Sun,

$$a_{Sun} / a_{Earth} = (2.0 \times 10^{30} \text{ kg} / 5.98 \times 10^{24} \text{ kg})$$
$$\times (6.378 \times 10^6 \text{ m} / 7.0 \times 10^8 \text{ m})^2$$
$$= (3.34 \times 10^5) (9.11 \times 10^{-3})^2$$
$$= 2.8 \times 10^1 = 28$$

(c) Jupiter (M_J = 318 $M_\oplus$, R_J = 11.2 $R_\oplus$)

For Jupiter, at the equator,

$$a_{Jupiter} / a_{Earth} = (318 \, M_\oplus / 1.0 \, M_\oplus)(1 \, R_\oplus / 11.2 \, R_\oplus)^2$$
$$= 2.54$$

Note: Appendix Table A3-3 lists the surface gravity as 2.74 times that of the Earth. This value was calculated using the average radius of Jupiter of $R_J = 68{,}700$ km $= 10.77 \, R_\oplus$, which would give a surface gravitational acceleration

$$a_{Jupiter} / a_{Earth} = (318 \, M_\oplus / 1.0 \, M_\oplus)(1 \, R_\oplus / 10.77 \, R_\oplus)^2 = 2.74$$

1-5. What are the perihelion and aphelion speeds of Mercury? What are the perihelion and aphelion distances of this planet? Compute the product vr (speed times distance) at each of these two points and interpret your result physically.

Using Equations 1-17a and 1-17b for the speed at perihelion and aphelion

$$v_{perihelion} = (2\pi a / P)[(1+e)/(1-e)]^{1/2}$$
$$v_{aphelion} = (2\pi a / P)[(1-e)/(1+e)]^{1/2}$$

From Appendix Table A3-1, for Mercury:

$a = 0.387$ AU $= 57.9 \times 10^6$ km

$P = 87.96$ days $= 7.60 \times 10^6$ s

$e = 0.206$

so at perihelion

$$v_{perihelion} = [2\pi (57.9 \times 10^6 \text{ km}) / (7.60 \times 10^6 \text{ s})]$$
$$\times [(1 + 0.206)/(1 - 0.206)]^{1/2}$$
$$= (47.9 \text{ km/s}) \times (1.232)$$
$$= 59.0 \text{ km/s}$$

and at aphelion

$$v_{aphelion} = [2\pi (57.9 \times 10^6 \text{ km}) / (7.60 \times 10^6 \text{ s})]$$
$$\times [(1 - 0.206)/(1 + 0.206)]^{1/2}$$
$$= (47.9 \text{ km/s}) \times (0.811)$$
$$= 38.9 \text{ km/s}$$

The perihelion and aphelion distances are (see Figure 1-7),

$$r_{perihelion} = a(1-e) = 57.9 \times 10^6 \text{ km} (1 - 0.206) = 4.60 \times 10^7 \text{ km}$$
$$r_{aphelion} = a(1+e) = 57.9 \times 10^6 \text{ km} (1 + 0.206) = 6.98 \times 10^7 \text{ km}$$

Computing the product of the velocity and radius at perihelion (p) and aphelion (a):

$$v_p r_p = (59.0 \text{ km/s})(4.60 \times 10^7 \text{ km}) = 2.71 \times 10^9 \text{ km}^2/\text{s}$$
$$v_a r_a = (38.9 \text{ km/s})(6.98 \times 10^7 \text{ km}) = 2.71 \times 10^9 \text{ km}^2/\text{s}$$

This confirms Kepler's Second Law (Law of Equal Areas), and also (from Equation 1-18) that angular momentum is conserved: $L/m = rv = $ constant.

1-6. Find the relative position of the center of mass for
(a) the Sun-Jupiter system
Using Figure 1-14, the distance (r_1 and r_2) of the center of mass from the center of the two bodies of mass m_1 and m_2, respectively, is given by Equation 1-21:

$$m_1 r_1 = m_2 r_2$$

where the distance between the two bodies is

$$a = r_1 + r_2$$

giving Equation 1-22:

$$r_1 = a\, m_2 / (m_1 + m_2) = a / [(m_1 / m_2) + 1]$$

For the Sun-Jupiter system:

$$r_{Sun-CM} = 778.3 \times 10^6 \text{ km} / [(2 \times 10^{30} \text{ kg} / 1.9 \times 10^{27} \text{ kg}) + 1]$$

$$= 7.783 \times 10^8 \text{ km} / (1053 + 1)$$

$$= 7.4 \times 10^5 \text{ km} \approx 10.6\, R_{Sun}$$

Since the Sun's radius is $\approx 7 \times 10^4$ km, the center of mass of the Sun-Jupiter system is ≈ 10.6 solar radii from the center of the Sun.

(b) the Earth-Moon system

$$r_{Earth-CM} = 3.84 \times 10^5 \text{ km} / [(5.98 \times 10^{24} \text{ kg} / 0.073 \times 10^{24} \text{ kg}) + 1]$$

$$= 3.84 \times 10^5 \text{ km} / (82 + 1)$$

$$= 4.7 \times 10^3 \text{ km} \approx 0.74\, R_\oplus$$

Since the Earth's radius is $\approx 6.4 \times 10^3$ km, the center of mass of the Earth-Moon system is ≈ 0.74 Earth radii from the center of the Earth -- thus within the Earth.

1-7. A television satellite is in circular orbit about the Earth, with a sidereal period of exactly 24 hours. What is the distance from the Earth's surface of such a satellite? (*Hint:* Use Kepler's laws.) If the satellite appears stationary to an earthbound observer, what is the orientation of its orbital plane?
Use Newton's form of Kepler's Third Law (Equation 1-24)

$$P^2 = 4\pi^2 a^3 / [G(m_1 + m_2)]$$

$$a^3 = G(m_1 + m_2)\, P^2 / (4\pi^2)$$

The satellite has negligible mass compared to the Earth,

$$a^3 = (6.67 \times 10^{-11} \text{ N m}^2/\text{kg}^2)(5.98 \times 10^{24} \text{ kg})(8.64 \times 10^4 \text{ s})^2 / (4\pi^2)$$

$$= 7.54 \times 10^{22} \text{ m}^3$$

$$a = 4.23 \times 10^7 \text{ m} = 4.23 \times 10^4 \text{ km}$$

If the satellite appears stationary to an Earth-bound observer, its orbit must be in the same plane as the earth's equator. Otherwise the satellite would move North-South with a 24 hour period. The satellite is in geosynchronous orbit.

1-8. Using orbital data for Titan (Appendix 3), find the mass of Saturn.
We can use Newton's form of Kepler's Third Law (Equation 1-24):
$$P^2 = 4\pi^2 a^3 / [G(m_1 + m_2)]$$
If Titan has negligible mass compared with Saturn
$$M_{Saturn} = 4\pi^2 a^3 / (G P^2)$$
Using the values from Appendix Table A3-6,
$$P_{Titan} = 15.945 \text{ days} = 1.378 \times 10^6 \text{ s}$$
$$a_{Titan} = 1.222 \times 10^6 \text{ km} = 1.222 \times 10^9 \text{ m}$$
we get
$$M_{Saturn} = 4\pi^2 (1.222 \times 10^9 \text{ m})^3 /$$
$$[(6.67 \times 10^{-11} \text{ N m}^2/\text{kg}^2)(1.378 \times 10^6 \text{ s})^2]$$
$$= 5.69 \times 10^{26} \text{ kg}$$
This agrees with the value of Saturn's mass given in Appendix Table A3-3.

1-9. A stone is released from rest at the Moon's orbit and falls toward Earth. What is the stone's speed when it is 192,000 km from the center of Earth?
From Equations 1-30 and 1-31,
$$(KE + PE)_a = (KE + PE)_b$$
so
$$(1/2) m_s v_{sa}^2 - G m_s m_E / r_a = (1/2) m_s v_{sb}^2 - G m_s m_E / r_b$$
where the subscripts s and E refer to the stone and Earth, respectively; and subscripts a and b refer to the initial and later times. The change in the kinetic energy of the Earth is negligible.
So
$$v_{sa}^2 - 2 G m_E / r_a = v_{sb}^2 - 2 G m_E / r_b$$
and since $v_{sa} = 0$,
$$v_{sb}^2 = 2 G m_E [(1/r_b) - (1/r_a)]$$
$$= 2 (6.67 \times 10^{-11} \text{ N m}^2/\text{kg}^2)(5.98 \times 10^{24} \text{ kg}) \times$$
$$[(1 / 1.92 \times 10^8 \text{ m}) - (1 / 3.84 \times 10^8 \text{ m})]$$
$$= (7.98 \times 10^{14} \text{ m}^3/\text{s}^2)(2.60 \times 10^{-9} \text{ m}^{-1})$$
$$= 2.08 \times 10^6 \text{ m}^2/\text{s}^2$$
$$v_{sb} = 1.44 \times 10^3 \text{ m/s} = 1.44 \text{ km/s}$$

1-10. An object is observed from the Earth to have a synodic period of 1.5 years. What are the two possible values for the semi-major axis of the object's orbit?
The asteroid could have an orbit exterior or interior to the Earth's orbit, although the latter is less likely since asteroids generally orbit with semi-major axes greater than 1 AU. Using the equations from Section 1-1A, we can solve for the orbital (sidereal) period of the asteroid

$1/S_{asteroid} = 1/P_{Earth} - 1/P_{asteroid}$ (superior orbit - external to Earth's)

and $1/S_{asteroid} = 1/P_{asteroid} - 1/P_{Earth}$ (inferior orbit - internal to Earth's)

where S is the synodic period, and P is the sidereal period.

The semi-major axis can be found using Kepler's Third (Harmonic) Law [Section 1-2A]

$$P^2 = k a^3$$

where P is the orbital period and a is the semi-major axis. The constant, k, is defined as 1 when units of years and AU are used for P and a, respectively.

For the superior orbit,

$1/1.5 = 1/1.0 - 1/P_{asteroid}$

$1/P_{asteroid} = 1 - 2/3 = 1/3$

$P_{asteroid} = 3$ years

The orbital period of 3 years corresponds to a semi-major axis of

$a = P^{2/3} = 3^{2/3} = 2.08$ AU

For the inferior orbit,

$1/1.5 = 1/P_{asteroid} - 1/1.0$

$1/P_{asteroid} = 1 + 2/3 = 5/3$

$P_{asteroid} = 0.6$ years

The orbital period of 0.6 years corresponds to a semi-major axis of

$a = P^{2/3} = 0.6^{2/3} = 0.71$ AU

1-11. Using orbital data for the Earth found in Appendix 3, estimate the mass of the Sun. Does the mass of the Earth matter significantly in this calculation?

Use Newton's form of Kepler's Third (Harmonic) Law (Equation 1-24):

$$P^2 = 4\pi^2 a^3 / G (M_\odot + m_\oplus) .$$

The mass of the Earth ($m_\oplus = 6 \times 10^{24}$ kg) is so much less (by a factor of more than 10^5) than the mass of the Sun ($M_\odot = 2 \times 10^{30}$ kg) that we can ignore it in the calculations, so

$$P^2 \approx 4\pi^2 a^3 / G (M_\odot)$$

So,

$$M_\odot = 4\pi^2 a^3 / G P^2$$

The orbital period, P, of the Earth is 1 year = 3.16×10^7 s, the semi-major axis of its orbit is 1.496×10^{11} m.

$M_\odot = 4\pi^2 (1.496 \times 10^{11} \text{ m})^3 / (6.67 \times 10^{-11} \text{ m}^3 \text{ kg}^{-1} \text{ s}^{-2}) (3.16 \times 10^7 \text{ s})^2$

$= 1.99 \times 10^{30}$ kg

1-12. Compare the mutual gravitational force between you and the following two objects:
(a) another person with mass 100 kg located 1 m from you
Using Newton's Law of Universal Gravitation (Equation 1-15)
$$F_{grav} = GMm/r^2$$
Since we are going to compare two gravitational forces, we can leave our mass as a variable.
$$F_{person-me} = (6.67 \times 10^{-11} \text{ m}^3 \text{ kg}^{-1} \text{ s}^{-2})(100 \text{ kg})(m_{me})/(1 \text{ m})^2$$
$$= 6.7 \times 10^{-9} \, m_{me} \text{ N} \quad \quad [\text{for } m_{me} \text{ expressed in kg}]$$

(b) Mars at opposition
Use Appendices A3-1 and A3-3 for the orbital parameters and mass of Mars. Mars' orbit has a semi-major axis of ≈ 1.5 AU, so at opposition Mars is about
$$1.5 - 1.0 = 0.5 \text{ AU} \approx 0.5 \times (1.5 \times 10^{11} \text{ m}) = 0.75 \times 10^{11} \text{ m distant, so}$$
$$F_{Mars-me} = (6.67 \times 10^{-11} \text{ m}^3 \text{ kg}^{-1} \text{ s}^{-2})(0.64 \times 10^{24} \text{ kg})(m_{me})$$
$$/ (0.75 \times 10^{11} \text{ m})^2$$
$$= 7.6 \times 10^{-9} \, m_{me} \text{ N} \quad \quad [\text{for } m_{me} \text{ expressed in kg}]$$

Comment on your result.
The ratio of the force of Mars on me to the force of a 100 kg person 1 m from me is
$$F_{Mars-me} / F_{person-me} = (7.6 \times 10^{-9} \, m_{me} \text{ N}) / (6.7 \times 10^{-9} \, m_{me} \text{ N}) \approx 1.1$$
Mars and the person have nearly the same mutual gravitational force with me.

1-13. (a) Venus has a maximum elongation of 47°. What is its distance from the Sun in astronomical units?
Using Figure 1-5A, we see that the distance of any inferior planet from the Sun is
$$r = 1 \text{ AU} \times \sin(\text{greatest elongation}).$$
For a maximum elongation for Venus of 47°,
$$r_V = 1 \text{ AU} \times \sin(47°) = 0.73 \text{ AU}$$

(b) Mars has a synodic period of 779.9 days and a sidereal period of 686.98 days. On February 11, 1990, Mars had an elongation of 43° West. The elongation of Mars 687 days later, on December 30, 1991, was 15° West. What is the distance of Mars from the Sun in astronomical units?
The two dates of observation are one Martian sidereal period apart, so Mars has returned to its original position in its orbit. The Earth has revolved just less than twice around the Sun, as shown in the figure

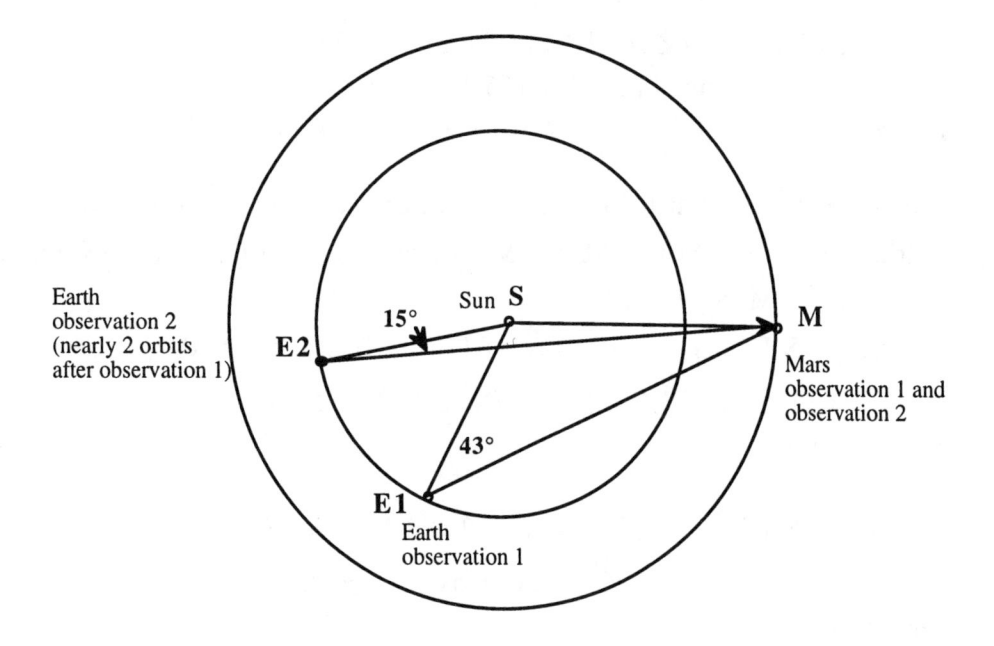

[Figure: Diagram showing Earth's orbit (inner circle) and Mars' orbit (outer circle) around the Sun S. Points E1 (Earth observation 1) and E2 (Earth observation 2, nearly 2 orbits after observation 1) on Earth's orbit, with 15° angle marked at Sun between E2 and M direction, and 43° angle at E1. Point M labeled "Mars observation 1 and observation 2" on Mars' orbit.]

The Earth has moved 687 / 365.26 = 1.881 orbital circumferences, or 677°, in the 687 days it takes Mars to complete one orbit. The angle ∠E1-S-E2 is thus 720° - 677° = 43°. Using the law of cosines (Appendix A9-1C), and the Earth-Sun distance as 1 AU, we can determine the distance E1-E2

$$(E1\text{-}E2)^2 = (E1\text{-}S)^2 + (E2\text{-}S)^2 - 2(E1\text{-}S)(E2\text{-}S)\cos(\angle E1\text{-}S\text{-}E2)$$
$$= 1^2 + 1^2 - 2(1)(1)\cos(43°) = 0.540$$
$$E1\text{-}E2 = 0.73 \text{ AU}$$

Since ΔE1-S-E2 is an isosceles triangle, and ∠E1-S-E2 = 43°, angles ∠S-E1-E2 and ∠S-E2-E1 are (180° - 43°) / 2 = 68.5°

By inspecting the figure, we can determine that
$$\angle M\text{-}E1\text{-}E2 = 68.5° + 43° = 111.5°$$
$$\angle M\text{-}E2\text{-}E1 = 68.5° - 15° = 53.5°$$
so consequently
$$\angle E1\text{-}M\text{-}E2 = 180° - 111.5° - 53.5° = 15°$$

We now know the three angles and one side of the triangle E1-M-E2, from which we can calculate the Earth-Mars distances using the law of sines (Appendix A9-1C)

$$E1\text{-}M / E1\text{-}E2 = \sin(\angle M\text{-}E2\text{-}E1) / \sin(\angle E1\text{-}M\text{-}E2)$$
$$= \sin(53.5°) / \sin(15°) = 3.11$$
$$E1\text{-}M = 3.11 \, (E1\text{-}E2) = 3.11 \, (0.73 \text{ AU}) = 2.27 \text{ AU}$$

and

$$E2\text{-}M / E1\text{-}E2 = \sin(\angle M\text{-}E1\text{-}E2) / \sin(\angle E1\text{-}M\text{-}E2)$$
$$= \sin(111.5°) / \sin(15°) = 3.59$$
$$E2\text{-}M = 3.59 \, (E1\text{-}E2) = 3.59 \, (0.73 \text{ AU}) = 2.62 \text{ AU}$$

Now that we have the Earth-Mars and Earth-Sun distances, we can return to the original triangles ΔE1-S-M and ΔE2-S-M, and use the law of cosines to solve for the semi-major axis of Mars' orbit (S-M).

$$(S\text{-}M)^2 = (S\text{-}E1)^2 + (E1\text{-}M)^2 - 2(S\text{-}E1)(E1\text{-}M)\cos(\angle S\text{-}E1\text{-}M)$$
$$= (1)^2 + (2.27)^2 - 2(1)(2.27)\cos(43°) = 2.83$$
$$S\text{-}M = 1.68$$

and

$$(S\text{-}M)^2 = (S\text{-}E2)^2 + (E2\text{-}M)^2 - 2(S\text{-}E2)(E2\text{-}M)\cos(\angle S\text{-}E2\text{-}M)$$
$$= (1)^2 + (2.62)^2 - 2(1)(2.62)\cos(15°) = 2.80$$
$$S\text{-}M = 1.67$$

As expected, to within the errors due to round-off during different steps in the calculations, the answers agree. The semi-major axis of Mars' orbit from this calculation is 1.67 AU, in reasonable agreement with the value of 1.52 AU given in Appendix A3-1. In actuality, the orbit of Mars is eccentric, so the values for the Sun-Mars distance obtained with this technique would show variations depending on where in the orbit Mars was. These different values led Kepler to conclude that Mars (and hence the other planets) had elliptical rather than circular orbits.

1-14. Using astronomical units as the unit of length, years as the time, and the mass of the Sun as the unit of mass, the value for k in Kepler's third law is 1.0. In these units, what is the value of Newton's constant of gravitation G?

Use Newton's form of Kepler's Third (Harmonic) Law (Equation 1-24):

$$P^2 = 4\pi^2 a^3 / G(M_1 + m_2)$$

For a planet orbiting the Sun (where $M_\odot \gg m_{planet}$), this reduces to

$$P^2 = (4\pi^2 / G)(a^3 / M_\odot)$$

If the planet is Earth, P = 1 yr, a = 1 AU, so

$$(1 \text{ yr})^2 = (4\pi^2 / G)((1 \text{ AU})^3 / (1 \, M_\odot))$$
$$G = 4\pi^2 \, [\text{AU}^3 / M_\odot \, \text{yr}^2]$$

Note, normally the units of G are expressed as $N \, m^2 / kg^2$, which is equivalent to $m^3 / kg \, s^2$; so the unit $\text{AU}^3 / M_\odot \, \text{yr}^2$ is dimensionally correct.

Chapter 2: The Solar System in Perspective

2-1. (a) Describe the apparent path of the Sun across the sky of Mercury during one solar "day," as seen by an observer at the planet's equator.

To understand the motion of the Sun as viewed from Mercury, we need to consider the rotation and revolution periods of Mercury and to a lesser extent the eccentricity of Mercury's orbit. The rotation axis of Mercury is nearly aligned with its revolution axis (Appendix Table A3-2), so we do not have to consider seasonal effects of the rising and setting points for the Sun or its seasonal path across the sky. For an observer on Mercury's equator, the Sun would rise due East, transit at the zenith, and set due West. Mercury's sidereal rotation period is 2/3 its sidereal orbital period, so it rotates one and a half times per revolution. The Mercury "day" takes two orbital periods (176 Earth days) -- Mercury rotates three times during two revolutions.

Use Figure 2-5 as a reference. At position 1 it is noon and the Sun is directly overhead as seen from Mercury's equator. The Sun would move slowly westward, setting when Mercury was at aphelion one half orbital period later (4). Although Mercury will have completed one sidereal rotation after completing two-thirds of an orbit (5), due to the revolution of Mercury, the "time" will be only a little after sunset. When Mercury is again at perihelion (7), Mercury will have rotated one and a half times so it will be midnight. Sunrise will occur when Mercury is again at aphelion. During the latter half of this second revolution, the Sun will rise perpendicular to the horizon, reaching the zenith (noon) when Mercury is once again at perihelion. Thus, one Mercurian "day" occurs every two sidereal revolutions and three sidereal rotations.

An interesting effect occurs due to the high eccentricity of Mercury's orbit. From Kepler's Law of Equal Areas (Second Law), we see that Mercury's orbital velocity varies (greatest when Mercury at perihelion, least at aphelion). The rotational speed of Mercury does not change, so the orbital speed varies relative to the rotational speed. The motion of the Sun across Mercury's sky would not be uniform. Indeed, for a period of a few Earth days once each orbit (at aphelion) the Sun would appear to move Eastward relative to the background stars, before resuming its normal Westwardly motion. If this occurred at a point on Mercury's surface where the Sun had just risen (or set), an observer would see two sunrises (or sunsets)!

(b) Describe the seasons of Uranus, giving their durations where appropriate.

Uranus' rotational axis is tilted 98° to its revolution axis; the orbital period of Uranus is ≈ 84 Earth years (see Figure 2-4B). Due to the high axial tilt, each pole of Uranus points nearly towards (and away from) the Sun during part of its orbit. Each season would last 21 years. Since the Uranian rotation period is short (≈ 17 Earth hours; Appendix Table A3-2), the Sun would daily move in a circular path in the Uranian sky. The yearly motion of the Sun would result in a small day-to-day variation in this path (as is the case for Earth), although much more extreme than what we witness on Earth. For an observer near a Uranian pole, the Sun would be nearly overhead during the beginning of "summer"; moving in a circle with an angular radius of 8°. As summer progresses, the Sun's daily path would move lower in the sky, in an increasingly larger circle. The Sun would set about 21 years later, and would be below the horizon for about half the orbital period (rising again in spring, 63 years after "summer"). (Due to the axial tilt not being exactly 90°, the Sun would rise and set each Uranian day for several Earth years near times of spring and autumn.)

For an observer on the Uranian equator, the Sun would move in a small circular path (angular radius 8°) in the north on the first day of summer. Unlike the conditions at the pole, the Sun would rise and set each Uranian day. The circle of the Sun's daily path would increase in angular size during the next 21 years -- moving further from the north until autumn - when the Sun would rise in the East, transit nearly overhead, and set in the West. During the next 21 years the Sun's path would be in the southern sky, in decreasing size daily paths -- describing an 8° circle in the south at the beginning of winter. This motion is reversed during the latter 42 years of the Uranian orbit.

2-2. Consider the planets Uranus, Neptune, and Pluto. Show that the ratios of their orbital periods are approximately *commensurable*, that is, nearly fractions such as 3/2.

The ratio of the actual orbital periods (Appendix Table A3-1) are:

P(Neptune) / P(Uranus) = 164.8 years / 84.01 years = 1.96 ≈ 4/2 = 2/1
P(Pluto) / P(Neptune) = 248.6 years / 164.8 years = 1.51 ≈ 3/2
P(Pluto) / P(Uranus) = 248.6 years / 84.01 years = 2.96 ≈ 3/1

Thus for every Neptune revolution, Uranus revolves about the Sun nearly twice; for every two Pluto orbits, Neptune orbits three times; and for every Uranus orbit, Pluto orbits three times. The orbits are approximately commensurable.

2-3. Show that the two satellites of Mars, Phobos and Deimos, obey Kepler's third law. Deduce the mass of Mars from the orbits of these moons.

Newton's formulation of Kepler's Third Law is (Equation 1-24 or 2-10):

$$P^2 = 4\pi^2 a^3 / [G(M + m)]$$

In the case of Mars and its satellites, M » m.
From Appendix Tables A3-3 and A3-4,

$M_{Mars} = 0.64 \times 10^{24}$ kg $= 0.107\, M_{Earth}$

$P_{Phobos} = 0.32$ days $A_{Phobos} = 9 \times 10^3$ km

$P_{Deimos} = 1.26$ days $A_{Deimos} = 23 \times 10^3$ km

So, checking Kepler's Third Law:

$P_{Phobos}^2 / a_{Phobos}^3 =?\ P_{Deimos}^2 / a_{Deimos}^3$

$0.32^2 / 9^3 =?\ 1.26^2 / 23^3$

$1.4 \times 10^{-4} =?\ 1.3 \times 10^{-4}$

Thus, to the accuracy of the values used, Mars' satellites obey Kepler's Third Law.

Using Deimos to calculate the mass of Mars:

$$P^2 = 4\pi^2 a^3 / [G(M)]$$

$M_{Mars} = 4\pi^2 (2.3 \times 10^7 \text{ m})^3 /$

$[(6.67 \times 10^{-11} \text{ N m}^2/\text{kg}^2) \times (1.26 \text{ days} \times 8.64 \times 10^4 \text{ s/day})^2]$

$$M_{Mars} = 6.1 \times 10^{23} \text{ kg}$$

This is near the value for the mass of Mars given in Appendix Table A3-3 (the error is due to the approximate values used for the period and semi-major axis of Deimos' orbit).

2-4. (a) How much does a spaceship having a mass of 10^3 kg weigh on the Earth's equator? At its poles? (*Hint:* Consider centripetal force.)

The weight of a spaceship of mass 10^3 kg will modified by the gravitational acceleration (depends on radius of the planet) and the centrifugal effect caused by the Earth's rotation (see also Problem 3-6). This latter effect is numerically equal to the centripetal force (Equation 1-14):

$$|F_{cent}| = m\,a = m\,v^2/r = m\,(2\pi r/P)^2/r = 4\pi^2\,m\,r/P^2$$

For the Earth's equator:

$$|F_{cent\text{-}\oplus eq}| = 4\pi^2\,(1.0 \times 10^3 \text{ kg})\,(6.378 \times 10^6 \text{ m})/(8.64 \times 10^4 \text{ s})^2$$
$$= 3.37 \times 10^1 \text{ N} = 33.7 \text{ N}$$

For the Earth's pole, the centripetal force is zero.

Finding the gravitational acceleration (see Equation 1-15):

$$g = F_{grav}/m = GM/r^2$$

so at the Earth's equator,

$$g_{\oplus eq} = (6.67 \times 10^{-11} \text{ N m}^2/\text{kg}^2)\,(5.98 \times 10^{24} \text{ kg})/(6.378 \times 10^6 \text{ m})^2$$
$$= 9.81 \text{ m}/\text{s}^2$$

Due to the Earth's rotation, its oblateness is: $\varepsilon = 0.0034$ (Appendix Table A3-2). The polar radius is (see Section 2-1A):

$$r_{\oplus pole} = (1-\varepsilon)\,r_{\oplus eq}$$
$$= (1 - 0.0034)\,(6.378 \times 10^6 \text{ m}) = 6.356 \times 10^6 \text{ m}$$

so $\quad g_{\oplus pole} = (6.67 \times 10^{-11} \text{ N m}^2/\text{kg}^2)\,(5.98 \times 10^{24} \text{ kg})/(6.356 \times 10^6 \text{ m})^2$
$$= 9.87 \text{ m}/\text{s}^2$$

The weight, W, of the spacecraft is: $W = F_{grav} - |F_{cent}| = m\,g - |F_{cent}|$. So

$$W_{\oplus pole} = (1.0 \times 10^3 \text{ kg}) \times (9.87 \text{ m}/\text{s}^2) - 0 = 9{,}870 \text{ N}$$
$$W_{\oplus equator} = (1.0 \times 10^3 \text{ kg}) \times (9.81 \text{ m}/\text{s}^2) - 33.7 \text{ N} = 9{,}776 \text{ N}$$

(b) Remembering Jupiter's rapid rotation, determine how much this craft will weigh at the equator of Jupiter's surface and at the poles.

For Jupiter's equator (using $P_J = 9^h 50^m = 3.54 \times 10^4$ s):

$$|F_{cent\text{-}Jeq}| = 4\pi^2\,(1.0 \times 10^3 \text{ kg})\,(6.87 \times 10^7 \text{ m})/(3.54 \times 10^4 \text{ s})^2$$

$$= 2.16 \times 10^3 \text{ N} = 2,160 \text{ N}$$
For Jupiter's pole, the centripetal force is zero.

Finding the gravitational acceleration for Jupiter's equator,
$$g_{Jeq} = (6.67 \times 10^{-11} \text{ N m}^2/\text{kg}^2)(1.90 \times 10^{27} \text{ kg})/(6.87 \times 10^7 \text{ m})^2$$
$$= 2.69 \times 10^1 \text{ m/s}^2 = 26.9 \text{ m/s}^2$$

Due to the Jupiter's rapid rotation, its oblateness is quite large: $\varepsilon = 0.062$. So:
$$r_{Jpole} = (1 - 0.062)(6.87 \times 10^7 \text{ m}) = 6.44 \times 10^7 \text{ m}$$
and $g_{Jpole} = (6.67 \times 10^{-11} \text{ N m}^2/\text{kg}^2)(1.90 \times 10^{27} \text{ kg})/(6.44 \times 10^7 \text{ m})^2$
$$= 3.06 \times 10^1 \text{ m/s}^2 = 30.6 \text{ m/s}^2$$

The weight, W, of the spacecraft is
$$W_{Jpole} = (1.0 \times 10^3 \text{ kg}) \times (30.6 \text{ m/s}^2) - 0 = 30,600 \text{ N}$$
$$W_{Jequator} = (1.0 \times 10^3 \text{ kg}) \times (26.9 \text{ m/s}^2) - 2,160 \text{ N} \approx 24,700 \text{ N}$$

2-5. Derive the formula that gives the synodic rotation period (solar day) of Venus with respect to the Earth. Verify the number quoted in the text.

The synodic rotation period of a planet with respect to the Earth is the time interval between two successive times when a longitude on the planetary surface faces the Earth. If the Earth and the planet did not orbit the Sun, the planet would appear to rotate at a rate $360°/D_{sid}$, where D_{sid} is the planet's sidereal rotation rate. However, the Earth is moving relative to the sidereal frame at a rate $360°/P_\oplus$, so the apparent rotation rate as seen from the Earth for an inferior planet is:
$$360°/D_{syn} = 360°/P_\oplus - 360°/D_{sid}.$$
For Venus, the sidereal rotation is retrograde with period $D_{sid} = -243$ days, so
$$1/D_{syn}[\text{days}] = 1/365 - 1/(-243) = 0.00274 + 0.00412 = 0.00686$$
$$D_{syn} = 146 \text{ days} \quad \text{--- this agrees with the value given in Section 2-1A}$$

2-6. In the 17th century, O. Roemer concluded that the speed of light is finite by observing that the satellite Jupiter I (Io) was occulted by the planet approximately 16.7 minutes *earlier* when Jupiter was in opposition than when it was near superior conjunction. Use this information to draw an approximate diagram and to calculate an approximate value for the speed of light (which Roemer did *not* do!).

From the sketch, we see that the difference between the distance that light travels when Jupiter is at superior conjunction compared to when it is at opposition is equal to twice the radius of the Earth's orbit. Thus, the time difference of 16.7 minutes corresponds to light traveling a distance of 2 AU. Calculating the speed of light:
$$c = d/v = (2 \text{ AU})(1.5 \times 10^{11} \text{ m/AU})/(1.0 \times 10^3 \text{ s})$$
$$= 3.0 \times 10^8 \text{ m/s} \quad \text{-- this is correct!}$$

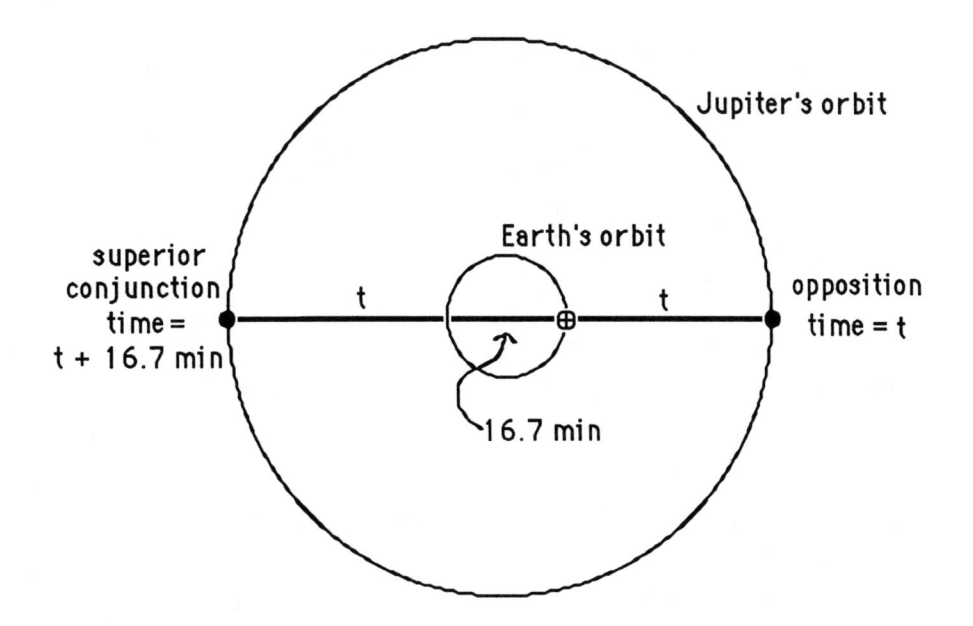

[Figure: Diagram showing Earth's orbit and Jupiter's orbit with superior conjunction time = t + 16.7 min on the left, opposition time = t on the right, and 16.7 min label near Earth's orbit.]

2-7. How far from the star would we have to be (in astronomical units) to find a substellar temperature comparable to that for the Earth for

(a) Rigel (surface temperature T = 12,000 K, radius R = 35 $R_\odot$)?

The subsolar temperature is given by (Equation 2-4):

$$T_{ss} = (R_*/r_p)^{1/2} T_*$$
$$r_p = R_* (T_*/T_{ss})^2$$

where R_* is the radius of the star, and r_p is the planet's orbital radius.

For Rigel, T_* = 12,000 K, R_* = 35 $R_\odot$ = 2.44 x 10^7 km = 0.163 AU, so if we want a subsolar temperature comparable to Earth (T ≈ 395 K),

$$r_p = 0.163 \text{ AU } (12,000/395)^2 = 150 \text{ AU}$$

(b) Barnard's star (T = 3,000 K, radius R = 0.5 $R_\odot$)?

For Barnard's star, T_* = 3,000 K, R_* = 0.5 $R_\odot$ = 3.48 x 10^5 km = 2.33 x 10^{-3} AU, so if we want a subsolar temperature comparable to Earth (T ≈ 395 K),

$$r_p = 2.33 \times 10^{-3} \text{ AU } (3,000/395)^2 = 0.13 \text{ AU}$$

2-8. Formaldehyde (H_2CO) has been discovered in interstellar space.

(a) Calculate its "mean" molecular speed for T = 280 K. Would our Moon retain this gas for billions of years?

The mass of a Formaldehyde molecule (H_2CO) is

$$(2 \times 1 + 12 + 16) \text{ amu} = 30 \, m_H = 30 \times (1.67 \times 10^{-27} \text{ kg}) = 5.0 \times 10^{-26} \text{ kg}.$$

The mean molecular speed is given by (Equation 2-7):

$$v_{rms} = (3kT/m)^{1/2}$$

$$= [(3)(1.38 \times 10^{-23} \text{ J/K})(280 \text{ K})/(5.0 \times 10^{-26} \text{ kg})]^{1/2}$$
$$= 4.8 \times 10^2 \text{ m/s} = 480 \text{ m/s} = 0.48 \text{ km/s}$$

The escape speed for the Moon is given by (Equation 2-8):
$$v_{esc} = (2GM/R)^{1/2}$$
$$= [2(6.67 \times 10^{-11} \text{ N m}^2/\text{kg}^2)(7.3 \times 10^{22} \text{ kg})/(1.74 \times 10^6 \text{ m})]^{1/2}$$
$$= 2.4 \times 10^3 \text{ m/s} = 2.4 \text{ km/s}$$

To retain an atmosphere, the escape speed should be greater than 10 times the mean molecular speed ($v_{esc} \geq 10\, v_{rms}$) (Section 2-1D). Since this is not the case ($v_{esc} \approx 5\, v_{rms}$) we would not expect the Moon to retain a formaldehyde atmosphere.

Alternately, we can argue that since from Equation 2-9, we know that the molecule will be retained by a body if
$$T \leq GMm/(100\, k R).$$
For the Moon,
$$T \leq (6.67 \times 10^{-11} \text{ N m}^2/\text{kg}^2)(7.3 \times 10^{22} \text{ kg})(5.0 \times 10^{-26} \text{ kg})/$$
$$[(100)(1.38 \times 10^{-23} \text{ J/K})(1.74 \times 10^6 \text{ m})]$$
$$\approx 100 \text{ K}$$

Since the Moon is nearly four times warmer than this ($T \approx 386$ K; see Figure 2-8 or Section 4-4B), Formaldehyde would escape from the Moon.

(b) Would Saturn's satellite Titan retain formaldehyde? (For Titan, radius $\approx$ 2600 km, mass $\approx 10^{23}$ kg.)

For Titan:
$$T \leq (6.67 \times 10^{-11} \text{ N m}^2/\text{kg}^2)(1.0 \times 10^{23} \text{ kg})(5.0 \times 10^{-26} \text{ kg})/$$
$$[(100)(1.38 \times 10^{-23} \text{ J/K})(2.6 \times 10^6 \text{ m})]$$
$$\approx 93 \text{ K}$$

The surface temperature of Titan is $\approx$ 94K (Figure 2-8 or Section 7-1C), so it would be possible (just barely) for Titan to retain some Formaldehyde in its atmosphere.

2-9. Consider a comet with an aphelion distance of 5×10^4 AU and an orbital eccentricity of 0.995.

(a) What are the perihelion distance and orbital period?

For an elliptical orbit of eccentricity, e, and semi-major axis, a, the perihelion distance is given by (see Section 1-2B)
$$r_p = a - ae = a(1-e)$$
and the aphelion distance is given by
$$r_a = a + ae = a(1+e).$$
For $r_a = 5 \times 10^4$ AU $= 7.5 \times 10^{15}$ m, and $e = 0.995$,
$$a = r_a/(1+e) = (5 \times 10^4 \text{ AU})/(1+0.995)$$
$$= 2.5 \times 10^4 \text{ AU} = 3.8 \times 10^{15} \text{ m}$$

$$r_p = a(1-e) = (2.5 \times 10^4 \text{ AU})(1-0.995)$$
$$= 1.25 \times 10^2 \text{ AU} = 1.9 \times 10^{13} \text{ m}$$

For the orbital period, use Newton's form of Kepler's Third Law (Equation 1-24 or 2-10), expressing distances in AU and revolution periods in years and assuming the $M_{comet} \ll M_{Sun}$:

$$P = a^{3/2} = (2.5 \times 10^4 \text{ AU})^{3/2}$$
$$= 4.0 \times 10^6 \text{ years} = 1.3 \times 10^{14} \text{ s}$$

The comet has a revolution period of 4 million years -- a once in a lifetime experience!

(b) What is the comet's speed at perihelion and at aphelion?

The orbital speed is given by (Equation 1-17):

$$v_p = \{2\pi a / P\}[(1+e)/(1-e)]^{1/2}$$
$$= \{2\pi (3.8 \times 10^{15} \text{ m})/(1.3 \times 10^{14} \text{ s})\}[(1+0.995)/(1-0.995)]^{1/2}$$
$$= 3.7 \times 10^3 \text{ m/s}$$

$$v_a = \{2\pi a / P\}[(1-e)/(1+e)]^{1/2}$$
$$= \{2\pi (3.8 \times 10^{15} \text{ m})/(1.3 \times 10^{14} \text{ s})\}[(1-0.995)/(1+0.995)]^{1/2}$$
$$= 9.2 \text{ m/s}$$

Equation 1-34 could also have been used to solve this problem:

$$v^2 = G(m_1 + m_2)(2/r - 1/a)$$

where a is the semi-major axis and r is either the perihelion or aphelion distance. The speed at any distance from the Sun could be calculated with this more general equation.

(c) What is the escape speed from the Solar System at the comet's aphelion? What do you conclude from this result?

The escape speed at the aphelion distance is given by (Equation 2-8):

$$v_{esc} = (2 G M_\odot / r_a)^{1/2}$$
$$= [2 (6.67 \times 10^{-11} \text{ N m}^2/\text{kg}^2)(2.0 \times 10^{30} \text{ kg})/(7.5 \times 10^{15} \text{ m})]^{1/2}$$
$$= 1.9 \times 10^2 \text{ m/s} = 190 \text{ m/s}$$

So $v_a < v_{esc}$ [9.2 m/s < 190 m/s].

The comet does not escape the solar system, since it's in a closed orbit -- as we expect. But it would not require much of a perturbation on the comet to give it sufficient speed to change its orbit or to escape the Solar System.

2-10. Draw a diagram to explain why some meteor showers are consistent from year to year whereas others are spectacular on occasion and feeble at other times.

A meteor shower occurs when the earth crosses the orbit of a swarm of meteoroids. If the swarm is evenly distributed throughout the orbit, then the number of meteors seen would be consistent from year to year. If the swarm is unevenly distributed, then some years we would see more meteors than in other years. If there is a single dense clump of meteoroids, then the interval between intense showers would be equal to the orbital period of the meteoroids.

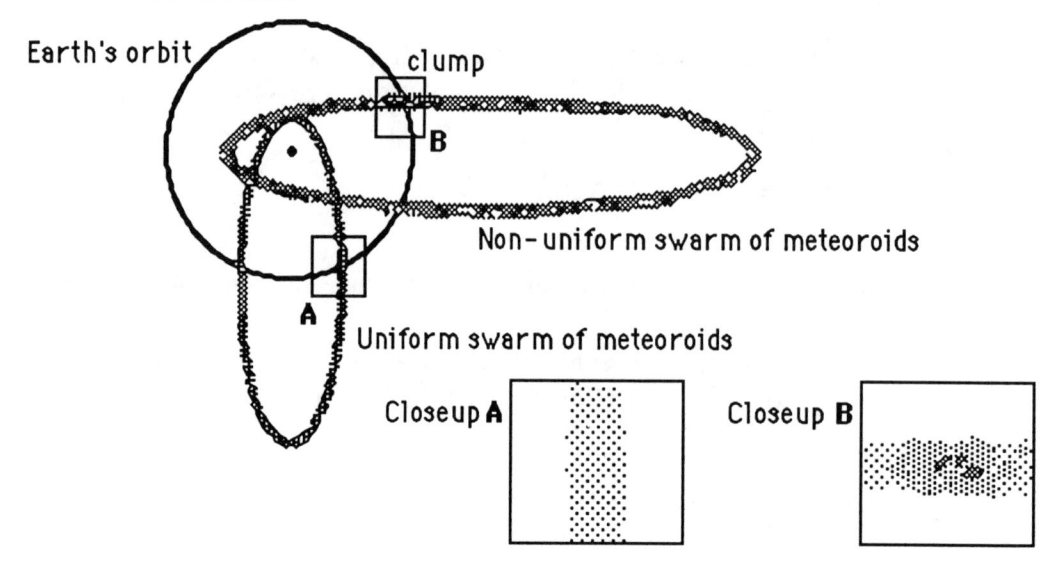

2-11. The albedo of Venus is about 0.77 because of the cloudy atmosphere. What would the noontime temperature be? (The measured temperature is 750 K.)
Neglecting the Greenhouse Effect, but not the albedo effect, the noontime temperature is given by (Equation 2-5a):
$$T_p = (1 - A)^{1/4} (R_\odot / 2 r_p)^{1/2} T_\odot \approx (279 \text{ K}) (1 - A)^{1/4} (r_p)^{-1/2}$$
where A is the albedo of the planet, r_p is the distance of the planet from the Sun (in AU), and $R_\odot$ and $T_\odot$ are the radius and temperature of the Sun. For Venus,
$$T_{Venus} = (279 \text{ K}) (1 - 0.77)^{1/4} (0.723)^{-1/2} = 227 \text{ K}$$
The much higher observed temperature of 750 K is due to the Greenhouse Effect!

2-12. (a) At noon, Mercury's surface temperature is roughly 700 K; at midnight, 125 K. Calculate the peak wavelength at which the surface emits at noon and at midnight.
Using Wien's Displacement Law (Equation 2-2), the wavelength of maximum emission is $\lambda_{max} = (2898 \text{ μm}) / T$, where the temperature is in K.
For Mercury,
$$\lambda_{max-noon} = (2898 \text{ μm}) / (700 \text{ K}) = 4.1 \text{ μm}$$
$$\lambda_{max-midnight} = (2898 \text{ μm}) / (125 \text{ K}) = 23 \text{ μm}$$

(b) Calculate the energy output per square meter of the surface at midnight and noon.
The energy output per square meter of surface is given by (Equation 2-3):
$$F = \sigma T^4 = (5.67 \times 10^{-8} \text{ W/m}^2 \text{ K}^4) T^4.$$
$$F_{Mercury-noon} = (5.67 \times 10^{-8} \text{ W/m}^2 \text{ K}^4) (700 \text{K})^4 = 1.4 \times 10^4 \text{ W/m}^2$$
$$F_{Mercury-midnight} = (5.67 \times 10^{-8} \text{ W/m}^2 \text{ K}^4) (125 \text{ K})^4 = 1.4 \times 10^1 \text{ W/m}^2$$

2-13. The Earth, Venus, and Mars all have carbon dioxide in their atmospheres. Find the ratio of the root mean square speed to escape speed for each. Make a statement about the retention of carbon dioxide for these planets.

The root mean squared speed is given by (Equation 2-7):
$$v_{rms} = (3kT/m)^{1/2}$$
and the escape speed is given by (Equation 2-8):
$$v_{esc} = (2GM/R)^{1/2}$$
so the ratio of the root mean squared speed to the escape speed is
$$v_{rms}/v_{esc} = [(3kTR)/(2GMm)]^{1/2}$$
For carbon dioxide (CO_2), with mass,
$$m = (12 + 2 \times 16) m_H = 44 m_H = 44 \times 1.67 \times 10^{-27} \text{ kg} = 7.35 \times 10^{-26} \text{ kg}$$
we have
$$v_{rms}/v_{esc} = \{[(3)(1.38 \times 10^{-23} \text{ J/K}) T R]$$
$$/ [(2)(6.67 \times 10^{-11} \text{ N m}^2/\text{kg}^2)(M)(7.35 \times 10^{-26} \text{ kg})]\}^{1/2}$$
$$= 2.05 \times 10^6 (TR/M)^{1/2}$$
where T is in K, R in m, and M in kg.

From Appendix Table A3-3,

for Venus, $M = 4.87 \times 10^{24}$ kg, $R = 6.05 \times 10^6$ m, $T = 700$ K
$$v_{rms}/v_{esc} = 2.05 \times 10^6 [(700)(6.05 \times 10^6)/(4.87 \times 10^{24})]^{1/2}$$
$$= 6.0 \times 10^{-2} = 0.060$$

for Earth, $M = 5.98 \times 10^{24}$ kg, $R = 6.38 \times 10^6$ m, $T = 250 - 300$ K
$$v_{rms}/v_{esc} = 2.05 \times 10^6 [(275)(6.38 \times 10^6)/(5.98 \times 10^{24})]^{1/2}$$
$$= 3.5 \times 10^{-2} = 0.035$$

for Mars, $M = 0.64 \times 10^{24}$ kg, $R = 3.39 \times 10^6$ m, $T = 210 - 300$ K
$$v_{rms}/v_{esc} = 2.05 \times 10^6 [(250)(3.39 \times 10^6)/(0.64 \times 10^{24})]^{1/2}$$
$$= 7.5 \times 10^{-2} = 0.075$$

A gas is retained by a planet if $v_{rms}/v_{esc} \leq 0.1$ (Section 2-1D). Thus we see that carbon dioxide will be retained by all these planets.

2-14. What is the approximate orbital period for a cometary nucleus in the Oort cloud?

If the Oort cometary cloud (see Section 2-2D) is at a distance of 50,000 AU, the orbital period can be estimated by Kepler's Third Law (Equation 1-24 or 2-10):
$$P^2 = 4\pi^2 a^3 / [G(M+m)]$$
Expressing the period, P, in years, and the semi-major axis, a, in AU:
$$P = a^{3/2}.$$
For the Oort cloud
$$P = (5 \times 10^4)^{3/2} = 1.1 \times 10^7 \text{ years} = 11 \text{ million years}$$

2-15. Calculate the blackbody equilibrium temperature of a fast-rotating asteroid with a radius of 100 km and an albedo of 0.5 in an orbit between Mars and Jupiter of semimajor axis 2.8 AU.

The blackbody temperature of a rotating asteroid is given by (Equation 2-5a):

$$T_p = (1 - A)^{1/4} (R_\odot / 2 r_p)^{1/2} T_\odot \approx (279 \text{ K}) (1 - A)^{1/4} (r_p)^{-1/2}$$

where A is the albedo of the asteroid, r_p is the distance of the asteroid from the Sun (in AU), and $R_\odot$ and $T_\odot$ are the radius and temperature of the Sun.

$$T_{asteroid} = (279 \text{ K}) (1 - 0.5)^{1/4} (2.8)^{-1/2}$$
$$\approx 140 \text{ K}$$

2-16. Roughly speaking, the spin angular momentum of a sphere is MVR, where V is the equatorial velocity. (For a homogeneous sphere, it is 2/5 MVR.) Make an approximate calculation of the spin and orbital angular momenta of Jupiter and compare them with the spin angular momentum of the Sun.

As specified in the problem, take the spin angular momentum of a sphere as

$$L \approx M V R = M (2 \pi R / P) R = 2 \pi M R^2 / P$$

(In actuality, it is less than this. For a uniform sphere $L = 0.8 \pi M R^2 / P$.)

For Jupiter

$$L_{spin} \approx 2 \pi (1.9 \times 10^{27} \text{ kg}) (6.9 \times 10^7 \text{ m})^2 / (3.5 \times 10^4 \text{ s})$$
$$\approx 1.6 \times 10^{39} \text{ kg m}^2 / \text{s}$$

and for the Sun, $P \approx 27$ days $\approx 2.3 \times 10^6$ s,

$$L_{spin} \approx 2 \pi (2.0 \times 10^{30} \text{ kg}) (7.0 \times 10^8 \text{ m})^2 / (2.3 \times 10^6 \text{ s})$$
$$\approx 2.7 \times 10^{42} \text{ kg m}^2 / \text{s}$$

The orbital angular momentum of Jupiter (for $r = 7.8 \times 10^{11}$ m; $P_{orbital}$ = 11.86 years = 3.75×10^8 s) is

$$L_{orbital} \approx M v r \approx 2 \pi M r^2 / P$$
$$\approx 2 \pi (1.9 \times 10^{27} \text{ kg}) (7.8 \times 10^{11} \text{ m})^2 / (3.75 \times 10^8 \text{ s})$$
$$\approx 1.9 \times 10^{43} \text{ kg m}^2 / \text{s}$$

Inspection indicates that the orbital angular momentum of Jupiter dominates the solar spin angular momentum; and that the spin angular momentum of Jupiter is insignificant.

2-17. (a) What is the semimajor axis of the least-energy elliptical orbit of a space probe from Earth to Venus? [See also Problem 2-23, which is a similar problem, but for travel to Mars.]

The least energy elliptical orbit will be one with aphelion coinciding with the Earth's orbit and perihelion coinciding with Venus' orbit. The semi-major axis is

$$a = (a_{Venus-Sun} + a_{Earth-Sun}) / 2 = (0.723 \text{ AU} - 1.000 \text{ AU}) / 2$$
$$= 0.862 \text{ AU} = 1.29 \times 10^{11} \text{ m}$$

and the orbital eccentricity (Section 1-2B) is given by

$$r_{aphelion} = a(1+e)$$
$$1.000 \text{ AU} = 0.862 \text{ AU}(1+e)$$
$$e = 0.16$$

(b) Relative to the Earth, what is the velocity of such a probe at the Earth's orbit?

The speed of the space probe at the Earth's orbit (aphelion) is given by Equation 1-28 where $m_{Sun} \gg m_{satellite}$ (Note, Equations 1-17 could be used if the orbital period was calculated from Kepler's Third Law):

$$v^2 = G(m_{Sun} + m_{satellite})[(2/r) - (1/a)]$$
$$= (6.67 \times 10^{-11} \text{ N m}^2/\text{kg}^2)(1.99 \times 10^{30} \text{ kg}) \times$$
$$[2/(1.50 \times 10^{11} \text{ m}) - 1/(1.29 \times 10^{11} \text{ m})]$$
$$= (1.33 \times 10^{20} \text{ m}^3/\text{s}^2)(5.58 \times 10^{-12} \text{ m}^{-1})$$
$$= 7.42 \times 10^8 \text{ m}^2/\text{s}^2$$
$$v = 2.72 \times 10^4 \text{ m/s} = 27.2 \text{ km/s}$$

The Earth's speed at 1 AU is

$$v^2 = G(m_{Sun} + m_{Earth})[(2/r) - (1/a)]$$
$$= (1.33 \times 10^{20} \text{ m}^3/\text{s}^2) \times [2/(1.50 \times 10^{11} \text{ m}) - 1/(1.50 \times 10^{11} \text{ m})]$$
$$= 8.87 \times 10^8 \text{ m}^2/\text{s}^2$$
$$v = 2.98 \times 10^4 \text{ m/s} = 29.8 \text{ km/s}$$

Thus, the satellite has a speed of 2.6 km/s (slower) relative to the Earth at the Earth's orbit (1 AU).

(c) When the probe reaches Venus, what is the velocity relative to that planet?

The speed of the space probe at the Venus' orbit (perihelion = 0.723 AU = 1.08×10^{11} m) is (using Equation 1-28):

$$v^2 = (1.33 \times 10^{20} \text{ m}^3/\text{s}^2) \times [2/(1.08 \times 10^{11} \text{ m}) - 1/(1.29 \times 10^{11} \text{ m})]$$
$$= (1.33 \times 10^{20} \text{ m}^3/\text{s}^2)(1.08 \times 10^{-11} \text{ m}^{-1})$$
$$= 1.44 \times 10^9 \text{ m}^2/\text{s}^2$$
$$v = 3.79 \times 10^4 \text{ m/s} = 37.9 \text{ km/s}$$

Venus' speed at 0.723 AU is

$$v^2 = (1.33 \times 10^{20} \text{ m}^3/\text{s}^2) \times [2/(1.08 \times 10^{11} \text{ m}) - 1/(1.08 \times 10^{11} \text{ m})]$$
$$= 1.23 \times 10^9 \text{ m}^2/\text{s}^2$$
$$v = 3.51 \times 10^4 \text{ m/s} = 35.1 \text{ km/s}$$

Thus, the satellite has a speed of 2.8 km/s (faster) relative to Venus at Venus' orbit (0.723 AU).

2-18. Suppose that a projectile has a burnout speed $1.44v$ (where $v_c/1.44 < v < v_c$) at a distance r from the Earth's center ($r > R_\oplus$). If the velocity vector points 45° above the local parallel to the Earth's surface, find the semimajor axis a, the sidereal period P, and the eccentricity e of the resulting elliptical orbit in terms of r, v, and constants. Can you also find the angle from burnout to the orbit's perigee? [*Hint:* Section 2-3C is very helpful.]

It is worthwhile to review Sections 1-5C and D and Sections 2-3B and C.

The burnout speed 1.44 v (= $\sqrt{2}$ v) is greater than the circular orbital speed, v_c, and less than the escape speed, 1.44 v_c,

$$v_c < 1.44\, v < 1.44\, v_c$$

so the projectile will go into an elliptical orbit about the Earth.

The semi-major axis, a, is given by (Equation 1-28):

$$v_{orbital}^2 = G\,(M_\oplus + m_{projectile})\,[(2/r) - (1/a)]]$$

$$(1.44\,v)^2 = G\,M_\oplus\,[(2/r) - (1/a)]$$

$$(2/r) - (1/a) = 2\,v^2 / (G\,M_\oplus)$$

$$1/a = (2/r) - [2\,v^2 / (G\,M_\oplus)] = (2\,G\,M_\oplus - 2\,v^2\,r) / (G\,M_\oplus\,r)$$

$$a = G\,M_\oplus\,r / (2\,G\,M_\oplus - 2\,v^2\,r) = r / [2 - 2\,v^2\,r / (G\,M_\oplus)]$$

Defining the dimensionless quantity

$$B \equiv G\,M_\oplus / (v^2\,r)$$

we get for the semi-major axis

$$a = r / (2 - 2\,B^{-1})$$

The orbital period is given by Kepler's Third Law (Equation 2-10):

$$P^2 = 4\,\pi^2\,a^3 / [G\,(M_\oplus + m_{projectile})]$$

$$= 4\,\pi^2\,\{r / [2 - 2\,v^2\,r / (G\,M_\oplus)]\}^3 / (G\,M_\oplus)$$

$$P = [2\,\pi / (G\,M_\oplus)^{1/2}]\,\{r / [2 - 2\,v^2\,r / (G\,M_\oplus)]\}^{3/2}$$

$$= [2\,\pi / (G\,M_\oplus)^{1/2}]\,[r / (2 - 2\,B^{-1})]^{3/2}$$

To determine the eccentricity is a little more complicated.

Since the burnout velocity vector points 45° above the local parallel to the Earth's surface, the radial speed is

$$v_r = v_{orbital}\,\sin(\phi) = 1.44\,v\,\sin(45°) = v$$

and the angular speed is

$$v_\theta = v_{orbital}\,\cos(\phi) = 1.44\,v\,\cos(45°) = v\,.$$

Using Equations 1-26a and 1-26b:

$$v_r = (2\,\pi\,a / P)\,(e\,\sin(\theta))\,(1 - e^2)^{-1/2}$$

$$v_\theta = (2\,\pi\,a / P)\,(1 + e\,\cos(\theta))\,(1 - e^2)^{-1/2}\,.$$

Evaluating the quantity (using the expression for P given in the previous step)

$$2\pi a / P = 2\pi [r/(2 - 2B^{-1})] / \{[2\pi/(GM_\oplus)^{1/2}][r/(2 - 2B^{-1})]^{3/2}\}$$
$$= (GM_\oplus)^{1/2} [r/(2 - 2B^{-1})]^{-1/2}$$
$$= (GM_\oplus)^{1/2} \{[2 - 2v^2 r/(GM_\oplus)]/r\}^{1/2}$$
$$= [(2GM_\oplus/r) - (2v^2)]^{1/2}$$
$$= v [2GM_\oplus/(v^2 r) - 2]^{1/2}$$
$$= v (2B - 2)^{1/2}$$

Thus, substituting into the equation for the radial speed
$$v = v (2B - 2)^{1/2} (e \sin(\theta)) (1 - e^2)^{-1/2}$$
$$e \sin(\theta) = (1 - e^2)^{1/2} (2B - 2)^{-1/2}$$
and the equation for the angular speed
$$v = v (2B - 2)^{1/2} (1 + e \cos(\theta)) (1 - e^2)^{-1/2}$$
$$1 + e \cos(\theta) = (1 - e^2)^{1/2} (2B - 2)^{-1/2}$$

Now for ease of computation, make the simplifying definitions:
$$X \equiv (2B - 2)$$
$$Y \equiv (1 - e^2)$$
so $e^2 = 1 - Y$.
The equations for the radial and angular speeds become
$$\sin(\theta) = Y^{1/2} X^{-1/2} / e = Y^{1/2} / (e X^{1/2})$$
$$\cos(\theta) = (Y^{1/2} X^{-1/2} - 1)/e = (Y^{1/2} - X^{1/2})/(e X^{1/2}).$$
Since, for any angle θ,
$$\sin^2(\theta) + \cos^2(\theta) = 1$$
$$Y/(e^2 X) + (Y - 2 Y^{1/2} X^{1/2} + X)/(e^2 X) = 1$$
$$2Y - 2 Y^{1/2} X^{1/2} + X = e^2 X = (1 - Y) X = X - XY$$
$$2Y - 2 Y^{1/2} X^{1/2} + XY = 0$$
Since the orbit is an ellipse, $Y \neq 0$ ($e \neq 1$) so we can divide through by Y
$$2 - 2 X^{1/2} Y^{-1/2} + X = 0$$
$$Y^{-1/2} = (2 + X)/(-2 X^{1/2})$$
$$Y = 4X/(2 + X)^2$$
$$1 - e^2 = 4X/(2 + X)^2$$
$$e^2 = [(2 + X)^2 - 4X]/(2 + X)^2 = (4 + 4X + X^2 - 4X)/(2 + X)^2$$
$$= (4 + X^2)/(2 + X)^2$$
$$e = (4 + X^2)^{1/2}/(2 + X)$$
$$= [4 + (2B - 2)^2]^{1/2}/[2 + (2B - 2)]$$

$$= (4 + 4 B^2 - 8 B + 4)^{1/2} / (2 B)$$
$$= 2 (2 - 2 B + 1 B^2)^{1/2} / (2 B)$$
$$= (1 - 2 B^{-1} + 2 B^{-2})^{1/2}$$
$$= \{1 - 2 [G M_\oplus / (v^2 r)]^{-1} + 2 [G M_\oplus / (v^2 r)]^{-2}\}^{1/2}$$

The angle from perigee to the orbit's burnout, θ, is given by (Equation 1-26a):
$$v = (2 \pi a / P) (e \sin(\theta)) (1 - e^2)^{-1/2}$$
noting that in Figure 1-15, θ should be shown as the angle from perigee (rather than apogee) to the position being considered on the ellipse -- see Figure 1-7.

$$\sin(\theta) = Y^{1/2} / (e X^{1/2})$$
$$\sin^2(\theta) = Y / (e^2 X) = (1 - e^2) / [e^2 (2 B - 2)]$$
$$= [1 - (1 - 2 B^{-1} + 2 B^{-2})] / [(1 - 2 B^{-1} + 2 B^{-2}) (2 B - 2)]$$
$$= [2 B^{-1} - 2 B^{-2}] / [(1 - 2 B^{-1} + 2 B^{-2}) (2 B - 2)]$$
$$= B^{-2} [2 B - 2] / [(1 - 2 B^{-1} + 2 B^{-2}) (2 B - 2)]$$
$$= B^{-2} / (1 - 2 B^{-1} + 2 B^{-2}) = 1 / [B^2 (1 - 2 B^{-1} + 2 B^{-2})]$$
$$= 1 / (B^2 - 2 B + 2)$$
$$\sin(\theta) = 1 / (B^2 - 2 B + 2)^{1/2}$$
$$= 1 / \{[G M_\oplus / (v^2 r)]^2 - 2 [G M_\oplus / (v^2 r)] + 2\}^{1/2}$$

Whew!

2-19. Compare the escape speed of a rocket launched from the Earth with the escape speed of one at a distance of 1 AU from the Sun (that is, the escape speed from the Solar System at the Earth's distance from the Sun).

The escape speed of a rocket from the surface of the Earth is given (see Section 2-3C) by:
$$v_{esc} = \sqrt{2} \, v_o = 1.44 \, v_o$$
where the circular speed at the Earth's surface is
$$v_o = (G M_\oplus / R_\oplus)^{1/2}$$
$$= [(6.67 \times 10^{-11} \text{ N m}^2 / \text{kg}^2) (6.0 \times 10^{24} \text{ kg}) / (6.38 \times 10^6 \text{ m})]^{1/2}$$
$$= 7.9 \times 10^3 \text{ m/s} = 7.9 \text{ km/s}$$
Thus, the escape speed from the Earth's surface is
$$v_{esc} = \sqrt{2} \, (7.9 \text{ km/s}) = 11 \text{ km/s}$$

The escape speed of the rocket from the Sun at the Earth's orbit is $\sqrt{2}$ times the circular orbital speed at a radius of 1 AU from the Sun -
$$v_o = (G M_\odot / r_{\odot\text{-}\oplus})^{1/2}$$
$$= [(6.67 \times 10^{-11} \text{ N m}^2 / \text{kg}^2) (2.0 \times 10^{30} \text{ kg}) (1.5 \times 10^{11} \text{ m})]^{1/2}$$

$$= 2.98 \times 10^4 \text{ m/s} = 29.8 \text{ km/s}$$

Note that this is the average orbital speed of the Earth.
Thus, the escape speed from the Sun at a distance of 1 AU is

$$v_{esc} = \sqrt{2} \, (29.8 \text{ km/s}) = 42 \text{ km/s}$$

The escape speed from the Sun for a spacecraft on the Earth's surface (1 AU from the Sun) is ≈ 4 times greater than the escape speed from the Earth's surface. However, since v_{earth} = 29.8 km/s, only an additional ≈ 12 km/s speed must be provided to the rocket to escape the Sun. Thus, in that sense, the escape speed from the Earth and from the Solar System at 1 AU from the Sun are nearly equal.

2-20. The inner (terrestrial) planets have relatively small amounts of hydrogen in their atmospheres, yet the outer (Jovian) planets are predominantly hydrogen. Taking the Earth and Jupiter as typical examples of each class, calculate the ratio of the root mean square speed to the escape velocity of hydrogen for both the Earth and Jupiter. How do these values affect the relative observed amounts of hydrogen in the atmosphere of each of these planets?

The root mean square speed for a gas of mass m at temperature T is (Equation 2-7):

$$v_{rms} = (3 \, k \, T / m)^{1/2}$$

and the escape speed from a planet of mass M and radius R is (Equation 2-8):

$$v_{esc} = (2 \, G \, M / R)^{1/2}$$

From Appendix Table A3-3, the mass, radius, and atmospheric temperature of Earth and Jupiter are:

$$M_\oplus = 5.98 \times 10^{24} \text{ kg}; \quad R_\oplus = 6.5 \times 10^6 \text{ m}; \quad T_\oplus \approx 300 \text{ K}$$

$$M_J = 1.9 \times 10^{27} \text{ kg}; \quad R_J = 6.87 \times 10^7 \text{ m}; \quad T_J \approx 150 \text{ K}$$

where the temperatures are the maximum observed temperatures, which will determine the maximum escape rate.

The mass of the hydrogen atom is (Appendix Table A7-2): $m_H = 1.67 \times 10^{-27}$ kg.

For hydrogen in Earth's atmosphere:

$$v_{rms} = [3 \, (1.38 \times 10^{-23} \text{ J/K}) \, (300 \text{ K}) / (1.67 \times 10^{-27} \text{ kg})]^{1/2}$$

$$= 2,730 \text{ m/s} = 2.73 \text{ km/s}$$

$$v_{esc} = [2 \, (6.67 \times 10^{-11} \text{ N m}^2 \text{ kg}^{-2}) \, (5.98 \times 10^{24} \text{ kg}) / (6.4 \times 10^6 \text{ m})]^{1/2}$$

$$= 11,200 \text{ m/s} = 11.2 \text{ km/s}$$

so the ratio of escape speed to mean hydrogen speed is

$$v_{esc} / v_{rms} = (11.2 \text{ km/s}) / (2.73 \text{ km/s}) = 4.1 \ll 10$$

For hydrogen in Jupiter's atmosphere:

$$v_{rms} = [3 \, (1.38 \times 10^{-23} \text{ J/K}) \, (150 \text{ K}) / (1.67 \times 10^{-27} \text{ kg})]^{1/2}$$

$$= 1,930 \text{ m/s} = 1.93 \text{ km/s}$$

$$v_{esc} = [2 \, (6.67 \times 10^{-11} \text{ N m}^2 \text{ kg}^{-2}) \, (1.9 \times 10^{27} \text{ kg}) / (6.87 \times 10^7 \text{ m})]^{1/2}$$

$$= 60{,}700 \text{ m/s} = 60.7 \text{ km/s}$$

so the ratio of escape speed to mean hydrogen speed is

$$v_{esc}/v_{rms} = (60.7 \text{ km/s})/(1.93 \text{ km/s}) = 31.5 \gg 10$$

In order for a planet to retain an atmosphere, the escape speed should be at least 10 times the root mean square speed for the gas. This is not true for hydrogen in the Earth's atmosphere, so Earth should have little of its primordial hydrogen; Jupiter's atmosphere should contain most of its primordial hydrogen.

2-21. From information given in the chapter, what is the approximate lifetime of Halley's Comet, assuming a constant mass loss rate? What does this lifetime tell us about the existence of the Oort Comet Cloud?

For each perihelion passage, Comet Halley loses 0.001 of its mass (0.1%) (Section 2-3D). Assuming a constant mass loss rate, Comet Halley will survive roughly 1000 perihelion passages, over a period of 76,000 years. This time is much less than the age of the Solar System (5 billion years); hence, there must be a source of new comets — the Oort Comet Cloud. Correcting the lifetime of a comet for the fact that the mass loss rate is not really constant will add a few more perihelion passages. The effect is, however, not large enough to increase the lifetime from 76,000 years to 5 billion years.

2-22. Comet Halley has an orbital period of 76 years and an orbital eccentricity of 0.967.
(a) What is the comet's perihelion distance? Aphelion distance?

Using Kepler's Third Law (Equations 1-24 or 2-10), we can determine the semi-major axis of the orbit of Comet Halley:

$$P^2 = a^3$$

where the orbital period P is in years and semi-major axis in AU.

$$a = P^{2/3} = (76)^{2/3} = 17.9 \text{ AU}$$

The perihelion and aphelion distances are given by (Section 1-2b)

$$r_{perihelion} = a(1-e) = 17.9 \text{ AU}(1 - 0.967) = 0.59 \text{ AU}$$
$$r_{aphelion} = a(1+e) = 17.9 \text{ AU}(1 + 0.967) = 35.2 \text{ AU}$$

Inspection of Appendix Table 3-1 indicates that Comet Halley's perihelion is inside the orbit of Venus (a = 0.72 AU), and the aphelion is outside the orbit of Neptune (a = 20.1 AU).

(b) What is the subsolar temperature on Comet Halley at perihelion? At aphelion?

The subsolar temperature is given by (Equation 2-4):

$$T_{ss} = (R_\odot/r_p)^{1/2} T_\odot \approx 394 \, (r_p)^{-1/2}$$

where r_p is the distance from the Sun, in AU.

For Comet Halley,

at perihelion: $T_{ss} \approx 394 \, (0.59)^{-1/2} \approx 510 \text{ K}$

at aphelion: $T_{ss} \approx 394 \, (35.2)^{-1/2} \approx 66 \text{ K}$

(c) The albedo of Comet Halley is 3%. What is the equilibrium blackbody temperature at perihelion? At aphelion?

When the albedo (and the rotation of the comet nucleus) is taken into account (Equation 2-5a), the equilibrium blackbody temperature is:

$$T_{eq} = (1 - A)^{1/4} (R_\odot / 2 r_p)^{1/2} T_\odot \approx 279 (1 - A)^{1/4} (r_p)^{-1/2}$$

For Comet Halley:

at perihelion: $T_{eq} \approx 279 (1 - 0.03)^{1/4} (0.59)^{-1/2} \approx 360$ K

at aphelion: $T_{eq} \approx 279 (1 - 0.03)^{1/4} (35.2)^{-1/2} \approx 47$ K

Most of the difference between the subsolar and equilibrium blackbody temperature is due to the rotation of the comet. The albedo is so small (3%), that most of the incoming energy is absorbed by the nucleus, heating the comet.

2-23. What would be the correct launch speed for a least energy orbit to Mars? How long would it take a probe to reach Mars from the Earth?

[See also Problem 2-17, which is a similar problem, but for travel to Venus.]

Following Sections 2-3D and 1-5C, the least energy elliptical orbit will be one with perihelion coinciding with the Earth's orbit (1 AU = 1.50 x 10^{11} m) and aphelion coinciding with Mars' orbit (1.524 AU = 2.28 x 10^{11} m; Appendix Table A3-1). The semi-major axis is

$$a = (a_{\text{Earth-Sun}} + a_{\text{Mars-Sun}}) / 2 = (1.000 \text{ AU} + 1.524 \text{ AU}) / 2$$
$$= 1.262 \text{ AU} = 1.89 \times 10^{11} \text{ m}$$

The speed of the space probe at the Earth's orbit (perihelion) is given by Equation 1-28 (the vis viva equation) where $m_{Sun} \gg m_{satellite}$ (Note, Equations 1-17 could be used if the orbital period was calculated from Kepler's Third Law):

$$v^2 = G (m_{Sun} + m_{satellite}) [(2 / r) - (1 / a)]$$
$$= (6.67 \times 10^{-11} \text{ N m}^2 / \text{kg}^2) (1.99 \times 10^{30} \text{ kg}) \times$$
$$[2 / (1.50 \times 10^{11} \text{ m}) - 1 / (1.89 \times 10^{11} \text{ m})]$$
$$= (1.33 \times 10^{20} \text{ m}^3 / \text{s}^2) (8.04 \times 10^{-12} \text{ m}^{-1})$$
$$= 1.07 \times 10^9 \text{ m}^2 / \text{s}^2$$
$$v = 3.27 \times 10^4 \text{ m} / \text{s} = 32.7 \text{ km} / \text{s}$$

The Earth's speed at 1 AU is

$$v^2 = G (m_{Sun} + m_{Earth}) [(2 / r) - (1 / a)]$$
$$= (1.33 \times 10^{20} \text{ m}^3 / \text{s}^2) \times [2 / (1.50 \times 10^{11} \text{ m}) - 1 / (1.50 \times 10^{11} \text{ m})]$$
$$= 8.87 \times 10^8 \text{ m}^2 / \text{s}^2$$
$$v = 2.98 \times 10^4 \text{ m} / \text{s} = 29.8 \text{ km} / \text{s}$$

Thus, the probe has a speed of 2.9 km / s (faster) relative to the Earth at the Earth's orbit (1 AU), or a speed of 32.7 km / s relative to the Sun.

The orbital period is given by Kepler's Third Law (Equation 2-10), which simplifies to:
$$P^2 = a^3$$
where the orbital period, P, is in years, semimajor axis, a, is in astronomical units.
$$P = a^{3/2}$$
$$= (1.262)^{3/2}$$
$$= 1.418 \text{ yr} = 518 \text{ days}$$
To reach Mars takes half this time,
$$t_{travel} = 0.709 \text{ yr} = 259 \text{ days}$$

Chapter 3: The Dynamics of the Earth

3-1. (a) How much does a sidereal clock gain (or lose) on a mean solar clock in 5 mean solar hours?
The sidereal clock runs faster than the solar clock. After 24 hours of mean solar time a sidereal clock would read $24^h\ 03^m\ 56^s$. After 5 hours of mean solar time the sidereal clock would gain: $(5/24) \times (3^m\ 56^s) = (5/24) \times (236^s) = 49^s$.

(b) What is the approximate sidereal time when it is noon apparent solar time on the following days: (i) the first day of spring, (ii) the first day of summer, (iii) April 21, (iv) January 2?

(i) The Sun is at a right ascension of 0^h at the vernal equinox, so at apparent solar noon the local sidereal time is 0^h.

(ii) The Sun is at a right ascension of 6^h at the summer solstice, so at apparent solar noon the local sidereal time is 6^h.

(iii) The Sun is at a right ascension of 2^h one month past the vernal equinox, so at apparent solar noon the local sidereal time is 2^h.

(iv) The Sun is about one third of a month (or 40^m in right ascension) past the winter solstice (right ascension 18^h), so at apparent solar noon the local sidereal time is $18^h 40^m$.

3-2. In terms of azimuth and altitude (Appendix 10), describe the Sun's daily path across the sky during every season of the year at the three locations. Use such descriptive terms as noon altitude, sunrise azimuth, sunset azimuth, and angle at which Sun meets horizon.
Note: A celestial globe can help in understanding these motions.
(a) the equator
At the equator, the Sun rises and sets vertically to the horizon. On the first day of spring and fall it rises at 90° azimuth, transits at the zenith, and sets at azimuth 270°. On the first day of summer it rises and sets 23.5° north of east and west, respectively, and transits 23.5° north of the zenith. On the first day of winter it rises and sets 23.5° south of east and west, respectively, and transits 23.5° south of the zenith.

(b) latitude 35° N
At latitude 35°N, the Sun rises and sets obliquely to the horizon. On the first day of spring and fall it rises at 90° azimuth, transits 35° south of the zenith, and sets at azimuth 270°. On the first day of summer it rises in the northeast (near azimuth 45°), sets in the northwest (near azimuth 315°), and transits 11.5° (= 35° - 23.5°) south of the zenith. On the first day of winter it rises in the southeast (near azimuth 135°), sets in the southwest (near azimuth 225°), and transits 57.5° (= 35° + 23.5°) south of the zenith.

(c) the north pole
At the north pole the sun's daily path is nearly parallel to the horizon during all seasons. It appears near the horizon on the first day of spring, and climbs higher in the sky until

the first day of summer when it reaches its maximum altitude of 23.5° above the horizon. It then slowly sinks, dropping below the horizon on the first day of autumn, and is not seen again until the first day of spring. (On the first day of winter, the Sun is 23.5° below the horizon.)

3-3. Cape Canaveral is at longitude 80°23' W and latitude 28°30' N. A rocket is launched from there due south, and it lands on the equator 10 min later. What is the longitude of impact?

The Coriolis displacement is equal to the Earth's rotational speed at the equator minus the rotational speed at the launch latitude, multiplied by the time of flight of the projectile.

$$d = [(v_{equator}) - (v_{equator} \times \cos(latitude))] \times \text{(time of flight)}$$
$$= [(0.46 \text{ km/s}) - (0.46 \text{ km/s} \times \cos(28.5°))] \times (10 \text{ min}) \times (60 \text{ s/min})$$
$$= (0.056 \text{ km/s}) \times (600 \text{ s})$$
$$= 33.4 \text{ km WEST}$$

The displacement in longitude is
$$[(33.4 \text{ km}) / (2\pi \times 6378 \text{ km})] \times 360° = 0.30° = 18.0' \text{ WEST}$$

The rocket impact is at
$$80°23' + 0°18' = 80°41'W$$

3-4. In discussing the Coriolis effect, we mentioned that a body, which undergoes a constant acceleration a, will travel a distance $s = at^2/2$ in a time t. Show that the speed of the body is proportional to time and that the body's acceleration is indeed a.

The velocity is the first derivative of the distance with respect to time:
$$v = ds/dt = d[(1/2)at^2]dt = (1/2)a \times d(t^2)/dt = (1/2)a \times (2t)$$
$$= at$$

The acceleration is the first derivative of the velocity with respect to time:
$$a = dv/dt = d(at)/dt = a \times d(t)/dt = a$$

3-5. The Earth's rotation also produces an aberration of starlight.
(a) What is the maximum value of this daily aberration?
For small angles (Equation 3-2):
$$\theta \approx \tan\theta = v/c.$$
where v is the rotational speed of the Earth, which is a maximum at the equator.
$$v = 2\pi R_\oplus / 24 \text{ hours}$$
$$= (2\pi \times 6378 \text{ km}) / (86,400 \text{ s}) = 0.464 \text{ km/s}$$

The maximum value of this daily aberration is
$$\theta = (0.464 \text{ km/s}) / (300,000 \text{ km/s}) = 1.55 \times 10^{-6} \text{ radians} = 0.32''$$

The total daily aberration of starlight is twice this value, or 0.64".
[Compare this with the maximum annual aberration of $2 \times 20.49'' \approx 41''$ for a star at the ecliptic pole. See Problem 3-15a.]

(b) Where on the Earth is this effect maximum?
This effect is a maximum at the equator, where the speed of rotation is the greatest.

(c) For what stars (location on the celestial sphere) is the effect maximum?

Similar to the annual aberration orbits due to the Earth's revolution about the Sun, the rotation of the Earth causes stars at the celestial poles to move in a circle of angular radius 0.32", stars on the celestial equator to move in a line of half-length 0.32", and stars at intermediate declinations (δ) to move in an ellipse of semi-major axis 0.32" and semi-minor axis (0.32") x sin (δ).

3-6. (a) If the centripetal acceleration from the Earth's rotation is $\omega^2 R_\oplus$ at the equator and if $\omega = 2\pi/P$, where P is one day, then by what percentage is a person's weight reduced as he or she walks from the north pole to the equator? Ignore the Earth's oblateness.

See also Problem 2-4 for a slightly different approach to this problem.

At the equator the centripetal acceleration is

$$a = \omega^2 R = (4\pi^2/P^2) R$$
$$= [4\pi^2 \times 6{,}378 \text{ km} \times 10^3 \text{ m/km}] / (86{,}400 \text{ s/day})^2$$
$$= 0.034 \text{ m/s}^2$$

The person's weight is reduced by the ratio of the centripetal acceleration to the gravitation's acceleration:

$$a_{centrip}/g = (0.034 \text{ m/s}^2)/(9.8 \text{ m/s}^2) = 0.0035 = 0.35\%$$

For a person at the pole, the centripetal acceleration is zero.

A person at the equator weighs 0.35% less than at the pole due to centripetal acceleration.

(b) Now ignore the Earth's rotation and find by what percentage a person's weight increases as he or she walks from the equator to pole on the oblate Earth.

The gravitational acceleration $g \propto 1/R^2$.

The equatorial radius is greater than the polar radius (see Section 3-2C):

$$R_{equator} = (1 + 1/298.3) R_{pole} = 1.00335 R_{pole}$$
$$(R_{equator})^2 = 1.0067 (R_{pole})^2$$

The person's weight is increased by 0.67% as he or she walks from the equator to pole on the oblate Earth.

(c) Compare your results from (a) and (b) and combine them to deduce how the *effective g* varies from the Earth's poles to the equator.

The difference in the Earth's radius has a greater effect than the centripetal acceleration on a person's weight -- both effects decrease the weight of a person at the equator relative to at the poles. A person at the equator weighs 0.35% + 0.67% = 1.02% less than he or she would at the poles.

So, a person who weighs 50 kg on the north pole would weigh 49.5 kg at the equator. It's more effective to diet for a few weeks if one wants to lose weight!

3-7. The Earth's orbital speed is approximately 30 km/s. A star emits a spectral line at wavelength λ_e = 517.3 nm (1 nm = 10^{-9} m). Over what amplitude does this wavelength oscillate as the Earth orbits the

Sun when the star is located at the ecliptic (celestial latitude $\beta = 0°$)?

The oscillation in wavelength is due to the Doppler effect, and is greatest for a star on the ecliptic (Equation 3-4):

$$\Delta\lambda / \lambda = v / c$$

$$\begin{aligned}\Delta\lambda &= \lambda \times (v/c) \\ &= (517.3 \text{ nm}) \times [(30 \text{ km/s}) / (300,000 \text{ km/s})] \\ &= 0.0517 \text{ nm}\end{aligned}$$

Since the Earth approaches and recedes from this star during its orbit, the amplitude of the star's wavelength change is

$$2\Delta\lambda = 0.1034 \text{ nm}$$

3-8. In Section 3-4A, we computed the Moon's differential tidal forces at the Earth's surface, ignoring the motion of the Earth-Moon system about its center of mass every month. Choose a coordinate system centered upon this center of mass and rotating eastward with the angular speed ω (due to this sidereal monthly motion) and include the centripetal acceleration at the Earth's surface to deduce the correct dependence of the tidal acceleration at the Earth's surface. Ignore the daily rotation of the Earth. (*Hint:* The Earth-Moon center of mass is located within the Earth.)

Ignoring the latitude dependence, the centripetal acceleration is

$$a_{centripetal} = \omega^2 R$$

where R is the distance of the center of the Earth from the center of mass:

$$R = (M_m / M_\oplus) d$$

The distance from this center to the Earth's surface is: $R_\oplus - R \cos(l)$,

where l is the longitude from the sublunar point.

The combined acceleration is

$$a \approx 2 G M_m R_\oplus \cos\phi / d^3 + \omega^2 [R_\oplus - (M_m / M_\oplus) d \cos(l)]$$

where $\omega = 4\pi^2 / P^2$

with P being the sidereal month.

3-9. On a large piece of graph paper, plot *to scale* the distances of the Roche limit and orbits of the innermost planetary satellites and rings for (i) the Earth, (ii) Mars, (iii) Jupiter, (iv) Saturn, and (v) Uranus. Assume that $\rho_M = \rho_m$ in every case. Write a brief statement summarizing your results.

The Roche limit is given by (Equation 3-9):

$$d \approx 2.44 (\rho_M / \rho_m)^{1/3} R$$

where d is the distance between the center of the bodies, and R is the radius of the primary. If we assume that $\rho_M = \rho_m$, then the Roche limit is given by:

$$d/R \approx 2.44$$

Using the Tables in Appendix 3, and information in Chapter 7:
- (i) Earth - Moon: $d/R = 384,000 / 6,378 \approx 60$
- (ii) Mars - Phobos: $d/R = 9,000 / 3,394 \approx 2.7$
- (iii) Jupiter - ring: $d/R = 128,500 / 68,700 \approx 1.9$

                Jupiter - J3 (small satellite): $d/R \approx 1.8$

                Jupiter - Amalthea: $d/R \approx 2.5$

(iv)   Saturn - A ring: $d/R = 140{,}000 / 57{,}550 \approx 2.4$

        Saturn - Atlas (small satellite): $d/R \approx 2.3$

        Saturn - Epimetheus (and Janus): $d/R \approx 2.5$

(v)   Uranus - ε ring: $d/R = 51{,}000 / 25{,}050 \approx 2.0$

        Uranus - Miranda: $d/R = 130{,}100 / 25{,}050 \approx 5.2$

Conclusion: All rings are inside the Roche limit, all major satellites are outside the Roche limit (a few small satellites with diameters < 40 km are within the Roche limit).

3-10. Compare the tidal forces the Moon exerts on the Earth (at perigee) with those the Sun exerts on the Earth (at perihelion) and those Venus exerts on the Earth (at closest approach).

The tidal force of a body (of mass M, distance d) on the Earth is given by (Equation 3-8):

$$F_{tidal} \propto M/d^3$$

For the Moon, $M_{Moon} \approx 0.012\, M_\oplus$,

    and at perigee, $d = 384{,}000\,(1 - 0.055)$ km $= 0.363 \times 10^6$ km.

For the Sun, $M_{Sun} \approx 3.33 \times 10^5\, M_\oplus = 333{,}000\, M_\oplus$,

    and at perihelion, $d = 149.6 \times 10^6\,(1 - 0.017)$ km $= 147 \times 10^6$ km.

For Venus $M_{Venus} \approx 0.815\, M_\oplus$,

    and at closest approach, $d \approx (149.6 - 108.2) \times 10^6$ km $\approx 41.4 \times 10^6$ km.

The ratio of the tidal forces of these bodies on the Earth is:

$F_{Moon}/F_{Sun}\ \ = (0.012/333{,}000) / (0.363 \times 10^6 / 147 \times 10^6)^3 \approx 2.4$

$F_{Moon}/F_{Venus} = (0.012/0.815) / (0.363 \times 10^6 / 41.4 \times 10^6)^3\ \ \approx 22{,}000$

$F_{Sun}/F_{Venus}\ \ = (333{,}000/0.815) / (147 \times 10^6 / 41.4 \times 10^6)^3 \approx 9{,}100$

The Moon exerts the greatest tidal force on the Earth, with the Sun only $1/2.4 = 42\%$ as strong; Venus is significantly weaker (only $\approx 1/100\ \%$ that of the Sun).

3-11. Assume that the Earth and Moon are spherical and that the Moon orbits the Earth in a circle. Calculate the *spin* angular momenta of the Earth and Moon and compare these with the *orbital* angular momentum of the Moon. A spherical mass of uniform density has a spin angular momentum of (2/5)MVR, where V is the equatorial velocity and R the radius. The sum of these momenta must be a constant for the Earth-Moon system (if we ignore external torques). From the rate of loss of the Earth's spin angular momentum from tidal friction, estimate the rate at which the Moon moves radially away from the Earth.

The spin angular momentum of a spherical body is given by:

$$L_{spin} \equiv 0.4\, M\, V\, R = 0.4\, M\,(2\pi R/P)\, R = 0.8\, \pi\, M\, R^2/P$$

The orbital angular momentum is given by:

$$L_{orbit} = M\, V\, D = M\,(2\pi D/P)\, D = 2\pi\, M\, D^2/P$$

See Appendix 3 for the masses, sizes, distances, and periods in the above equations. The Earth's distance from the center of mass of the Earth-Moon system is given by (Equation 1-21):

$$D_\oplus = (M_m / M_\oplus) D_m$$
$$= [(7.35 \times 10^{22} \text{ kg}) / (5.98 \times 10^{24} \text{ kg})] (3.84 \times 10^8 \text{ m})$$
$$= 4.72 \times 10^6 \text{ m}$$

We need to consider the Moon's and the Earth's rotational (spin) and orbital momentum.

$$L_{\oplus\text{-rot}} = 0.8 \, \pi \, (5.98 \times 10^{24} \text{ kg}) (6.378 \times 10^6 \text{ m})^2 / (8.64 \times 10^4 \text{ s})$$
$$= 7.08 \times 10^{33} \text{ kg m}^2 / \text{s}$$
$$L_{m\text{-rot}} = 0.8 \, \pi \, (7.35 \times 10^{22} \text{ kg}) (1.738 \times 10^6 \text{ m})^2 / (2.36 \times 10^6 \text{ s})$$
$$= 2.36 \times 10^{29} \text{ kg m}^2 / \text{s}$$
$$L_{m\text{-orbit}} = 2 \, \pi \, (7.35 \times 10^{22} \text{ kg}) (3.84 \times 10^8 \text{ m})^2 / (2.36 \times 10^6 \text{ s})$$
$$= 2.89 \times 10^{34} \text{ kg m}^2 / \text{s}$$
$$L_{\oplus\text{-orbit}} = 2 \, \pi \, (5.98 \times 10^{24} \text{ kg}) (4.72 \times 10^6 \text{ m})^2 / (2.36 \times 10^6 \text{ s})$$
$$= 3.55 \times 10^{32} \text{ kg m}^2 / \text{s}$$

The sum of the momenta is:

$$L_{total} = L_{\oplus\text{-rot}} + L_{m\text{-rot}} + L_{m\text{-orbit}} + L_{\oplus\text{-orbit}} = 3.63 \times 10^{34} \text{ kg m}^2 / \text{s}$$

which must remain a constant.

The Earth's rotation slows by (Section 3-4B):

$$\Delta P = -0.002 \text{ s / century} = -2.0 \times 10^{-5} \text{ s / yr} = -6.3 \times 10^{-13} \text{ s / s}$$

This causes a decrease in the Earth's spin angular momentum

$$dL_\oplus / dt = 0.8 \, \pi \, M R^2 \, [(1/P^2)(dP/dt)] = L_\oplus [(1/P)(dP/dt)]$$

To conserve angular momentum, the orbital angular momentum must increase by this amount. The change in the Moon's spin angular momentum is insignificant, as is the Earth's orbital angular momentum --

$$L_{m\text{-rot}} \ll L_{\oplus\text{-orbit}} < L_{\oplus\text{-rot}} < L_{m\text{-orbit}}$$

Now, from Kepler's Third Law (Equation 1-24)

$$P^2 \approx (4\pi^2 / G M_\oplus) D^3$$
$$P = [2\pi / (G M_\oplus)^{1/2}] D^{3/2}$$
$$1 / P = [(G M_\oplus)^{1/2} / (2\pi)] D^{-3/2}$$

we get

$$L_{orbit} = 2\pi M_m D^2 / P = (G M_\oplus)^{1/2} M_m D^{1/2}$$

or

$$dL_{orbit}/dt = (GM_\oplus)^{1/2} M_m [(1/2) D^{-1/2} (dD/dt)]$$
$$= 0.5 L_{orbit} (1/D)(dD/dt)$$

From the conservation of angular momentum
$$dL_{\oplus\text{-rot}}/dt + dL_{m\text{-orbit}}/dt = 0$$
or
$$-L_{\oplus\text{-rot}}[(1/P_{\oplus\text{-rot}})(dP_{\oplus\text{-rot}}/dt)] = 0.5 L_{m\text{-orbit}}(1/D_m)(dD_m/dt)$$
which can be rewritten as
$$(dD_m/dt) = -(dP_{\oplus\text{-rot}}/dt) \times (L_{\oplus\text{-rot}}/L_{m\text{-orbit}}) \times (2D_m/P_{\oplus\text{-rot}})$$
$$= -(-2.0 \times 10^{-5}\text{ s/yr})$$
$$\times [(7.08 \times 10^{33}\text{ kg m}^2/\text{s})/(2.89 \times 10^{34}\text{ kg m}^2/\text{s})]$$
$$\times [2 \times (3.84 \times 10^8\text{ m})/(8.64 \times 10^4\text{ s})]$$
$$= (2.0 \times 10^{-5}\text{ s/yr}) \times (2.45 \times 10^{-1})(8.89 \times 10^3\text{ m/s})$$
$$= 4.4 \times 10^{-2}\text{ m/yr} \approx 4\text{ cm/yr}$$

The Moon is thus moving away from the Earth at a rate of ≈ 4 cm / yr due to tidal forces which are slowing the Earth's rotation.

3-12. The Moon will move away from the Earth until it no longer lags the tidal bulges, and the angular momentum transfer will stop. Computer calculations indicate that this will occur at an Earth-Moon distance of 6.45×10^5 km. Calculate the Moon's orbital period then.

Using Kepler's Third Law (Equation 1-24) for the Earth-Moon system:
$$P^2 \propto a^3$$
so
$$(P_{new}/P_{current})^2 = (d_{new}/d_{current})^3$$
$$(P_{new}/27.3\text{ days})^2 = (6.45 \times 10^5\text{ km}/3.84 \times 10^5\text{ km})^3 = 4.74$$
$$P_{new} = (4.74)^{0.5} \times 27.3\text{ days} = 59.4\text{ days}$$

3-13. Compare the solar insolation at noon in Albuquerque, New Mexico (latitude about 35° N), for the day of the summer solstice and the day of the winter solstice.

If A is the solar energy falling on a unit area of the Earth's surface when the Sun is at the zenith, the energy falling on a unit area when the Sun is at an altitude of θ is given by: A × sin(θ). At Albuquerque, the celestial equator is at an altitude of 55° (90° - 35°) at the meridian. On the summer solstice, the Sun is at an altitude of 78.5° (= 55° + 23.5°) at noon; on the winter solstice, the Sun is at an altitude of 31.5° (= 55° - 23.5°). The ratio of the solar insolation on the first day of summer to that on the first day of winter is: sin(78.5°) / sin(31.5°) = 1.88. Each unit area of the Earth in Albuquerque receives 1.88 times more energy at noon on the summer solstice than on the winter solstice.

3-14. An astronomer from Cullowhee, North Carolina (latitude 35° N, longitude 83° W) wants to visit a colleague in Hamilton, New York (latitude 43° N, longitude 75.5° W) (in order to convince him to finish writing this textbook instructors' manual). He hops into a plane, but finds on taking off that the weather is overcast and that it is impossible to navigate on the basis of landmarks. The compass is also broken, so he is able to navigate only by flying above the clouds in the direction of the North Star. He relies on the Coriolis effect to allow for east-west motion while aiming his plane due north.

(a) Will the Coriolis effect operate in the correct direction?

The plane will land East of the position due North of the starting location due to the Coriolis force. The deflection is to the right in the northern hemisphere. This is in the correct direction, since Hamilton is East of Cullowhee.

(b) If so, at what average speed must he fly in order to be at the longitude of Hamilton when he has reached the proper latitude?

[This solution is a simplistic solution. A more rigorous solution is beyond the scope of how the text treats the Coriolis force.]

At a latitude of 43°, the target metropolis (Hamilton) moves with a speed given by:

$$v = 2\pi R_E \cos(\text{lat}) / \text{day} = 2\pi (6.38 \times 10^3 \text{ km}) \cos(43°) / (8.64 \times 10^4 \text{ s})$$
$$= 0.339 \text{ km/s}$$

while at latitude 35°, the origin (Cullowhee) moves with a speed given by:

$$v = 2\pi R_E \cos(\text{lat}) / \text{day} = 2\pi (6.38 \times 10^3 \text{ km}) \cos(35°) / (8.64 \times 10^4 \text{ s})$$
$$= 0.380 \text{ km/s}$$

The difference in the speeds is 0.041 km/s.

To land at the correct location, the Earth needs to rotate so that the plane goes from 83° longitude to 75.5° (or 7.5°). At latitude 43°, this corresponds to a distance of

$$d = 2\pi R_E \cos(\text{lat}) (7.5°/360°)$$
$$= 2\pi (6.38 \times 10^3 \text{ km}) \cos(43°) (0.0208)$$
$$= 611 \text{ km}$$

The time for Hamilton to move this relative distance is

$$t = d/v = (611 \text{ km}) / (0.041 \text{ km/s}) = 14{,}900 \text{ s} = 248 \text{ min} = 4.14 \text{ hr}$$

The pilot has 248 minutes to travel 8° of latitude North. 8° corresponds to a distance of approximately

$$d = 2\pi (6.38 \times 10^3 \text{ km})(8°/360°) = 891 \text{ km}$$

The pilot would have to fly at a speed of 891 km / (4.14 hr) = 215 km/hr to use the Coriolis force alone to meet his destination. This speed is attainable by a private plane. The astronomer will not have to phone his colleague (except to tell him he's arriving soon with a pack of Frontier rolls!).

3-15. (a) Because of the eccentricity of the Earth's orbit, the magnitude of the aberration of starlight due to the Earth's orbital motion is not constant. Determine the aberration at perihelion and aphelion.

The amplitude of the annual aberration varies when Earth's speed changes due to the orbital eccentricity. Earth's orbital speed can be calculated from Equations 1-17a and b:

$$v_{per} = (2\pi a/P)[(1+e)/(1-e)]^{1/2}$$
$$= (2\pi(1.496 \times 10^8 \text{ km})/(3.156 \times 10^7 \text{ s}))[(1+0.0167)/(1-0.0167)]^{1/2}$$

$$= 30.29 \text{ km/s}$$

$$v_{ap} = (2\pi a / P) [(1-e)/(1+e)]^{1/2}$$
$$= (2\pi(1.496 \times 10^8 \text{ km})/(3.156 \times 10^7 \text{ s})) [(1 - 0.0167)/(1 + 0.0167)]^{1/2}$$
$$= 29.29 \text{ km/s}$$

The aberration of starlight is given by Equation 3-2:

$$\theta \approx \tan\theta = v/c.$$

where, here, v is the orbital speed of the Earth.

The maximum value of this yearly aberration occurs at perihelion, when the orbital speed is greatest

$$\theta_{per} = (30.29 \text{ km/s}) / (3.0 \times 10^5 \text{ km/s}) = 1.010 \times 10^{-4} \text{ radians}$$
$$= 20.83''$$

and the minimum aberration occurs at aphelion

$$\theta_{ap} = (29.29 \text{ km/s}) / (3.0 \times 10^5 \text{ km/s}) = 9.763 \times 10^{-5} \text{ radians}$$
$$= 20.14''$$

The yearly range in aberration of starlight is the sum of these values, or 40.97". Recall, in problem 3-5 that the total daily aberration is only 0.64".

(b) Determine the aberration of starlight as seen from Mars at perihelion and aphelion.

Mars' orbital speed can be calculated from Equations 1-17a and b:

$$v_{per} = (2\pi (1.524 \text{ AU} \times 1.496 \times 10^8 \text{ km/AU}) /$$
$$(1.881 \text{ Earth yr} \times 3.156 \times 10^7 \text{ s/Earth yr})) [(1+0.093)/(1 - 0.093)]^{1/2}$$
$$= 26.49 \text{ km/s}$$

$$v_{ap} = (2\pi (1.524 \text{ AU} \times 1.496 \times 10^8 \text{ km/AU}) /$$
$$(1.881 \text{ Earth yr} \times 3.156 \times 10^7 \text{ s/Earth yr})) [(1 - 0.093)/(1+ 0.093)]^{1/2}$$
$$= 21.98 \text{ km/s}$$

The maximum value of the yearly aberration occurs at perihelion

$$\theta_{per} = (26.49 \text{ km/s}) / (3.0 \times 10^5 \text{ km/s}) = 8.830 \times 10^{-5} \text{ radians}$$
$$= 18.21''$$

and the minimum aberration occurs at aphelion

$$\theta_{ap} = (21.98 \text{ km/s}) / (3.0 \times 10^5 \text{ km/s}) = 7.327 \times 10^{-5} \text{ radians}$$
$$= 15.11''$$

The yearly range in aberration of starlight is 33.32".

3-16. In the early 1980s the planets were all located on the same side of the Sun, with a maximum angular separation of roughly 90° as seen from the Sun. This rough "alignment" was sufficient to make possible the Voyager spacecraft grand tour. Some people claimed that this planetary alignment would produce destructive earthquakes, triggered by the cumulative tidal effects of all the planets acting together. Very few scientists took this prediction seriously! To understand why, compute the maximum tidal effects on the Earth produced by Jupiter (the most massive planet) and Venus (the closest). Compare these tidal effects to those caused by the Moon each month.

Problem 3-10 also dealt with the tidal forces of different objects on the Earth, where it was found that the tidal force of the Moon was much greater than that of Venus. Here we will look at the tidal force of the most massive planet, Jupiter, on the Earth. The tidal force of a body (of mass M, distance d) is given by (Equation 3-8):

$F_{tidal} \propto M/d^3$.

For the Moon, $M_{Moon} \approx 0.012\ M_\oplus$,

and at perigee, d = 384,000 (1 - 0.055) km = 0.363×10^6 km.

For Jupiter, $M_{Jupiter} \approx 318\ M_\oplus$,

closest approach, d = (149.6 × 10^6 km / AU) (5.2 - 1 AU) = 629 × 10^6 km.

For Venus $M_{Venus} \approx 0.815\ M_\oplus$,

closest approach, d ≈ (149.6 - 108.2) × 10^6 km ≈ 41.4 × 10^6 km.

The ratio of the tidal forces of these bodies on the Earth is:

$F_{Moon}/F_{Jupiter}$ = (0.012 / 318) / (0.363 × 10^6 / 629 × 10^6)^3 ≈ 196,000

F_{Moon}/F_{Venus} = (0.012 / 0.815) / (0.363 × 10^6 / 41.4 × 10^6)^3 ≈ 22,000

The Moon exerts the greatest tidal force on the Earth, with the closest planet, Venus, being 10s of thousands times weaker; and most massive planet, Jupiter, being 100s of thousands of times weaker! It is clear that the effects of both Venus and Jupiter (and the other planets) are completely negligible (even if they align) compared to the effect of the Moon. Also evidence exists against the tidal effects due to a planetary alignment -- the alignment occurred and nothing happened!

3-17. In Problem 10, you compared the magnitude of tidal forces of the Sun and Moon on the Earth. Are there special alignments of the Sun, Moon, and Earth that would result in greater than average high tides? Smaller than average low tides? Explain.

Yes, there are alignments of the Sun, Moon, and Earth that result in both greater than average high tides and smaller than average low tides.

During New Moon and Full Moon (when the Earth, Moon and Sun are in a straight line) the Sun and Moon work together so that the total tidal force is greater. Hence the tides are more extreme: higher at high tide and lower at low tide.

During the First and Third Quarter phases (when the Sun and Moon are 90° apart as seen from the Earth) the tidal forces from the Moon and Sun work to partially cancel each other out. They don't exactly cancel because the Moon produces a greater tidal force than the Sun (see Problem 3-10). If they exactly canceled there would be no tides during these phases. They don't cancel, but the net tidal force is less than average. Hence the tides are less extreme. The high tides are not as high and the low tides are not as low.

There are also effects due to the non-circular orbits of the Moon about the Earth and the Earth about the Sun. The tides are greater when the Moon is at perigee (and smaller when at apogee); the tides are also greater when the Earth is at perihelion (and smaller when at aphelion). The greatest tides would occur when both the Earth is at perihelion (in early January) and the Moon is at perigee; the smallest tides would occur when both the Earth is at aphelion and the Moon is at apogee.

There are also minor effects due to the lunar orbit being inclined to the ecliptic. There are also latitude dependent effects due to the declination changes of the Sun and Moon - the tides would be greater if the Moon and/or Sun were close to the zenith compared to when they are lower in the sky at transit.

3-18. The precession cycle of the Earth's axis is about 26,000 years. Estimate the average torque applied to the Earth. (*Hint:* As in Problem 11, you need to calculate the Earth's spin angular momentum.)

As in Problem 3-11, the Earth's spin angular momentum of a spherical body is given by:

$$L_{spin} \equiv 0.4\,M\,V\,R = 0.4\,M\,(2\pi\,R/P)\,R = 0.8\,\pi\,M\,R^2/P$$
$$= 0.8\,\pi\,(5.98 \times 10^{24}\,\text{kg})\,(6.378 \times 10^6\,\text{m})^2 / (8.64 \times 10^4\,\text{s})$$
$$= 7.08 \times 10^{33}\,\text{kg m}^2/\text{s}$$

Defining the precession angular velocity

$$\Omega = 2\pi / (26{,}000\,\text{yr})\,(3.16 \times 10^7\,\text{s/yr})$$
$$= 7.65 \times 10^{-12}\,\text{s}^{-1}$$

As in Section 3-4C (see also Section 1-5A and Prelude Section P1-7), if we apply Newton's 2nd Law to rotational motion, the torque (rate of change of angular momentum) is

$$N = dL/dt = L\,\Omega$$
$$= (7.08 \times 10^{33}\,\text{kg m}^2/\text{s})\,(7.65 \times 10^{-12}\,\text{s}^{-1})$$
$$= 5.4 \times 10^{22}\,\text{kg m}^2/\text{s}^2$$

Chapter 4: The Earth-Moon System

4-1. (a) An Apollo spaceship heads for the Moon. At what distance between the Earth and the Moon will the ship experience no net gravitational acceleration?

The spaceship will experience no net acceleration when the gravitational force (Equation 1-15) on the ship by the Earth equals that on the ship by the Moon, but in opposite directions:

$$G M_\oplus M_s / d_{\oplus s}^2 = G M_m M_s / d_{ms}^2$$

where the subscript $\oplus$ refers to the Earth, m to the Moon, and s to the spaceship.

$$M_\oplus / d_{\oplus s}^2 = M_m / d_{ms}^2$$
$$(d_{ms} / d_{\oplus s})^2 = M_m / M_s = 0.0123$$
$$d_{ms} / d_{\oplus s} = 0.111$$

Now, $d_{ms} = d_{\oplus m} - d_{\oplus s}$, so

$$d_{\oplus m} - d_{\oplus s} = 0.111 \, d_{\oplus s}$$
$$d_{\oplus s} = d_{\oplus m} / 1.111 = 0.90 \, d_{\oplus m}$$
$$= 3.84 \times 10^5 \text{ km} / 1.111 = 3.46 \times 10^5 \text{ km}$$

(b) How long will it take the ship to circumnavigate the Moon in a circular orbit 50 km above the lunar surface?

The spaceship's orbital period is given by Kepler's Third Law (Equation 1-24):

$$P^2 = 4 \pi^2 a^3 / [G (M_m + M_s)]$$
$$= 4 \pi^2 (1.738 \times 10^6 \text{ m} + 5.0 \times 10^4 \text{ m})^3 /$$
$$[(6.67 \times 10^{-11} \text{ N m}^2 / \text{kg}^2) (7.3 \times 10^{22} \text{ kg})]$$
$$= 4.63 \times 10^7 \text{ s}^2$$
$$P = 6.81 \times 10^3 \text{ s} = 113 \text{ min} = 1 \text{ hr } 53 \text{ min}$$

4-2. Using the length of the sidereal month (27.322^d) and the periods of the Earth's revolution (365.26^d), the regression of the nodes (18.6 years), and the precession of the Moon's perigee (8.85 years), compute the lengths of

(a) the synodic month

Following the argument for sidereal and synodic periods for planetary motions as seen from the Earth (Section 1-1A),

$$1 / \text{synodic month} = (1 / \text{sidereal month}) - (1 / \text{year})$$
$$= (1 / 27.322 \text{ days}) - (1 / 365.26 \text{ days})$$
$$= 3.386 \times 10^{-2} \text{ days}^{-1}$$
synodic month = 29.531 days

(b) the nodical month

The nodical month relates to the regression of the nodes

$$1 / \text{nodical month} = (1 / \text{sidereal month}) + (1 / \text{regression of the nodes})$$

where the plus sign is due to the regression of the nodes being westward while the sidereal motion is eastward.

$$1 / \text{nodical month} = (1 / 27.322 \text{ days}) + 1 / [(18.6 \text{ yr}) \times (365.26 \text{ days} / \text{yr})]$$
$$= 3.675 \times 10^{-2} \text{ days}^{-1}$$
$$\text{nodical month} = 27.212 \text{ days}$$

(c) the anomalistic month

The anomalistic month relates to the precession of the Moon's perigee

$$1 / \text{anomalistic month} = (1 / \text{sidereal month}) - (1 / \text{precession of perigee})$$
$$= (1 / 27.322 \text{ days}) - 1 / [(8.85 \text{ yr}) \times (365.26 \text{ days} / \text{yr})]$$
$$= 3.6292 \times 10^{-2} \text{ days}^{-1}$$
$$\text{anomalistic month} = 27.555 \text{ days}$$

Each of these results agree with the values given in Section 4-2A.

4-3. Assume that the Moon's orbit is circular and lies in the ecliptic plane. Find the difference in the solar attraction on the Moon's orbit at opposition and inferior conjunction; compare this with the gravitational attraction from the Earth. Can you now understand why the lunar orbit is not a simple ellipse?

If we assume the Moon's orbit is circular, with radius 3.84×10^8 m and the Earth's orbit is circular with radius 1.496×10^{11} m, then gravitational force of the Sun on the Moon at opposition is (Equation 1-15):

$$F = G M_\odot M_m / r^2$$
$$= (6.67 \times 10^{-11} \text{ N m}^2 / \text{kg}^2) (1.99 \times 10^{30} \text{ kg}) (7.3 \times 10^{22} \text{ kg}) /$$
$$(1.496 \times 10^{11} \text{ m} + 3.84 \times 10^8 \text{ m})^2$$
$$= 4.31 \times 10^{20} \text{ N}$$

and at inferior conjunction

$$F = (6.67 \times 10^{-11} \text{ N m}^2 / \text{kg}^2) (1.99 \times 10^{30} \text{ kg}) (7.3 \times 10^{22} \text{ kg}) /$$
$$(1.496 \times 10^{11} \text{ m} - 3.84 \times 10^8 \text{ m})^2$$
$$= 4.35 \times 10^{20} \text{ N}$$

while the gravitational force of the Earth on the Moon is

$$F = (6.67 \times 10^{-11} \text{ N m}^2 / \text{kg}^2) (5.98 \times 10^{24} \text{ kg}) (7.3 \times 10^{22} \text{ kg}) /$$
$$(3.84 \times 10^8 \text{ m})^2$$
$$= 1.97 \times 10^{20} \text{ N}$$

The difference in the solar gravitational attraction on the Moon at inferior conjunction is 0.04×10^{20} N greater than at opposition, which corresponds to a difference of 2% of the Earth's gravitational attraction on the Moon. This significant variable solar attraction causes the Moon's orbit to be a non-perfect ellipse.

4-4. (a) If the Moon had a core radius of $R_m/10$ and the remainder of its interior had a uniform density of ρ = 3000 kg/m^3, what must be the uniform density of the core?

The Moon's average density is 3370 kg / m^3 (Section 4-3B). In this simplified lunar model the mass (= density times volume) of the core plus the mass of the remainder of the interior equals the mass of the entire Moon:

$$[(4/3) \pi (R_m / 10)^3] \rho_c + [(4/3) \pi (R_m)^3 - (4/3)\pi (R_m / 10)^3] \rho_i$$
$$= [(4/3) \pi (R_m)^3] <\rho>$$

$(R_m / 10)^3 \rho_c + [(R_m)^3 - (R_m / 10)^3] \rho_i = (R_m)^3 <\rho>$

$0.001 \rho_c + 0.999 \rho_i = <\rho>$

$0.001 \rho_c + (0.999) (3000 \text{ kg} / m^3) = 3370 \text{ kg} / m^3$

$0.001 \rho_c = 373 \text{ kg} / m^3$

$\rho_c = 3.73 \times 10^5 \text{ kg} / m^3$

This value is a bit high due to the simple model used in this calculation.

(b) Compare the Earth's tidal forces on the Moon at the perigee and apogee of the Moon's orbit; comment on your results.

The Moon is 363,263 km from the Earth at perigee and 405,547 km at apogee (Section 4-1). The differential tidal force on the Moon by the Earth is given by (Equation 3-8):

$$dF = - (2 G M_\oplus m_m / R^3) dR$$

where R is the Earth-Moon separation and dR is the diameter of the Moon (3476 km).
At perigee

$dF = - 2 (6.67 \times 10^{-11} \text{ N } m^2 / kg^2) (5.98 \times 10^{24} \text{ kg}) (7.3 \times 10^{22} \text{ kg}) \times$
$(3.48 \times 10^6 \text{ m}) / (3.63 \times 10^8 \text{ m})^3$
$= -4.2 \times 10^{18} \text{ N}$

and at apogee

$dF = - 2 (6.67 \times 10^{-11} \text{ N } m^2 / kg^2) (5.98 \times 10^{24} \text{ kg}) (7.3 \times 10^{22} \text{ kg}) \times$
$(3.48 \times 10^6 \text{ m}) / (4.06 \times 10^8 \text{ m})^3$
$= -3.0 \times 10^{-18} \text{ N}$

Thus the differential tidal force of the Earth on the Moon at perigee is ≈40% stronger than at apogee (thus stretching the Moon more). This variable differential tidal force results in a change in the global shape of the Moon during its orbit.

4-5. The mean kinetic energy per molecule in a gas at temperature T is $mv^2/2 = 3kT/2$, where $k = 1.38 \times 10^{-23}$ J/K. A particle with a vertical speed v at the Earth's surface will rise to a height $h = v^2/2g$ before it falls back to the Earth.

(a) Show that the characteristic height for a molecule of mass m at temperature T is $h = 3kT/2mg$.
The mean speed of a molecule at temperature T is (see Equation 2-7):

$v^2 = 3 k T / M.$

Thus the height that it will reach in the atmosphere is

$h = v^2 / (2 g)$
$= (3 k T / M) / (2 g)$
$= 3 k T / (2 g M)$

(b) At T = 250 K, compute the characteristic heights for nitrogen (N_2), oxygen (O_2), carbon dioxide (CO_2), and hydrogen (H_2). What does this tell you about the *compositional* structure of the Earth's atmosphere?

The masses of these four molecules are (see Appendix 5):

$N_2 = (2 \times 14) \times (1.67 \times 10^{-27} \text{ kg}) = 4.68 \times 10^{-26}$ kg

$O_2 = (2 \times 16) \times (1.67 \times 10^{-27} \text{ kg}) = 5.34 \times 10^{-26}$ kg

$CO_2 = (12 + 2 \times 16) \times (1.67 \times 10^{-27} \text{ kg}) = 7.35 \times 10^{-26}$ kg

$H_2 = (2 \times 1) \times (1.67 \times 10^{-27} \text{ kg}) = 3.34 \times 10^{-27}$ kg

The characteristic height is given by:

$h = 3 (1.38 \times 10^{-23} \text{ J/K}) (250 \text{ K}) / [2 (9.8 \text{ m/s}^2) M]$

$= (5.28 \times 10^{-22} \text{ kg m}) / M$

so

$h(N_2) = (5.28 \times 10^{-22} \text{ kg m}) / (4.68 \times 10^{-26} \text{ kg})$

$= 1.13 \times 10^4$ m $= 11.3$ km

$h(O_2) = (5.28 \times 10^{-22} \text{ kg m}) / (5.34 \times 10^{-26} \text{ kg})$

$= 9.9 \times 10^3$ m $= 9.9$ km

$h(CO_2) = (5.28 \times 10^{-22} \text{ kg m}) / (7.35 \times 10^{-26} \text{ kg})$

$= 7.2 \times 10^3$ m $= 7.2$ km

$h(H_2) = (5.28 \times 10^{-22} \text{ kg m}) / (3.34 \times 10^{-27} \text{ kg})$

$= 1.6 \times 10^5$ m $= 160$ km

Thus hydrogen molecules will be distributed from low to high altitudes in the Earth's atmosphere and can readily escape the atmosphere; nitrogen molecules are found at a higher height than oxygen, which in turn is higher than carbon dioxide -- though each of these three molecules do not readily escape the Earth.

(c) How does this calculation differ from that of the scale height done in the chapter?

The scale height, H, calculated in Section 4-5B is a factor of 2/3 less than the characteristic height. The characteristic height is a measure of the height to which a molecule can rise (on the average) whereas the scale height is a measure of the atmospheric pressure falloff with height.

4-6. (a) If a star emits the same intensity of radiation at all visible wavelengths, what will be its apparent color at the Earth's surface?

(See the end of Section 4-5B for a discussion of material relevant to this question.)

As a result of Rayleigh scattering of light by molecules in the Earth's atmosphere ($I_{scat} \propto 1/\lambda^4$), blue light (shorter wavelengths) is scattered more than red light (longer wavelengths). Scattering by dust particles also scatters blue light more than red light, but the wavelength dependence is less pronounced ($I_{scat} \propto 1/\lambda$). Thus, if a star emits the same intensity of radiation at all wavelengths, the star will appear red at the Earth's surface.

(b) Explain why the Sun appears flattened (like an ellipse) at sunset.
The light from the Sun is refracted as it passes through the Earth's atmosphere (see Figure 4-19) such that light appears higher in the sky than it would be without the atmospheric refraction. The refraction is greater at lower altitudes (where the light must pass through more atmosphere), so that portions of the Sun at lower altitudes appear to be elevated from the "true position" by an amount larger than the upper portion of the Sun -- thus the Sun appears as a flattened disk.

4-7. Show that the circular period P (in seconds) for a charge of uniform magnetic field B does *not* depend on the radius of the orbit. Evaluate this period for a proton moving at speed $v = 10^7$ m/s in a magnetic field of 10^{-4} T.
The radius of the orbit of a charged particle moving in the presence of a transverse magnetic field, of strength B, is (Equation 4-5):
$\quad r = m v / (q B)$.
The period, P, of the orbit is given by
$\quad P = 2 \pi r / v$
$\quad = 2 \pi [m v / (q B)] / v$
$\quad = 2 \pi m / (q B)$.
The period is independent of the radius of the orbit, and also independent of the velocity.
For a proton: $m = 1.67 \times 10^{-27}$ kg and $q = 1.60 \times 10^{-19}$ C. For $B = 10^{-4}$ T :
$\quad P = 2 \pi (1.67 \times 10^{-27} \text{ kg}) / [(1.60 \times 10^{-19} \text{ C}) (10^{-4} \text{ T})]$
$\quad = 6.6 \times 10^{-4}$ s

4-8. Draw a diagram and use the Lorentz force law to explain why electrons in the radiation belts drift *eastward* owing to the decrease in the magnetic field strength with distance from the Earth.
An electron will spiral in a clockwise direction along the magnetic field lines as seen by an observer looking in the direction of the magnetic field (see Figure 4-24B). The circular orbit expands when the electron moves into a weaker magnetic field region. Because the Earth's magnetic field decreases in strength with distance from the Earth, and because the electron is moving eastward when it is furthest from the Earth in its "circular" orbit around the magnetic field lines, the electron will move further east each orbit than it will move west. Thus the electron will drift eastward in the radiation belts. (A proton would likewise drift westward since its circular orbit would be counterclockwise.)

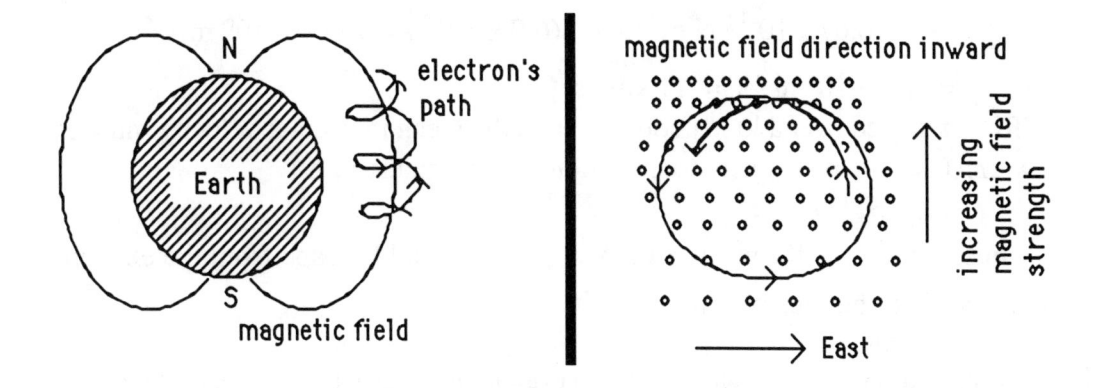

4-9. Find the radius of the Earth's core relative to the total radius if the core density is 10,000 kg/m^3, the mantle density is 4500 kg/m^3, and the average density is 5500 kg/m^3.

In this simplified Earth model, the mass (= density times volume) of the core plus the mass of the mantle equals the mass of the entire Earth:

$$[(4/3) \pi (R_c)^3] \rho_c + [(4/3) \pi (R_\oplus)^3 - (4/3) \pi (R_c)^3] \rho_m = [(4/3) \pi (R_\oplus)^3] <\rho>$$

$$(R_c)^3 \rho_c + [(R_\oplus)^3 - (R_c)^3] \rho_i = (R_\oplus)^3 <\rho>$$

$$(R_c / R_\oplus)^3 \rho_c + [1 - (R_c / R_\oplus)^3] \rho_i = <\rho>$$

$$(R_c / R_\oplus)^3 (10{,}000 \text{ kg}/\text{m}^3) + [1 - (R_c / R_\oplus)^3] (4{,}500 \text{ kg}/\text{m}^3) = 5{,}500 \text{ kg}/\text{m}^3$$

$$(10{,}000 - 4{,}500)(R_c / R_\oplus)^3 = 5{,}500 - 4{,}500$$

$$(R_c / R_\oplus)^3 = 1{,}000 / 5{,}500 = 0.18$$

$$R_c / R_\oplus = 0.57$$

In this model, the core radius of the Earth is 57% the radius of the Earth.

4-10. In the Earth's exosphere, the temperature can reach 2000 K. Estimate the lifetime of water vapor here by comparing its mean velocity with the Earth's escape velocity.

The mean molecular speed is given by (Equation 2-7):

$$v_{rms} = (3 k T / M)^{1/2}$$

$$= [3 (1.38 \times 10^{-23} \text{ J/K})(2000 \text{ K}) / M]^{1/2}$$

$$= (8.28 \times 10^{-20} \text{ J} / M)^{1/2}$$

The mass of a water molecule is

$$M(H_2O) = (2 \times 1 + 16) \times (1.67 \times 10^{-27} \text{ kg}) = 3.01 \times 10^{-26} \text{ kg}$$

so

$$v_{rms} = [(8.28 \times 10^{-20} \text{ J}) / (3.01 \times 10^{-26} \text{ kg})]^{1/2}$$

$$= 1.6 \times 10^3 \text{ m/s} = 1.6 \text{ km/s}$$

The escape speed (Equation 2-8) from the Earth is

$$v_{esc} = (2 G M_\oplus / r)^{1/2}$$

The exosphere begins at an altitude of h ≈ 700 km (Figure 4-18), or r ≈ 7000 km, so

45

$$v_{esc} = [2\,(6.67 \times 10^{-11}\,N\,m^2/kg^2)\,(6.0 \times 10^{24}\,kg)\,/\,(7.0 \times 10^6\,m)]^{1/2}$$
$$= 1.1 \times 10^4\,m/s = 11\,km/s$$

To be retained over a long period of time, the mean molecular velocity must be less than $\approx 1/10$ the escape velocity (see Section 2-1D; Equation 2-9)

$$v_{rms} \leq 0.1\,v_{esc}\,.$$

Thus, since v_{rms} for water $\approx 0.15\,v_{esc}$, water could escape from the exosphere over a timescale of billions of years.

4-11. Use the equation of hydrostatic equilibrium to estimate the central pressure of the Moon and compare it to the Earth's central pressure.

We follow the derivation in section 4-3A. The equation of hydrostatic equilibrium (Equation 4-1) is

$$dP/dr = -\rho(r)\,(G\,M/r^2)\,.$$

If we assume that the density is constant, $\langle\rho\rangle$, throughout the volume, integrating the equation of hydrostatic equilibrium gives

$$P_c = (2/3)\,\pi\,G\,\langle\rho\rangle^2\,R^2$$

For the Moon,

$$P_c \approx (2/3)\,\pi\,(6.67 \times 10^{-11}\,N\,m^2/kg^2)\,(3370\,kg/m^3)^2\,(1.74 \times 10^6\,m)^2$$
$$\approx 4.8 \times 10^9\,N/m^2 \approx 4.8 \times 10^4\,atm\,.$$

In the text, Section 4-3A, the central pressure of the Earth was estimated using this technique to be 1.7×10^6 atm, or 35 times that of the Moon. We would clearly expect a structural difference between the cores of the Earth and Moon.

4-12. (a) Use the strength of the magnetic field at the Earth's surface to estimate the strength of the Van Allen belts.

The Earth's dipole magnetic field strength decreases as $1/r^3$ (Section 4-6B). The surface magnetic field is $\approx 0.4 \times 10^{-4}$ T (Section 4-6A). The inner Van Allen belt lies between 1 and 2 $R_\oplus$ and the outer belt between 3 and 4 $R_\oplus$ (Figure 4-21). The inner belt has a magnetic field strength (use $r \approx 1.5\,R_\oplus$)

$$B \approx (0.4 \times 10^{-4}\,T)/(1.5)^3 \approx 1 \times 10^{-5}\,T$$

and the outer belt has a magnetic field strength (use $r \approx 3.5\,R_\oplus$)

$$B \approx (0.4 \times 10^{-4}\,T)/(3.5)^3 \approx 9 \times 10^{-7}\,T.$$

(b) Use this estimate to calculate the radius of curvature of a 50-MeV proton in the belts.

A 50-MeV proton has a velocity

$$E = (1/2)\,m\,v^2$$
$$(50 \times 10^6\,eV)\,(1.6 \times 10^{-19}\,J/eV) = (1/2)\,(1.67 \times 10^{-27}\,kg)\,v^2$$

$$v^2 = 9.6 \times 10^{15} \text{ m}^2/\text{s}^2$$
$$v \approx 10^8 \text{ m/s}$$

The radius of curvature of the orbit of a charged particle moving in the presence of a magnetic field is given by (Equation 4-5):

$$r = mv/(qB).$$
$$= (1.67 \times 10^{-27} \text{ kg})(10^8 \text{ m/s})/[(1.6 \times 10^{-19} \text{ C})(10^{-5} \text{ T})]$$
$$\approx 1 \times 10^5 \text{ m} = 100 \text{ km}$$

4-13. Demonstrate that a dipolar field decreasing as $1/r^3$ beats out an atmospheric density falloff that is exponential.

The Earth's atmospheric density falloff is exponential, with a scale height (1/e) of ≈8 km (Section 4-5B):

$$\rho(h) \approx (1 \text{ atm}) e^{-(h/8 \text{ km})}.$$

At 1/30 of an Earth radius (≈200 km) from the surface, the density has fallen drastically,

$$\rho(640 \text{ km}) \approx (1 \text{ atm}) e^{-(200 \text{ km}/8 \text{ km})} \approx 10^{-11} \text{ atm}.$$

The dipole magnetic field strength ($B \propto 1/r^3$) has fallen only

$$1/(1.033)^3 \approx 0.9 \qquad \text{--- to 90\% of the surface magnetic field.}$$

The magnetic field thus dominates over the atmosphere at heights a few hundred km above the surface of the Earth.

4-14. Assume that a rock sample contains ^{40}Ar and ^{40}K in the ratio ^{40}Ar/^{40}K = 3. How old is the rock? What assumptions do you have to make in order to calculate an age?

The half life, τ, for the reaction ^{40}K --> ^{40}Ar is 1.3×10^9 years (Section 4-4A). The abundance of radioactive ^{40}K will decrease in time according to Equation 4-3:

$$n/n_o = e^{-0.693 t/\tau}$$

Following the argument presented in the text (Section 4-4A after Equation 4-3), if the ratio ^{40}Ar/^{40}K = 3 today, then a majority of the potassium has decayed, and

$$K_{original} = K_{now} + K_{decayed} = K_{now} + 3 K_{now} = 4 K_{now}$$

so

$$n(K_{now})/n(K_{original}) = 1/4$$

then

$$n/n_o = 1/4 = e^{-0.693 t/\tau}$$
$$t = -[\ln(1/4)/0.693](1.3 \times 10^9 \text{ years})$$
$$= 2.6 \times 10^9 \text{ years} = 2.6 \text{ billion years}$$

Assumptions made include that the original rock contained no argon, so that all the argon present today resulted from the decay of potassium. It also assumes that no argon has escaped from the rock after being formed in the potassium decay.

4-15. Assuming that the average Sun rises at 6 A.M., crosses the meridian at noon, and sets at 6 P.M., what are the average times for the Moon's rising, crossing the meridian, and setting for each of the lunar phases: new, waxing crescent, first quarter, waxing gibbous, full, waning gibbous, third quarter, and waning crescent. Assume each of the phases is 45° apart in elongation, and ignore complicating factors.

A table can best demonstrate the systematics in the lunar motion relative to the local horizon during a cycle through its phases. The required times of rising, transit and setting can be computed by considering the elongation of the various phases, using the conversion 15° = 1 hour of right ascension. Elongations can be found from Figure 4-4.

phase	elongation	rising time	meridian crossing	setting time
new	0°	6 am	noon	6 pm
waxing crescent	45° E	9 am	3 pm	9 pm
first quarter	90° E	noon	6 pm	midnight
waxing gibbous	135° E	3 pm	9 pm	3 am
full	180°	6 pm	midnight	6 am
waning gibbous	135° W	9 pm	3 am	9 am
third quarter	90° W	midnight	6 am	noon
waning crescent	45° W	3 am	9 am	3 pm

4-16. You and a friend decide the determine the radius of the Earth. You synchronize watches; then your friend drives 50 miles due west, at latitude 40°. Each of you determines the time when the Sun lies due south — on the meridian. Your friend observes the Sun to be on her meridian 140 s after you observe the Sun on your meridian. What is your estimate of the Earth's radius?

We can convert the observed time delay in the meridian crossing of the Sun to a difference in longitude of the observers, by noting that 24 hours of longitude corresponds to 86,400 seconds of time:

$$[140 \text{ s} / (8.64 \times 10^4 \text{ s})] \times (360°) = 0.583° = 35 \text{ arc-minutes}.$$

The ratio of the observed time delay to 24 hours is equal to the ratio of distance between observers to the circumference of the Earth at the observers' latitude (= the equatorial circumference times the cosine of observers' latitude):

$$\Delta t / 24 \text{ hrs} = d / (2 \pi R_\oplus \cos(\text{lat}))$$

$$140 \text{ s} / (8.64 \times 10^4 \text{ s}) = 50 \text{ km} / (2 \pi R_\oplus \cos(40°))$$

$$R_\oplus = 6.41 \times 10^3 \text{ km} = 6,410 \text{ km}$$

4-17. The summit of Mauna Kea, Hawaii (4200 m above sea level), is considered one of the world's premier observing sites. It is especially good as an infrared site because of the low water vapor and carbon dioxide content in the atmosphere above this altitude. (Water and carbon dioxide have many infrared wavelength absorption bands.) However, working at these altitudes can be hard on astronomers who are not acclimated to the lower abundance of molecular oxygen.

(a) To understand both of these effects, calculate the scale height of H_2O, CO_2, and O_2 in the Earth's atmosphere.

The scale height is given by (Section 4-5B):

$$H = kT / gm$$

where m is the mean molecular weight of a molecule. Take $T \approx 270$ K as the average temperature (Figure 4-18).

For H_2O, of atomic mass $2 \times 1 + 16 = 18$, the scale height is

$$H = (1.38 \times 10^{-23} \text{ J/K})(270 \text{ K}) / (9.8 \text{ m/s}^2)(18 \times 1.67 \times 10^{-27} \text{ kg})$$
$$= 1.26 \times 10^4 \text{ m} \approx 12.6 \text{ km}$$

For CO_2, of atomic mass $12 + 2 \times 16 = 44$, the scale height is

$$H = (1.38 \times 10^{-23} \text{ J/K})(270 \text{ K}) / (9.8 \text{ m/s}^2)(44 \times 1.67 \times 10^{-27} \text{ kg})$$
$$= 5.2 \times 10^3 \text{ m} \approx 5.2 \text{ km}$$

For O_2, of atomic mass $2 \times 16 = 32$, the scale height is

$$H = (1.38 \times 10^{-23} \text{ J/K})(270 \text{ K}) / (9.8 \text{ m/s}^2)(32 \times 1.67 \times 10^{-27} \text{ kg})$$
$$= 7.1 \times 10^3 \text{ m} \approx 7.1 \text{ km}$$

(b) What percentage of each of these gases in the atmosphere is below 4200 m? (Assume the content decreases uniformly for each gas.)

The atmospheric pressure at altitude h, relative to the pressure at sea level, $h = h_o$, is given by (Section 4-5B):

$$P(h) / P(h_o) = \exp^{-h/H}$$

For H_2O at an altitude of 4.2 km,

$$P(\text{Mauna Kea}) / P(\text{sea level}) = \exp^{-4.2 \text{ km} / 12.6 \text{ km}} = 0.72$$

For CO_2, $\quad P(\text{Mauna Kea}) / P(\text{sea level}) = \exp^{-4.2 \text{ km} / 5.2 \text{ km}} = 0.45$

For O_2, $\quad P(\text{Mauna Kea}) / P(\text{sea level}) = \exp^{-4.2 \text{ km} / 7.1 \text{ km}} = 0.55$

Therefore, on Mauna Kea, the water and carbon dioxide contents are reduced to $\approx 3/4$ and $\approx 2/5$ of their sea level values (the water content in the atmosphere is not uniform; Mauna Kea actually lies above the major cloud layers), while the oxygen is just over 1/2 of the sea level value.

4-18. What mass would the Moon need to have (relative to the Earth) so that the center of mass of the Earth-Moon system falls just at the Earth's surface? In what sense is the Earth-Moon system *not* a "double planet"?

[This problem is similar to Problem 1-6.]

Using Figure 1-14, the distance (r_1 and r_2) of the center of mass from the center of the two bodies of mass m_1 and m_2, respectively, is given by Equation 1-21:

$$m_1 r_1 = m_2 r_2$$

where the distance between the two bodies is

$$a = r_1 + r_2$$
giving Equation 1-22:
$$r_1 = a\, m_2 / (m_1 + m_2) = a / [(m_1/m_2) + 1].$$

For the Earth-Moon system:
$$r_{Earth\text{-}CM} = 3.84 \times 10^5 \text{ km} / [(5.98 \times 10^{24} \text{ kg} / 0.073 \times 10^{24} \text{ kg}) + 1]$$
$$= 3.84 \times 10^5 \text{ km} / (82 + 1)$$
$$= 4.7 \times 10^3 \text{ km} \approx 0.74\, R_\oplus$$

Since the Earth's radius is $\approx 6.38 \times 10^3$ km, the center of mass of the Earth-Moon system is ≈ 0.74 Earth radii from the center of the Earth -- thus within the Earth. We thus have to make the Moon more massive to move the center of mass further from the Earth's center.

For the center of mass to fall just at the Earth's surface, we want the Moon's mass, $m_{Moon'}$, to be given by:

$$r_{Earth\text{-}CM'} = a / [(m_{Earth}/m_{Moon'}) + 1] = 1\, R_\oplus = 6.38 \times 10^3 \text{ km}$$

$$3.84 \times 10^5 \text{ km} / [(5.98 \times 10^{24} \text{ kg} / m_{Moon'}) + 1] = 6.38 \times 10^3 \text{ km}$$

$$5.98 \times 10^{24} \text{ kg} / m_{Moon'} = (3.84 \times 10^5 \text{ km}) / (6.38 \times 10^3 \text{ km}) - 1$$
$$= 5.92 \times 10^1$$

$$m_{Moon'} = 1.01 \times 10^{23} \text{ kg} = 1.38\, m_{Moon} = 1.7 \times 10^{-2}\, m_{Earth}$$

The Earth-Moon system is not a double planet in the sense that the center of mass of the system is inside the earth.

4-19. The main constituent of the Moon's atmosphere is neon. How long would be the lifetime of neon gas on the Moon's surface at noon? *Hint:* Refer to Section 2-1D. Note that if $v_{esc}/v_{rms} = 1$, the lifetime is a few years; for 4, several thousand years; for 5, a few hundred million years; for > 10, several billion years.
[See Problems 2-13, 2-20, and 4-10 for similar problems for other planets and gases.]
The root mean square speed for a gas of mass m at temperature T is (Equation 2-7):
$$v_{rms} = (3kT/m)^{1/2}$$
and the escape speed from a planet of mass M and radius R is (Equation 2-8):
$$v_{esc} = (2GM/R)^{1/2}$$
From Appendix Table A7-1, the mass and radius of the Moon are:
$$M_m = 7.3 \times 10^{22} \text{ kg}$$
$$R_m = 1.738 \times 10^6 \text{ m}$$
and the maximum surface temperature is (Section 4-4B): $T_m \approx 380$ K.

The mass of a neon atom is (Appendix Tables A5-1 and A7-2):
$$m_{Ne} = 20.2 \text{ amu} = 20.2 \, (1.66 \times 10^{-27} \text{ kg}) = 3.35 \times 10^{-26} \text{ kg}$$

For neon in Moon's atmosphere:
$$v_{rms} = [3 \, (1.38 \times 10^{-23} \text{ J/K}) (380 \text{ K}) / (3.35 \times 10^{-26} \text{ kg})]^{1/2}$$
$$= 685 \text{ m/s} = 0.685 \text{ km/s}$$
$$v_{esc} = [2 \, (6.67 \times 10^{-11} \text{ N m}^2 \text{ kg}^{-2}) (7.3 \times 10^{22} \text{ kg}) / (1.738 \times 10^6 \text{ m})]^{1/2}$$
$$= 2{,}370 \text{ m/s} = 2.37 \text{ km/s}$$
so the ratio of escape speed to mean neon speed is
$$v_{esc} / v_{rms} = (2.37 \text{ km/s}) / (0.685 \text{ km/s}) \approx 3.5 \ll 10.$$

In order for a planet to retain an atmosphere, the escape speed should be at least 10 times the root mean square speed for the gas. This is not true for neon in the Moon's atmosphere. According to the rules given in the question, neon should have a lifetime of several hundred to a few thousand years in the Moon's atmosphere.

Chapter 5: The Terrestrial Planets: Mercury, Venus, and Mars

5-1. (a) Calculate and compare Mercury's surface temperature at noon at perihelion and at aphelion.
The perihelion, r_p, and aphelion, r_a, distances are (see Equation 1-3):
$$r_p = a(1-e) \text{ and } r_a = a(1+e).$$
For Mercury, $a = 0.387$ AU and $e = 0.206$ (Appendix Table A3-1).
$$r_p = 0.387 \text{ AU } (1 - 0.206) = 0.307 \text{ AU}$$
$$r_a = 0.387 \text{ AU } (1 + 0.206) = 0.467 \text{ AU}$$
The subsolar temperature (Equation 2-5b) applies since Mercury has no appreciable atmosphere and it rotates slowly:
$$T_{ss} \approx (1-A)^{1/4} (R_\odot / r_p)^{1/2} T_\odot \approx 394 \text{ K } (1-A)^{1/4} (r_p)^{-1/2}$$
where r_p is the planet's distance from the Sun in AU. The albedo of Mercury is 0.06 (Appendix Table A3-3), so
$$T_{ss-p} = 394 \text{ K } (1 - 0.06)^{1/4} (0.307)^{-1/2} = 700 \text{ K} \quad \text{at perihelion}$$
$$T_{ss-a} = 394 \text{ K } (1 - 0.06)^{1/4} (0.467)^{-1/2} = 568 \text{ K} \quad \text{at aphelion}$$
A 132 K difference in surface temperature exists between perihelion and aphelion (but, regardless, both temperatures are very high!)

(b) At what wavelength would Mercury's thermal emission peak on the dayside? On the nightside?
The wavelength of maximum radiation is given by Wien's Displacement Law (Equation 2-2):
$$\lambda_{max} = 2898 \text{ μm} / T[K]$$
Using the daytime (700 K) and nighttime (100 K) temperatures given in Section 5-1B (see also Problem 2-12, where we used $T_{night} \approx 125$ K),
$$\lambda_{max-day} = 2898 \text{ μm} / 700 = 4.1 \text{ μm}$$
$$\lambda_{max-night} = 2898 \text{ μm} / 100 = 29 \text{ μm}$$

5-2. At a frequency of 10 GHz, calculate the difference (from rotation) between the Doppler shift of a radio signal bounced off one side of Mercury and that of a signal bounced off the other side.
The Doppler shift from planetary rotation is given by (Section 5-1A)
$$\Delta\lambda / \lambda_o = \Delta\nu / \nu = 2\pi R / (P c)$$
where $\Delta\lambda$ is the shift between the center and one limb of the planet.
For a rotation period of $P = 58.6$ days $= 5.06 \times 10^6$ s :
$$\Delta\nu / \nu = 2\pi (2.43 \times 10^6 \text{ m}) / [(5.06 \times 10^6 \text{ s})(3.0 \times 10^8 \text{ m/s})]$$
$$= 1.0 \times 10^{-8}$$
$$\Delta\nu = (1.0 \times 10^{-8})(10 \times 10^9 \text{ Hz}) = 100 \text{ Hz}$$
The difference in the Doppler shift between the approaching and receding limbs of Mercury is thus twice this amount, or 200 Hz at a frequency of 10 GHz.

5-3. Assume Mercury has a satellite whose composition is the same as Mercury's and whose mass is 1% of the mass of Mercury. How closely must it orbit Mercury to remain bound against the tidal force of the Sun?

The instability limit of a satellite in orbit about a parent body of mass M_1 against the perturbation of another massive body, M_2, is given by Equation 3-13 for $M_1 \ll M_2$:

$$d = (M_1 / 2 M_2)^{1/3} D$$

where d is the distance between the satellite and the parent body, and D is the distance to the perturber.

Using $M_1 = M_{Mercury} = 0.33 \times 10^{24}$ kg, and $M_2 = M_\odot = 2.0 \times 10^{33}$ kg, and the semi-major axis of Mercury's orbit $D = 5.8 \times 10^7$ km:

$$d = (0.33 \times 10^{24} \text{ kg} / 2 \times 2.0 \times 10^{30} \text{ kg})^{1/3} D$$
$$= 4.35 \times 10^{-3} D = (4.35 \times 10^{-3})(5.8 \times 10^7 \text{ km})$$
$$= 2.5 \times 10^5 \text{ km} = 100 R_{Mercury}$$

Thus if the satellite is more than 100 Mercury radii from Mercury, its orbit will be disrupted by the tidal force of the Sun.

5-4. Calculate the length of the solar day on Venus. Watch out for the retrograde rotation!
From Section 5-1A, the relationship between the synodic period of rotation (solar day), S, sidereal period of rotation, P, and orbital period, T, of a planet is:
$$1/S = 1/P - 1/T$$
Venus' rotation is retrograde (so $P < 0$) with $P = -243$ days, and the orbital period is 224.7 days, so
$$1/S = 1/(-243 \text{ days}) - 1/(224.7 \text{ days})$$
$$= (-4.12 \times 10^{-3} - 4.45 \times 10^{-3}) \text{ days}^{-1} = 8.57 \times 10^{-3} \text{ days}^{-1}$$
$$S = -117 \text{ days}$$
The solar day on Venus is 117 days, with the Sun rising in the West (retrograde).

5-5. (a) Calculate the maximum radar Doppler shift (at 10 GHz) due to rotation from Venus and Mars.
As in Problem 5-2, the Doppler shift from planetary rotation is given by (Section 5-1A)
$$\Delta\lambda / \lambda_o = \Delta v / v = 2 \pi R / (P c)$$
where $\Delta\lambda$ is the shift between the center and one limb of the planet.

For Venus, with a rotation period of $P = 243$ days $= 2.1 \times 10^7$ s
$$\Delta v / v = 2 \pi (6.05 \times 10^6 \text{ m}) / [(2.1 \times 10^7 \text{ s})(3.0 \times 10^8 \text{ m/s})]$$
$$= 6.0 \times 10^{-9}$$
$$\Delta v = (6.0 \times 10^{-9})(10 \times 10^9 \text{ Hz}) = 60 \text{ Hz}$$
The difference in the Doppler shift between the approaching and receding limbs of Venus is thus twice this amount, or 120 Hz at a frequency of 10 GHz.

For Mars, with a rotation period of $P = 24^h 37^m 23^s = 8.86 \times 10^4$ s

$$\Delta\nu/\nu = 2\pi (3.39 \times 10^6 \text{ m}) / [(8.86 \times 10^4 \text{ s})(3.0 \times 10^8 \text{ m/s})]$$
$$= 8.0 \times 10^{-7}$$
$$\Delta\nu = (8.0 \times 10^{-7})(10 \times 10^9 \text{ Hz}) = 8,000 \text{ Hz} = 8 \text{ KHz}$$

The difference in the Doppler shift between the approaching and receding limbs of Mars is thus twice this amount, or 16 KHz at a frequency of 10 GHz. This is more than 100 times larger than that observed for Venus, due to Mars' faster rotation.

(b) What accuracy in timing of radio signals must be obtained to detect the minimum height differences on Venus?

To detect a height difference Δh, the accuracy, Δt, needed in timing of radio signals is the time for radio waves to travel a distance of $2\Delta h$,

$$\Delta t = 2\Delta h / c$$

To detect height differences of 1 km, timing must be accurate to:

$$\Delta t = (2 \times 1.0 \text{ km}) / (3.0 \times 10^5 \text{ km/s}) = 6.7 \times 10^{-6} \text{ s}$$

To detect height differences of 10 m would require $\Delta t = 6.7 \times 10^{-8}$ s.

5-6. At what distance from the center of Venus would you expect a magnetic field strength equivalent to that of the Earth's Van Allen belts? Do the same calculations for Mars. What do you conclude?

As discussed in Section 4-6B, a magnetic dipole field strength varies as $1/r^3$. The Earth's magnetic field is 0.4×10^{-4} T at the surface (see Section 4-6A). The Earth's Van Allen belts are at 1-2 $R_\oplus$ for the inner belt and 3-4 $R_\oplus$ for the outer belt (Figure 4-21). The strength of the magnetic field in the Van Allen belt (assume $R = 3 R_\oplus$) is $\approx 0.4 \times 10^{-4}$ T $/ (3)^3 \approx 1.5 \times 10^{-6}$ T.

From the text (Section 5-2D), Venus' surface magnetic field (if any exists) is less than 10^{-4} that of the Earth, while Mars (Section 5-3E) has a magnetic field of 60×10^{-9} T at the surface. The upper limit to Venus's surface magnetic field is thus (since $R_V \approx R_\oplus$):

$$B_{Vsurface} \approx (10^{-4})(0.4 \times 10^{-4} \text{ T}) = 0.4 \times 10^{-8} \text{ T}$$

Equating Venus' magnetic field at a distance R from its center to that of the Earth's Van Allen belt ($3 R_\oplus$):

$$0.4 \times 10^{-8} \text{ T} / (R/R_V)^3 = 1.5 \times 10^{-6} \text{ T}$$
$$(R/R_V)^3 = 2.67 \times 10^{-3}$$
$$R = 0.14 R_V \quad \text{-- well within the planet}$$

Likewise, for Mars' magnetic field:

$$60 \times 10^{-9} \text{ T} / (R/R_M)^3 = 1.5 \times 10^{-6} \text{ T}$$
$$(R/R_M)^3 = 4.0 \times 10^{-2}$$
$$R = 0.34 R_M \quad \text{-- within the interior of Mars}$$

Thus for both Venus and Mars, we would have to be within the planet to find the magnetic field strength equal to that of the Earth's Van Allen belts. It is thus not surprising that there are no radiation belts surrounding either of these terrestrial planets.

5-7. Compare the appearance of the Earth-Moon system viewed from Mars with that viewed from Mercury. From Mars, the Earth-Moon system would be an inferior "double planet" system and therefore would always appear close to the Sun. The Earth and Moon would go through phases (just as Venus shows phases as seen from the Earth). From Mercury, the Earth-Moon system would be a superior system. Using the information in Appendix Table A3-1, we can find the range in angular size of the Earth-Moon system as seen from Mercury and Mars, and the maximum elongation as seen from Mars.

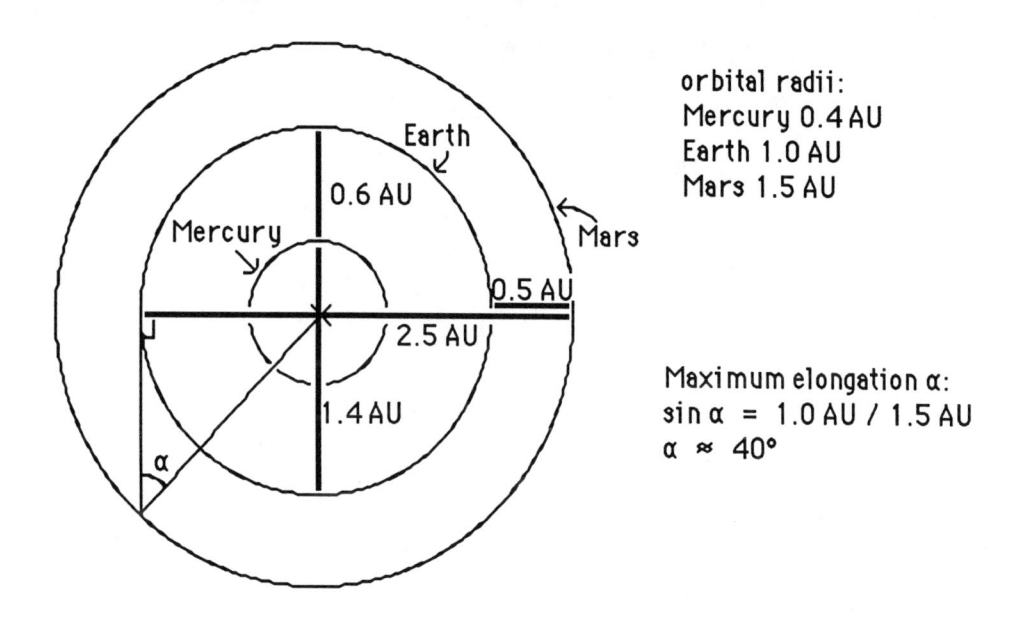

orbital radii:
Mercury 0.4 AU
Earth 1.0 AU
Mars 1.5 AU

Maximum elongation α:
sin α = 1.0 AU / 1.5 AU
α ≈ 40°

The radius of the Moon's orbit is 3.84×10^5 km ≈ 2.5×10^{-3} AU.
At closest approach to Mars (inferior conjunction with separation ≈ 0.5 AU), the Moon-Earth separation would be (at "New Earth-Moon")

2.5×10^{-3} AU / 0.5 AU = 5×10^{-3} radians ≈ 1/4 degree

while at superior conjunction (at "Full Earth-Moon") the separation would be

2.5×10^{-3} AU / 2.5 AU = 1×10^{-3} radians ≈ 1/20 degree ≈ 3 arc minutes.

The maximum elongation of the Earth-Moon ("Crescent Earth-Moon" phase) as seen from Mars is ≈ 40°.
As seen from Mercury, the Earth-Moon separation varies from

2.5×10^{-3} AU / 0.6 AU ≈ 4.2×10^{-3} radians ≈ 1/4 degree (opposition)

to 2.5×10^{-3} AU / 1.4 AU ≈ 1/10 degree (superior conjunction).

5-8. Calculate the Roche limit for Mars (with the planetary and satellite densities equal) and compare your results with the orbits of Deimos and Phobos, the moons of Mars.
The Roche limit is given by (Equation 3-9):

$$d = 2.44 \, (\rho_M / \rho_m)^{1/3} \, R$$

where ρ_M and ρ_m are the densities of the planet and satellite, R is the radius of the planet, and d is the distance between the planet's center and the satellite.

Considering the Mars system, and assuming that the planetary and satellite densities are equal,
$$d = 2.44 \, R = 2.44 \, (3.39 \times 10^3 \text{ km}) = 8.3 \times 10^3 \text{ km}$$
Phobos and Deimos (Appendix Table A3-4) are 9×10^3 km and 23×10^3 km from Mars, respectively. Both satellites are outside the Roche limit -- Phobos just outside.

5-9. Estimate the lifetime of CO_2 in the atmospheres of Mars and Venus.

[This problem is similar to Problems 2-13, 2-20, 4-10, and 4-19.]
Consider the ratios of the root mean squared speed and the escape speed:
$$v_{rms} = (3 \, k \, T / m)^{1/2} \quad \text{(Equation 2-7)}$$
$$v_{esc} = (2 \, G \, M / R)^{1/2} \quad \text{(Equation 2-8)}$$
so $\quad v_{rms} / v_{esc} = [(3 \, k \, T \, R) / (2 \, G \, M \, m)]^{1/2}$

For carbon dioxide (CO_2),
$$m = (12 + 2 \times 16) \, m_H = 44 \, m_H = 44 \times (1.67 \times 10^{-27} \text{ kg}) = 7.35 \times 10^{-26} \text{ kg}$$
We have
$$v_{rms} / v_{esc} = \{[(3) (1.38 \times 10^{-23} \text{ J/K}) \, T \, R] /$$
$$[(2) (6.67 \times 10^{-11} \text{ N m}^2/\text{kg}^2) \, M \, (7.35 \times 10^{-26} \text{ kg})]\}^{1/2}$$
$$= 2.05 \times 10^6 \, (T \, R / M)^{1/2}$$
where T is in K, R in m, and M in kg.

For Venus, $M = 4.87 \times 10^{24}$ kg, $R = 6.05 \times 10^6$ m, $T = 700$ K, so
$$v_{rms} / v_{esc} = 2.05 \times 10^6 \, [(700) (6.05 \times 10^6) / (4.87 \times 10^{24})]^{1/2}$$
$$= 6.0 \times 10^{-2} = 0.060$$
For Mars, $M = 0.64 \times 10^{24}$ kg, $R = 3.39 \times 10^6$ m, $T = 210 - 300$ K, so
$$v_{rms} / v_{esc} = 2.05 \times 10^6 \, [(250) (3.39 \times 10^6) / (0.64 \times 10^{24})]^{1/2}$$
$$= 7.5 \times 10^{-2} = 0.075$$

For a gas to be retained indefinitely, it is necessary that $v_{rms} / v_{esc} \leq 0.10$, which is true here. Carbon dioxide will be retained in the atmosphere of both Venus and Mars.

5-10. Compare the scale heights of CO_2 in the atmospheres of Mars and Venus. What do you conclude?

The scale height is given by (Section 4-5B)
$$H = k \, T / g \, m$$
and the pressure at height h is
$$P(h) = P(h_o) \, e^{-(h/H)}$$

For carbon dioxide (CO_2), $m = (12 + 2 \times 16) m_H = 44 m_H = 7.35 \times 10^{-26}$ kg.
The surface gravity for Venus is (Appendix Table A3-3)
$$g_V = 0.91 g_\oplus = 0.91 (9.8 \text{ m/s}^2) = 8.9 \text{ m/s}^2$$
and the surface temperature is T = 700 K. Thus,
$$H_V = [(1.38 \times 10^{-23} \text{ J/K})(700 \text{ K})] / [(8.9 \text{ m/s}^2)(7.35 \times 10^{-26} \text{ kg})]$$
$$= 1.5 \times 10^4 \text{ m} = 15 \text{ km}$$
For Mars, $g_M = 0.39 g_\oplus = 3.8$ m/s^2, and T ≈ 250 K, so
$$H_M = [(1.38 \times 10^{-23} \text{ J/K})(250 \text{ K})] / [(3.8 \text{ m/s}^2)(7.35 \times 10^{-26} \text{ kg})]$$
$$= 1.2 \times 10^4 \text{ m} = 12 \text{ km}$$

The scale heights of CO_2 in the atmospheres of Venus and Mars are comparable. Thus the atmospheres' pressures should decrease with height at a comparable rate.

5-11. For a magnetic dipole, the field strength far from the dipole varies as $1/R^3$. At what distance from Mercury's center does that planet's magnetic field have the same strength as the Earth's at the Van Allen belts?

This problem is the same as Problem 5-6, but for the planet Mercury. In Section 5-1D, Mercury's magnetic field is given as 220×10^{-9} T (presumably at the surface), so equating Mercury's' magnetic field at a distance R from its center to that of the Earth's Van Allen belt (3 $R_\oplus$):

$$220 \times 10^{-9} \text{ T} / (R/R_M)^3 = 1.5 \times 10^{-6} \text{ T}$$
$$(R/R_M)^3 = 1.47 \times 10^{-1}$$
$$R = 0.53 R_M \quad \text{-- within the planet}$$

As for Problem 5-6, this answer depends on the radius chosen for the Van Allen belt.

5-12. Use hydrostatic equilibrium to compare the central pressures of Mercury, Venus, and Mars. (*Hint:* Call the surface pressure zero and use the average density.)

See also the solution to Problem 4-11, where the central pressure of the Moon was calculated.
The equation of hydrostatic equilibrium (Equation 4-1) is:
$$dP/dr = -\rho(r)(GM/r^2)$$
Do a one layer model, letting

$\rho(r) = <\rho> =$ constant, for each planet   (average density)
$r = R$ (planet radius)
$dr = 0 - R = -R$ (center - planet radius)
$dP = P_c - 0 = P_c$ (central pressure - surface pressure)

so the equation of hydrostatic equilibrium becomes
$$P_c = -<\rho>(GM/R^2)(-R) = <\rho>(GM/R^2) R$$

Noting that the surface gravity is: $g = (GM/R^2)$
$$P_c = <\rho> g R$$

For Mercury (from Appendix Table A3-3),
$$g = 0.38 \, g_\oplus = 0.38 \, (9.8 \, m/s^2) = 3.8 \, m/s^2$$
$$<\rho> = 5.5 \times 10^3 \, kg/m^3$$
$$R = 2.43 \times 10^6 \, m$$
so $P_c = (5.5 \times 10^3 \, kg/m^3)(3.8 \, m/s^2)(2.43 \times 10^6 \, m)$
$$= 5.1 \times 10^{10} \, N/m^2$$

For Venus,
$$g = 0.94 \, g_\oplus = 9.2 \, m/s^2, \, <\rho> = 5.2 \times 10^3 \, kg/m^3, \, R = 6.05 \times 10^6 \, m$$
so $P_c = 3.0 \times 10^{11} \, N/m^2$

For Mars,
$$g = 0.39 \, g_\oplus = 3.8 \, m/s^2, \, <r> = 3.9 \times 10^3 \, kg/m^3, \, R = 3.39 \times 10^6 \, m$$
so $P_c = 5.0 \times 10^{10} \, N/m^2$

So Venus has a central pressure similar to the Earth's, while Mars and Mercury have central pressures nearly an order of magnitude smaller.

5-13. The mean albedo of Mars is 0.16. What is the noon-time temperature of Mars at perihelion? At aphelion? Compare your results to the observed temperature range of 210 - 300K.

This problem is similar to Problem 5-1 in which the temperature of Mercury was considered, except, here, for the case of Mars we want to consider the equilibrium blackbody temperature rather than the subsolar temperature.

The perihelion, r_p, and aphelion, r_a, distances are (see Equation 1-3):
$$r_p = a(1-e) \text{ and } r_a = a(1+e)$$
For Mars, $a = 1.524$ AU and $e = 0.093$ (Appendix Table A3-1).
$$r_p = 1.524 \text{ AU } (1 - 0.093) = 1.382 \text{ AU}$$
$$r_a = 1.524 \text{ AU } (1 + 0.093) = 1.666 \text{ AU}$$

The equilibrium blackbody temperature (Equation 2-5a) applies for Mars since it has some atmosphere and it rotates sufficiently fast:
$$T_{eq} \approx (1-A)^{1/4} (R_\odot/2r_p)^{1/2} T_\odot \approx 279 \, K \, (1-A)^{1/4} (r_p)^{-1/2}$$
where r_p is the planet's distance from the Sun in AU. The albedo of Mars is 0.16 (Appendix Table 3-3), so
$$T_{eq-p} = 279 \, K \, (1 - 0.16)^{1/4} (1.382)^{-1/2} = 236 \, K \quad \text{at perihelion}$$
$$T_{eq-a} = 279 \, K \, (1 - 0.16)^{1/4} (1.666)^{-1/2} = 215 \, K \quad \text{at aphelion}$$
These values are lower than the upper temperatures observed on Mars. Because of

Mars' thin atmosphere, a more realistic estimate would be somewhere between the subsolar and equilibrium blackbody temperatures.

Calculating the subsolar temperature, noting that it would be scaled by 394 / 279 = 1.41 times larger rather than the equilibrium blackbody temperature, the perihelion and aphelion values would be 333 K and 303 K, respectively.

5-14. In the late 19th century, Percival Lowell claimed that he was able to observe extensive canal structure on Mars. However, other astronomers at the time were unable to see the canals. Evaluate these claims by considering the size of features visible on Mars and the angular resolution of Earth-based telescopes. For a good telescope, angular resolution is limited by the atmospheric "seeing" and not the telescope optics. On a night that is considered excellent seeing, it is possible to resolve structures to an angular size of 1 arcsec. [On the best sites on the best nights, 0.3 arcsec is possible.]

(a) What is the angular size of Mars at opposition?

Ignoring the eccentricity of the planets' orbits, the distance from Earth to Mars at opposition is roughly (Appendix Table A3-1)

$$228 \times 10^6 \text{ km} - 150 \times 10^6 \text{ km} = 78 \times 10^6 \text{ km}$$

An object of linear size d and a distance D would subtend an angle θ (in radians) given by

$$\theta \approx \tan \theta = d / D$$

where the small angle approximation is used since d « D. In arc seconds

$$\theta_{[\text{"}]} \approx 2.06 \times 10^5 \, d / D$$

Mars has a radius of $\approx$ 3,400 km, or diameter $\approx$6,800 km, so its angular size at opposition is

$$\theta = 2.06 \times 10^5 \, (6.8 \times 10^3 \text{ km} / 78 \times 10^6 \text{ km}) \text{ arcsec}$$
$$= 18\text{"}$$

(b) Valles Marineris is one of the larger features on the Martian surface, with dimensions 5000 km by 500 km. At opposition what are the angular dimensions of Valles Marineris? Should it be resolved from the Earth?

The angular size of Valles Marineris would be

$$\theta = 2.06 \times 10^5 \, (5.0 \times 10^3 \text{ km} / 78 \times 10^6 \text{ km}) \text{ arcsec} = 13\text{"}$$

$$\theta = 2.06 \times 10^5 \, (0.5 \times 10^3 \text{ km} / 78 \times 10^6 \text{ km}) \text{ arcsec} = 1.3\text{"}$$

It would be possible to see Valles Marineris if the color or albedo contrast was great enough.

(c) At opposition what would be the angular resolution required to detect something roughly the size of a 1-km-wide canal? Could Lowell have seen this size canal?

For a canal 1 km wide,

$$\theta = 2.06 \times 10^5 \, (1.0 \times 10^0 \text{ km} / 78 \times 10^6 \text{ km}) \text{ arcsec} = 0.0026\text{"}$$

A feature this size could not be seen from within the Earth's atmosphere. Even the Hubble Space Telescope could not resolve such a thin canal.

(d) What is the minimum size feature resolved with 1 arcsec seeing?
The minimum linear size resolved with 1 arcsec seeing is

$$d = D\,\theta[''] / 2.06 \times 10^5 = (78 \times 10^6 \text{ km})(1) / 2.06 \times 10^5$$
$$= 3.8 \times 10^2 \text{ km} = 380 \text{ km}$$

With 0.3 arcsec seeing, features ≈ 115 km in size could be resolved.

5-15. At what distance from the Sun would Mercury be pulled apart by tidal forces? How does this compare with Mercury's actual distance from the Sun?
[See also Problem 5-3, which deals with a satellite of Mercury.]
The Roche limit is given by (Equation 3-12)

$$d = 2.44\,(\rho_\odot / \rho_M)^{1/3}\,R_\odot$$

where $\rho_\odot$ and ρ_M are the densities of the Sun and Mercury, $R_\odot$ is the radius of the Sun, and d is the distance between the the Sun and Mercury.

The average density of the Sun is (using values in Appendix Table 7-1)

$$\rho_\odot = M_\odot / (4\pi R_\odot^3 / 3)$$
$$= 1.99 \times 10^{30} \text{ kg} / [4\pi (6.96 \times 10^8 \text{ m})^3 / 3] = 1.41 \times 10^3 \text{ kg/m}^3$$

The average density of Mercury (Appendix Table A3-3) is 5.4×10^3 kg/m³, so

$$d = 2.44\,(1.41 \times 10^3 \text{ kg/m}^3 / 5.4 \times 10^3 \text{ kg/m}^3)^{1/3}\,R_\odot$$
$$= 1.6\,R_\odot \approx 1.1 \times 10^6 \text{ km}$$

Mercury is well outside the Roche limit of the Sun, its semimajor axis being ≈ 58 million km from the Sun or over 50 times further away than the Roche limit.

5-16. (a) To understand the magnitude of the greenhouse effect on Venus, calculate the equilibrium blackbody and subsolar blackbody temperatures of Venus. Compare these temperatures to the observed temperature of 750 K. Make a statement about the importance of the greenhouse effect.
The equilibrium blackbody temperature for Venus (albedo 0.76 and orbital semimajor axis 0.72 AU, from Appendix Tables A3-1 and A3-3) is given by Equation 2-5a:

$$T_{eq} \approx (1-A)^{1/4}\,(R_\odot / 2r_p)^{1/2}\,T_\odot = 279 \text{ K}\,(1-A)^{1/4}\,(r_p)^{-1/2}$$
$$= 279 \text{ K}\,(1 - 0.76)^{1/4}\,(0.72)^{-1/2} = 230 \text{ K}$$

The subsolar temperature is given by Equation 2-5b

$$T_{ss} \approx (1-A)^{1/4}\,(R_\odot / r_p)^{1/2}\,T_\odot = 394 \text{ K}\,(1-A)^{1/4}\,(r_p)^{-1/2}$$
$$= 394 \text{ K}\,(1 - 0.76)^{1/4}\,(0.72)^{-1/2} = 325 \text{ K}$$

If we ignored the albedo, the subsolar temperature (Equation 2-4) would be

$$T_{ss} \approx (R_\odot / r_p)^{1/2}\,T_\odot = 394 \text{ K}\,(r_p)^{-1/2} = 394 \text{ K}\,(0.72)^{-1/2} = 464 \text{ K}$$

Since Venus' observed temperature is 750 K, much greater than even the subsolar temperature with albedo ignored, the greenhouse effect is quite significant on Venus.

(b) Do the same for the Earth. Comment on the relative importance of the greenhouse effect for the Earth compared to Venus.

The equilibrium blackbody temperature for Earth (albedo 0.4 and orbital semimajor axis 1.0 AU, from Appendix Tables A3-1 and A3-3) is given by Equation 2-5a:

$$T_{eq} \approx 279 \text{ K } (1 - 0.4)^{1/4} (1.0)^{-1/2} = 246 \text{ K}$$

The subsolar temperature is given by Equation 2-5b

$$T_{ss} \approx 394 \text{ K } (1 - 0.4)^{1/4} (1.0)^{-1/2} = 347 \text{ K}$$

If we ignored the albedo, the subsolar temperature (Equation 2-4) would be

$$T_{ss} \approx 394 \text{ K } (1.0)^{-1/2} = 394 \text{ K}$$

The average temperature of the Earth is on the order of 300 K, in the range between the calculated equilibrium blackbody and subsolar temperatures, so the greenhouse effect is less significant (at least at the present time!) on Earth than for Venus.

5-17. Estimate the masses of the cores of Mercury, Venus, and the Earth and compare them. Are there any surprises?

From Figure 5-3, Mercury has a Ni-Fe core of radius

$$R_{coreMercury} = 1.80 \times 10^3 \text{ m}.$$

From Figure 5-9 and Section 5-2B, Venus has a Ni-Fe core of radius

$$R_{coreVenus} = 0.4 \ (6.05 \times 10^6 \text{ m}) = 2.4 \times 10^6 \text{ m}.$$

From Section 4-3A, the Earth has a Ni-Fe core of radius (inner plus outer core)

$$R_{coreEarth} = (1.3 \times 10^6 \text{ m}) + (2.2 \times 10^6 \text{ m}) = 3.5 \times 10^6 \text{ m}.$$

The Earth's density is roughly 13,000 kg/m^3 for the inner core and 9,900 to 12,000 kg/m^3 for the outer core.

The planetary masses from Appendix Table A3-3:

$$M_{totalMercury} = 0.33 \times 10^{24} \text{ kg}$$

$$M_{totalVenus} = 4.87 \times 10^{24} \text{ kg}$$

$$M_{totalEarth} = 5.97 \times 10^{24} \text{ kg}$$

For the sake of an approximate calculation, assume the average density of $\rho \approx 12,000$ kg/m^3 for the Ni-Fe cores of the three planets. So,

$$M_{core} = \rho \ V_{core} = \rho \ (4/3) \ \pi \ (R_{core})^3$$

For Mercury,

$$M_{coreMercury} = (12,000 \text{ kg/m}^3) \ (4/3) \ \pi \ (1.80 \times 10^6 \text{ m})^3$$

$$= 2.9 \times 10^{23} \text{ kg} \qquad \text{-- or about 90\% of the planet mass}$$

For Venus,

$$M_{coreVenus} = (12,000 \text{ kg/m}^3) \ (4/3) \ \pi \ (2.40 \times 10^6 \text{ m})^3$$

$$= 6.9 \times 10^{23} \text{ kg} \qquad \text{-- or about 15\% of the planet mass}$$

For Earth,
$$M_{coreEarth} = (12{,}000 \text{ kg/m}^3)(4/3)\pi(3.50 \times 10^6 \text{ m})^3$$
$$= 2.2 \times 10^{24} \text{ kg} \qquad \text{-- or about 35\% of the planet mass}$$

It is surprising that Mercury's core is such a significant fraction of the planet's total mass.

5-18. At what altitude do you have to go up in the atmosphere of Venus in order to reach a pressure of 1 atm? See Problem 5-10 for a calculation of the scale height of CO_2 in the atmosphere of Venus, which was found to be $H = kT/gm = 15$ km. (See Section 4-5B for a discussion of scale height.)

The surface pressure of Venus is 95 atm (Section 5-2B). The atmospheric pressure has dropped to 1 atm at a height, h, given by (Section 4-5B):

$$P(h) = P(h_o) e^{-(h/H)}$$
$$1 \text{ atm} = (95 \text{ atm}) e^{-(h/15 \text{ km})}$$
$$e^{(h/15 \text{ km})} = 95$$
$$h/15 \text{ km} = 4.55$$
$$h = 68 \text{ km}$$

Venus' atmospheric pressure is 1 atm at a height of 68 km. This is just at the top of the upper cloud deck (see Figure 5-11).

Chapter 6: The Jovian Planets and Pluto

6-1. Determine the orbital periods of particles at the inner and outer edges of Saturn's rings. At what distance from the center of Saturn will a particle orbit the planet in $10^h 14^m$? Show that the inner particles of the rings rise in the west and set in the east of Saturn's sky and the outer particles rise in the east and set in the west. Is this result paradoxical? Explain.

Use Newton's form of Kepler's Third Law (Equation 1-24):

$$P^2 = 4\pi^2 a^3 / [G(M+m)].$$

From Section 7-1C, the inner and outer edges of Saturn's rings have orbital radii of 7.1×10^7 m and 1.4×10^8 m, respectively. Using $M = 5.69 \times 10^{26}$ kg, noting $M_{Saturn} \gg m_{ring\ particles}$, an inner ring particle will have a period given by

$$P^2 = 4\pi^2 (7.1 \times 10^7\ m)^3 / [(6.67 \times 10^{-11}\ N\ m^2/kg^2)(5.69 \times 10^{26}\ kg)]$$

$$= 3.72 \times 10^8\ s^2$$

$$P = 1.9 \times 10^4\ s = 5.4\ hr$$

and an outer ring particle has a period of

$$P^2 = 4\pi^2 (1.4 \times 10^8\ m)^3 / [(6.67 \times 10^{-11}\ N\ m^2/kg^2)(5.69 \times 10^{26}\ kg)]$$

$$= 2.85 \times 10^9\ s^2$$

$$P = 5.3 \times 10^4\ s = 14.8\ hr$$

(See Problem 7-13 for calculations using slightly different particle orbital radii.)

Particles with a period of $10^h 14^m = 3.68 \times 10^4$ s (the equatorial rotation period of Saturn), have an orbital radius of

$$a^3 = G(M+m)P^2/[4\pi^2]$$

$$a^3 = (6.67 \times 10^{-11}\ N\ m^2/kg^2)(5.69 \times 10^{26}\ kg)(3.86 \times 10^4\ s)^2/(4\pi^2)$$

$$= 1.43 \times 10^{24}\ m^3$$

$$a = 1.1 \times 10^8\ m$$ -- this is between the inner and outer edge of the ring system, so some ring particles will be Saturn-ocentric orbits.

The outer ring particles revolve (in a West to East direction) around Saturn in a longer time than it takes Saturn to rotate, so they will rise in the East and set in the West as seen from Saturn. The inner particles revolve around Saturn in a shorter time than it takes Saturn to rotate, so they will rise in the West and set in the East. Particles in the ring with orbital radius 1.1×10^8 m will remain in the same position in the sky as viewed from Saturn. This is not paradoxical because the ring system is not a solid object -- particles in the rings revolve around Saturn according to Kepler's Harmonic Law.

6-2. Show how the orbits of Uranus' moons appear from the Earth over a period of 100 years.

This problem is similar to Problem 2-1b -- which analyzes the seasons on Uranus (see also Figure 2-4B). The satellites of Uranus orbit in the equatorial plane. Since

Uranus' rotation axis is tilted 98° to its orbital axis, the poles of Uranus point nearly towards and away from the Sun during Uranus' orbit with a period 84 Earth years. As seen from the Earth, the orbits of the Uranian satellites will be seen face on (when a pole is pointed toward the Sun); one quarter of a Uranian orbit later (≈21 years) the satellites' orbits will be seen edge on; after half an orbit (≈42 years) the satellites' orbits will be seen face on (from the other side); and after three-quarters of an orbit (≈63 years) the satellites' orbits will again be near edge on. After 84 years the orbits will be once again face on; after which this pattern repeats. During the last part of the twentieth century, the satellites orbits are nearly face on -- so observations of the moons show them moving in circles in the plane of the sky.

6-3. Show that the moons of Neptune obey Kepler's third (harmonic) law and deduce the mass of Neptune. (*Hint:* Use appropriate units or ratios.)

This problem is like Problem 2-3 (which is for the moons of Mars).
Newton's formulation of Kepler's Third Law is (Equation 1-24):

$$P^2 = 4\pi^2 a^3 / [G(M+m)]$$

In the case of Neptune and its satellites, M » m.
From Appendix Tables 3-3 and A3-7,

$$M_{Neptune} = 103 \times 10^{24} \text{ kg} = 17.2 \, M_{Earth}$$

$$P_{Triton} = 5.88 \text{ days}; \quad a_{Triton} = 3.54 \times 10^5 \text{ km};$$

$$P_{Nereid} = 359 \text{ days}; \quad a_{Nereid} = 5.51 \times 10^6 \text{ km}.$$

So, checking Kepler's Third Law:

$$P_{Nereid}^2 / a_{Nereid}^3 =? \; P_{Triton}^2 / a_{Triton}^3$$

$$359^2 / (5.51 \times 10^6)^3 =? \; 5.88^2 / (3.54 \times 10^5)^3$$

$$7.7 \times 10^{-16} =? \; 7.8 \times 10^{-16}$$

To the accuracy of the values used, Neptune's satellites obey Kepler's Third Law.

Using Triton to calculate the mass of Neptune:

$$P^2 = 4\pi^2 a^3 / [G(M)]$$

$$M_{Neptune} = 4\pi^2 (3.54 \times 10^8 \text{ m})^3 /$$

$$[(6.67 \times 10^{-11} \text{ N m}^2 / \text{kg}^2) \times (5.88 \text{ days} \times 8.64 \times 10^4 \text{ s/day})^2]$$

$$M_{Neptune} = 1.02 \times 10^{26} \text{ kg}$$

This is near the value for the mass of Neptune (1.03×10^{26} kg) given above.

6-4. If Pluto's radius is 1120 km, what must its mass be to give the planet the same density as an icy moon of Saturn?

[Note, Appendix Table A3-3 gives the equatorial radius of Pluto as 1140 km.]

The icy moons of Saturn have a density of ≈ 1200 kg/m^3 (Appendix Table 3-6). If Pluto's radius is 1120 km, then to have a density the same as the icy moons of Saturn, it must have a mass:

$$M = <\rho> V = <\rho> (4/3) \pi R^3$$
$$= (1.20 \times 10^3 \text{ kg/m}^3)(4/3) \pi (1.12 \times 10^6 \text{ m})^3$$
$$= 7.1 \times 10^{21} \text{ kg} \approx 1.2 \times 10^{-3} M_\oplus$$

6-5. Jupiter has a strong magnetic field, about 10^{-5} T at a distance of 25×10^3 km from the surface (Pioneer 11). Estimate the size of Jupiter's magnetosphere and compare it with that of the Earth. Assume that the field is a dipole and that the solar wind pressure falls of as $1/R^2$ with distance R from the Sun. (*Hint:* Magnetic field pressure is proportional to the *square* of field intensity.)

Assume that a planet's magnetosphere ends when the magnetic field pressure balances the solar wind pressure: $P_{mag} = P_{sw}$.

We can compare the size of the magnetosphere of Jupiter with that of the Earth:
$$P_{mag-\oplus} / P_{mag-J} = P_{sw-\oplus} / P_{sw-J}$$

For a dipole magnetic field the field strength, B, falls off as $1/r^3$, where the magnetic pressure $P_{mag} \propto B^2$.

The solar wind pressure falls off with $1/R^2$, where R is the distance to the Sun. Since Jupiter is 5.2 AU from the Sun,
$$(B_\oplus)^2 / (B_J)^2 = P_{sw-\oplus} / [P_{sw-J} / (5.2)^2] = (5.2)^2$$
$$(B_\oplus) / (B_J) = 5.2$$

At a distance r_J from the center of Jupiter:
$$B_J = 10^{-5} \text{ T } [(69 \times 10^3 \text{ km} + 25 \times 10^3 \text{ km})/ r_J]^3$$
$$= 10^{-5} \text{ T } [(9.4 \times 10^4 \text{ km})/ r_J]^3$$

and at a distance $r_\oplus$ from the center of the Earth:
$$B_\oplus = (0.4 \times 10^{-4} \text{ T}) (R_\oplus / r_\oplus)^3$$

The Earth's magnetosphere extends to about 10 $R_\oplus$ (Figure 4-21), where the field strength is:
$$B_\oplus = (0.4 \times 10^{-4} \text{ T}) (1/10)^3 = 4 \times 10^{-8} \text{ T}$$

Thus, from above,
$$(4 \times 10^{-8} \text{ T}) / (B_J) = 5.2$$
$$B_J = 7.7 \times 10^{-9} \text{ T} \quad \text{-- at the outer extent of Jupiter's magnetosphere.}$$

Calculating the size of Jupiter's magnetosphere:
$$7.7 \times 10^{-9} \text{ T} = 10^{-5} \text{ T } [(9.4 \times 10^4 \text{ km})/ r_J]^3$$
$$[r_J / (9.4 \times 10^4 \text{ km})]^3 = 1.3 \times 10^3$$
$$r_J / (9.4 \times 10^4 \text{ km}) = 1.1 \times 10^1$$
$$r_J = 1.0 \times 10^6 \text{ km} = 15 R_J$$

Both the Earth's and Jupiter's magnetospheres are on the order of ten times larger than the respective planet.

6-6. Infrared observations indicate that Saturn gives off 2.8 times the energy it receives from the Sun for a total internal power loss of 2×10^{17} W. Assume that gravitational contraction releases this thermal energy. How much must Saturn shrink per year to account for this output?

Following the same argument as in the text (Section 6-1B) which was done for Jupiter, the gravitational potential energy of a spherical mass is

$$P.E. \approx G M^2 / R$$

so the energy loss rate from gravitational collapse is:

$$\text{energy loss} = d(P.E.)/dt = (-G M^2 / R^2) \, dR/dt$$

$$dR/dt = -[d(P.E.)/dt] \, R^2 / (G M^2)$$

For Saturn, use the equatorial radius in the calculation. For an internal energy power loss of 2×10^{17} W:

$$dR/dt = -(2 \times 10^{17} \text{ J/s}) (5.75 \times 10^7 \text{ m})^2 /$$
$$[(6.67 \times 10^{-11} \text{ N m}^2/\text{kg}^2) (5.69 \times 10^{26} \text{ kg})^2]$$
$$= -3.1 \times 10^{-11} \text{ m/s} \approx -1.0 \times 10^{-3} \text{ m/yr}$$

If the source of this excess heat is due to gravitational contraction (which is not certain), Saturn's radius shrinks ≈ 1 mm per yr.

6-7. Assume that Saturn's internal heat is left over from primordial contraction. Calculate the *maximum* bulk thermal conductivity the planet would need to retain enough of its internal energy to account for its present luminosity. Theoretical calculations indicate that Saturn's maximum luminosity was about 10^{20} W 4.5 billion years ago. The *thermal conductivity*, κ, is the flow of heat energy per unit time per unit area per unit temperature gradient (units: J/s · m · K), so $\kappa = -H/A(\Delta T/\Delta x)$ where H is the flow of heat energy (J/s), $\Delta T/\Delta x$ is the temperature gradient (K/m), and A is the surface area (m^2).

First consider Saturn today, where the energy loss due to the flow of heat energy from the center of the planet is:

$$H = -2 \times 10^{17} \text{ W} \quad \text{(Table 6-1)}$$

$T_{center} = 15{,}000$ K (Figure 6-9), and $T_{surface} \approx 100$ K

$$\Delta T = 15{,}000 \text{ K} - 100 \text{ K} \approx 1.5 \times 10^4 \text{ K}$$

$$\Delta x = R = 5.8 \times 10^7 \text{ m}$$

$$A = 4 \pi R^2 = 4.2 \times 10^{16} \text{ m}^2$$

so $\kappa = -(-2 \times 10^{17} \text{ W}) / \{(4.2 \times 10^{16} \text{ m}^2) [(1.5 \times 10^4 \text{ K}) / (5.8 \times 10^7 \text{ m})]\}$
$$= 1.9 \times 10^4 \text{ J/s/m/K}$$

[For comparison, in Section 7-6D a model for Jupiter's primordial gravitational contraction 4.5 billion years ago is presented, where Jupiter's internal temperature was $\approx 16{,}000$ K, surface temperature ≈ 1000 K, the planet's radius was ≈ 16 times the current size, and the energy loss rate $H = 10^{-2} L_\odot = 10^{-2} \times 3.9 \times 10^{26}$ W,

$$\kappa_{Jupiter} = (3.9 \times 10^{24} \text{ W}) / \{(4\pi (16 \times 7.1 \times 10^7 \text{ m})^2) \times$$
$$[(16{,}000 \text{ K} - 1000 \text{ K}) / (16 \times 7.1 \times 10^7 \text{ m})]\}$$
$$\approx 1.8 \times 10^{10} \text{ J/s/m/K}$$

This is a much higher value than exists presently.]

6-8. Spectroscopic observations suggest that Pluto is covered with icy frost and thus has a high albedo (0.5). The brightness of Pluto at opposition (38 AU from the Earth) is 2×10^{-17} as bright as the Sun (1 AU from the Earth). From these two observations, calculate a radius for Pluto.

The flux of reflected sunlight that we receive from Pluto (see Sections 2-1C and 7-2) is:

$$F_{Pluto} = [L_\odot / (4\pi D^2)] (\pi R^2) [A / (4\pi d^2)]$$

where D is the Sun-Pluto distance, d is the Earth-Pluto distance, R is the radius of Pluto, and A is Pluto's albedo.

The flux from the Sun as seen on the Earth is:

$$F_{Sun} = L_\odot / (4\pi r^2)$$

where r is the Sun-Earth distance.
Therefore,

$$F_{Pluto} / F_{Sun} = \{[L_\odot / (4\pi D^2)] (\pi R^2) [A / (4\pi d^2)]\} / \{L_\odot / (4\pi r^2)\}$$
$$= (A/4) [(R r) / (D d)]^2$$

Using the values given in the problem,

$$2 \times 10^{-17} = (0.5/4) \{[(R)(1 \text{ AU})] / [(39 \text{ AU})(38 \text{ AU})]\}^2$$
$$= 5.7 \times 10^{-8} (R)^2 \quad \text{-- where R is in AU}$$
$$R^2 = 3.5 \times 10^{-10} \text{ AU}^2$$
$$R = 1.9 \times 10^x \text{ AU} = 2.8 \times 10^6 \text{ m} = 2{,}800 \text{ km}$$

This result is larger than the accepted value for Pluto's radius (≈ 1140 km, Appendix Table A3-3).

6-9. Imagine that you are viewing an eclipse of Charon by Pluto.
(a) How could you use your observations to infer a diameter for Pluto?
Observing eclipses of Charon by Pluto (or of Pluto by Charon) can be used to determine their sizes, just as eclipsing binary stars are used to determine the sizes of stars (see Section 12-4A and Figures 12-7 and 12-9). To determine the diameter of Pluto and Charon, we need to observe three of the following: the onset of the eclipse (first contact, t_1), the beginning of total eclipse (second contact, t_2), the end of total eclipse (third contact, t_3), and the end of eclipse (fourth contact, t_4) -- see Figure 12-9.

For simplicity, assume that the orbits are circular, that the size of the orbit can be determined from observations of the system, and that the eclipse is a central eclipse - the orbital plane is not inclined to our line of sight. Assuming Pluto is the larger of the two bodies,

$$2 R_{Pluto} = v(t_3 - t_1) = v(t_4 - t_2)$$

where v is the orbital velocity.

For a circular orbit
$$v = 2\pi a / P$$
where P is the orbital period (which can be determined by observing the time interval between eclipses).
Similarly, the size of Charon can be found:
$$2 R_{Charon} = v(t_4 - t_3) = v(t_2 - t_1)$$
If the orbit is not circular or if the orbital plane is inclined to our line of sight, the calculations would be more complicated.

(b) By what percentage would the total brightness of the system dim at mid-eclipse?
If the eclipse is total - with Pluto eclipsing Charon, then the fractional decrease in brightness is

$$f = \text{(brightness of Charon)} / \text{(brightness of Pluto + brightness of Charon)}$$
$$= (\text{area}_{Charon} \times \text{Albedo}_{Charon}) / [(\text{area}_{Pluto} \times \text{Albedo}_{Pluto}) + (\text{area}_{Charon} \times \text{Albedo}_{Charon})]$$
$$= (R_C^2 A_C) / [(R_P^2 A_P) + (R_C^2 A_C)]$$

If Charon and Pluto have the same albedo,
$$f = R_C^2 / (R_P^2 + R_C^2)$$
Using the radii given in Section 6-5, $R_P = 1{,}120$ km and $R_C = 560$ km, the percentage of brightness dimming would be
$$f = (560^2) / (1{,}120^2 + 560^2) = 0.20$$
[Note: Appendix Tables A3-3 and A3-7 give $R_P = 1{,}140$ km and $R_C = 593$ km, which would give $f = 0.21$.]

6-10. Estimate the lifetime of methane in the atmospheres of Jupiter and Uranus.
This problem is nearly identical to Problems 2-13 and 5-9.
Consider the ratios of the root mean squared speed and the escape speed:
$$v_{rms} = (3kT/m)^{1/2} \quad \text{(Equation 2-7)}$$
$$v_{esc} = (2GM/R)^{1/2} \quad \text{(Equation 2-8)}$$
so $v_{rms}/v_{esc} = [(3kTR)/(2GMm)]^{1/2}$
For methane (CH_4),
$$m = (12 + 4 \times 1) m_H = 16 m_H = 16 \times (1.67 \times 10^{-27} \text{ kg}) = 2.67 \times 10^{-26} \text{ kg}$$
We have
$$v_{rms}/v_{esc} = \{[(3)(1.38 \times 10^{-23} \text{ J/K}) T R] / [(2)(6.67 \times 10^{-11} \text{ N m}^2/\text{kg}^2) M (2.67 \times 10^{-26} \text{ kg})]\}^{1/2}$$
$$= 3.41 \times 10^6 (TR/M)^{1/2}$$
where T is in K, R in m, and M in kg.

For Jupiter (Appendix Table A3-3),

$$M = 1.9 \times 10^{27} \text{ kg}, R = 6.87 \times 10^7 \text{ m}, T = 100\text{-}150 \text{ K, so}$$

$$v_{rms} / v_{esc} = 3.41 \times 10^6 \, [(150) \, (6.87 \times 10^7) / (1.9 \times 10^{27})]^{1/2}$$

$$= 7.9 \times 10^{-3} \approx 0.008$$

For Uranus, $M = 8.7 \times 10^{25}$ kg, $R = 2.5 \times 10^7$ m, $T = 58$ K, so

$$v_{rms} / v_{esc} = 3.41 \times 10^6 \, [(58) \, (2.5 \times 10^7) / (8.7 \times 10^{25})]^{1/2}$$

$$= 1.4 \times 10^{-2} = 0.014$$

For a gas to be retained indefinitely, it is necessary that $v_{rms} / v_{esc} \leq 0.10$, which is true here. Methane will be retained in the atmospheres of both Jupiter and Uranus.

6-11. Use the equation of hydrostatic equilibrium to compare the central pressures of Saturn and Uranus.
This problem is similar to Problem 5-12.
The equation of hydrostatic equilibrium (Equation 4-1) is

$$dP / dr = -\rho(r) \, (G M / r^2)$$

Do a one layer model, letting

$\rho(r) = \langle \rho \rangle$ = constant, for each planet   (average density)
$r = R$ (planet radius)
$dr = 0 - R = -R$ (center - planet radius)
$dP = P_c - 0 = P_c$ (central pressure - surface pressure)

so the equation of hydrostatic equilibrium becomes

$$P_c = -\langle \rho \rangle \, (G M / R^2) \, (-R) = \langle \rho \rangle \, (G M / R^2) \, R$$

Noting that the surface gravity is: $g = (G M / R^2)$,

$$P_c = \langle \rho \rangle \, g \, R$$

For Saturn (from Appendix Table A3-3),

$g = 1.17 \, g_\oplus = 1.17 \, (9.8 \text{ m} / \text{s}^2) = 11.5 \text{ m} / \text{s}^2$

$\langle \rho \rangle = 0.71 \times 10^3 \text{ kg} / \text{m}^3$

$R = 5.76 \times 10^7 \text{ m}$

so $P_c = (0.71 \times 10^3 \text{ kg} / \text{m}^3) \, (11.5 \text{ m} / \text{s}^2) \, (5.76 \times 10^7 \text{ m})$

$= 4.7 \times 10^{11} \text{ N} / \text{m}^2$

For Uranus,

$g = 0.94 \, g_\oplus = 9.2 \text{ m} / \text{s}^2$, $\langle \rho \rangle = 1.32 \times 10^3 \text{ kg} / \text{m}^3$, $R = 2.50 \times 10^7 \text{ m}$

so $P_c = 3.0 \times 10^{11} \text{ N} / \text{m}^2$

Compare the central pressure of these two Jovian planets to that calculated for the terrestrial planets (Problem 5-12).

6-12. Calculate the central pressure of Pluto and compare it to the central pressures of Jupiter, our Moon, and the Earth.
This problem is similar to Problems 4-11 and 5-12.

We follow the derivation in section 4-3A. The equation of hydrostatic equilibrium (Equation 4-1) is

$$dP/dr = -\rho(r)(GM/r^2).$$

If we assume that the density is constant, $\langle\rho\rangle$, throughout the volume, integrating the equation of hydrostatic equilibrium gives

$$P_c = (2/3)\pi G \langle\rho\rangle^2 R^2 = (1.4 \times 10^{-10})\langle\rho\rangle^2 R^2$$

Using values from Appendix Tables A3-3 and A3-4, for Pluto,

$$P_{cPluto} \approx (1.4 \times 10^{-10})(500 \text{ kg/m}^3)^2 (1.14 \times 10^6 \text{ m})^2$$
$$\approx 4.5 \times 10^7 \text{ N/m}^2 = 4.5 \times 10^7 \text{ Pa} = 4.5 \times 10^2 \text{ atm}.$$

For comparison,

$$P_{cEarth} \approx 1.7 \times 10^{11} \text{ N/m}^2 = 1.7 \times 10^6 \text{ atm} \quad \text{(Section 4-3A)}$$
$$P_{cJupiter} \approx 1.2 \times 10^{12} \text{ N/m}^2 = 1.2 \times 10^7 \text{ atm} \quad \text{(Section 6-1B)}$$
$$P_{cMoon} \approx 4.8 \times 10^9 \text{ N/m}^2 = 4.8 \times 10^4 \text{ atm} \quad \text{(Problem 4-11)}$$

Thus Pluto's central pressure is two orders of magnitude less than that of the Moon, and much less than the Earth and Jupiter.

6-13. Calculate the relative Doppler shift from the approaching and receding edges of Jupiter due to planetary rotation for a spectral line at rest wavelength $\lambda_o = 500$ nm.

The Doppler shift from planetary rotation is given by (Section 5-1A)

$$\Delta\lambda/\lambda_o = \Delta v/v_o = 2\pi R/(P c)$$

where $\Delta\lambda$ is the wavelength shift between the center and one limb of the planet. For Jupiter, with a rotation period $P = 9^h 50^m = 3.54 \times 10^4$ s at the equator,

$$\Delta\lambda = \{2\pi(7.14 \times 10^7 \text{ m})/[(3.54 \times 10^4 \text{ s})(3.0 \times 10^8 \text{ m/s})]\}\lambda_o$$
$$= 4.22 \times 10^{-5}\lambda_o = 4.22 \times 10^{-5} \times 500 \text{ nm}$$
$$= 2.1 \times 10^{-2} \text{ nm}$$

The difference in the Doppler shift between the approaching and receding limbs of Jupiter is twice this amount, or 4.2×10^{-2} nm.

6-14. (a) At what distance from Jupiter's surface is its magnetic field strength equal to the Earth's surface field strength? Assume both planets have dipole magnetic fields and consider field strengths at the equator.

The magnetic field strength of a dipole varies as $1/r^3$. At the surface, the magnetic field of Jupiter (Table 6-2) is 4.28×10^{-4} T, while that at the surface of the Earth is 0.31×10^{-4} T. For Jupiter's field strength at radius r to be equal to that at the Earth's surface,

$$B_{Jupiter-surface}/(r/R_{Jupiter})^3 = B_{Earth-surface}$$

70

$$r = (B_{\text{Jupiter-surface}} / B_{\text{Earth-surface}})^{1/3} R_{\text{Jupiter}}$$
$$= (4.28 \times 10^{-4} \text{ T} / 0.31 \times 10^{-4} \text{ T})^{1/3} R_{\text{Jupiter}}$$
$$= 2.40 \, R_{\text{Jupiter}}$$

Jupiter's magnetic field strength is equal to the Earth's at its surface at a distance of 2.40 Jupiter radii from the center of Jupiter, or 1.40 Jupiter radii from the surface.

(b) Assuming ideal dipole behavior (even in Jupiter's interior) what would be the magnetic field strength at one Earth radius from Jupiter's center? (*Note:* The ideal dipole assumption would not really be valid at this point because it is inside the metallic hydrogen layer that generates Jupiter's magnetic field. But use it anyway.)

Jupiter has a radius 11.2 times Earth radius (Appendix Table 3-3), so at the Earth radius distance from the center of Jupiter, an ideal dipole would have a field strength of

$$B_{\text{Earth radius}} = B_{\text{Jupiter surface}} (R_{\text{Jupiter}} / R_{\text{Earth}})^3$$
$$= 4.28 \times 10^{-4} \text{ T } (11.2)^3$$
$$= 0.60 \text{ T}$$

6-15. Qualitatively compare the characteristics (size, temperature, rotation, meteorological characteristics) of Jupiter's Great Red Spot with Neptune's Great Dark Spot.
Refer to Sections 6.1B and 6.5C for descriptions of these features.

	Jupiter's Great Red Spot	Neptune's Great Dark Spot
size	20,000 x 50,000 km	30,000 km
temperature	few degrees cooler than surrounding zone	--
rotation	counterclockwise	counterclockwise, few days
meteorological	high pressure	high pressure
discovery	1664 - from Earth	1989 - Voyager II spacecraft

6-16. Compare the magnitude of the tidal forces of Jupiter on Io to that of Saturn on Titan. Comment on your conclusion.
From Equation 3-8, the differential tidal force is given by

$$dF = -(2 \, G \, M \, m / R^3) \, dR$$

where R is the planet-moon separation, dR is the diameter of the moon experiencing the tidal force, and M is the mass of the planet.

From Appendix Tables A3-3 and A3-5,

$$M_{\text{Jupiter}} = 1900 \times 10^{24} \text{ kg}$$
$$M_{\text{Io}} = 4.7 \times 10^{-5} \, M_{\text{Jupiter}} = 8.9 \times 10^{22} \text{ kg}$$
$$R_{\text{Jupiter-Io}} = 4.22 \times 10^{8} \text{ m}$$
$$dR_{\text{Io}} = 3.64 \times 10^{6} \text{ m}$$

So, the tidal force of Jupiter on Io

$$dF = -2\,(6.67 \times 10^{-11}\text{ N m}^2\text{ kg}^{-2})\,(1900 \times 10^{24}\text{ kg})\,(8.9 \times 10^{22}\text{ kg}) \times$$
$$(3.64 \times 10^6\text{ m}) / (4.22 \times 10^8\text{ m})^3$$
$$dF = -1.1 \times 10^{21}\text{ N}$$

For Saturn-Titan, from Appendix Tables A3-3 and A3-6,

$M_{Saturn} = 569 \times 10^{24}$ kg

$M_{Titan} = 2.36 \times 10^{-4}\, M_{Saturn} = 1.34 \times 10^{23}$ kg

$R_{Saturn\text{-}Titan} = 1.22 \times 10^9$ m

$dR_{Titan} = 5.15 \times 10^6$ m

So, the tidal force of Saturn on Titan

$$dF = -2\,(6.67 \times 10^{-11}\text{ N m}^2\text{ kg}^{-2})\,(569 \times 10^{24}\text{ kg})\,(1.34 \times 10^{23}\text{ kg}) \times$$
$$(5.15 \times 10^6\text{ m}) / (1.22 \times 10^9\text{ m})^3$$
$$dF = -2.9 \times 10^{19}\text{ N}$$

Comparing the results, Io experiences a considerable tidal force, much larger than either Titan or our Moon experience. (Io experiences nearly forty times the tidal force from Jupiter that Titan feels from Saturn.) Consequently, it experiences considerable internal friction and heating. This internal heat is what drives the volcanic activity on Io. Otherwise, a body the size of Io would no longer be geologically active.

6-17. Estimate the location of the center of mass of the Pluto-Charon system.
[This problem is similar to Problems 1-6 and 4-18 for other systems.]
Using Figure 1-14, the distance (r_1 and r_2) of the center of mass from the center of the two bodies of mass m_1 and m_2, respectively, is given by Equation 1-21:

$$m_1 r_1 = m_2 r_2$$

where the distance between the two bodies is

$$a = r_1 + r_2$$

giving Equation 1-22:

$$r_1 = a\, m_2 / (m_1 + m_2) = a / [(m_1 / m_2) + 1].$$

For the Pluto-Charon system, from Appendix Tables A3-3 and A3-7,

$m_{Pluto} = 1 \times 10^{22}$ kg

$m_{Charon} = 1 \times 10^{21}$ kg,

$a = 1.96 \times 10^7$ m

$r_{Pluto} = 1.14 \times 10^6$ m

so

$$r_{Pluto\text{-}CM} = 1.96 \times 10^7\text{ m} / [(1 \times 10^{22}\text{ kg} / 1 \times 10^{21}\text{ kg}) + 1]$$
$$= 1.96 \times 10^7\text{ m} / (10 + 1)$$

$$= 1.8 \times 10^6 \text{ m} \approx 1.6 \, R_{Pluto}$$

The center of mass of the Pluto-Charon system is ≈1.6 Pluto radii from the center of Pluto -- thus outside of Pluto. Pluto-Charon is thus a true "double planet," unlike the Earth-Moon system (Problem 4-18).

6-18. Estimate the frequency of the peak of the synchrotron spectrum from Uranus. Take a typical electron energy as 1 MeV.

From Section 6-1C, the peak frequency for synchrotron emission is given by

$$f_{max} \approx (4.8 \times 10^{-4}) \, E^2 \, B \, \sin(\alpha) \text{ MHz}$$

where E is the electron energy in MeV, B is the magnetic field in Tesla, and α is the pitch angle.

[*Note:* This equation (with the constant as given), does not work for the example given in the text for Jupiter's magnetic field on page 106, unless the constant is off by a factor of 10^6. The constant should be 4.8×10^2 for the units given for the variables.] The equation should thus be

$$f_{max} \approx (4.8 \times 10^2) \, E_{[MeV]}^2 \, B_{[T]} \, \sin(\alpha) \text{ MHz}$$

For Uranus, assume a pitch angle $\alpha = 90°$, and use a value of $B = 0.23 \times 10^{-4}$ T from Table 6-2 and E = 1 MeV,

$$f_{max} \approx 4.8 \times 10^2 \, (1)^2 \, (0.23 \times 10^{-4}) \sin(90°) \text{ MHz}$$

$$\approx 1.1 \times 10^{-2} \text{ MHz} \approx 11 \text{ kHz}$$

This is a very (!) low value for the frequency of peak emission because a low value for the electron energy was selected. (Remember, from Section 6-1C, the emission peaks at ≈ 1 GHz for Jupiter).

Chapter 7: Small Bodies and the Origin of the Solar System

7-1. What effects do radiation pressure and the Poynting-Robertson effect have upon an artificial space probe of mean density 1000 kg/m^3 and radius 1 m? Justify your answer quantitatively and state your assumptions.

The radiation pressure force pushing the artificial space probe away from the Sun (Equation 7-2) is:

$$F_R = (\pi \sigma r^2 R_\odot^2 T_\odot^4 / c) / d^2$$

where r is the probe's radius and d is its distance from the Sun.
The Sun's attractive gravitational force on the probe (Equation 7-3) is:

$$F_G = G M_\odot (4 \pi r^3 \rho / 3) / d^2$$

where ρ is the probe's density.
The ratio of the outward radiation pressure to the inward gravitational force is, after substituting the appropriate numerical values (see Equation 7-4):

$$F_R / F_G = 5.78 \times 10^{-5} / (\rho r)$$

where ρ is in units of kg / m^3 and r is in m.
For the probe:

$$F_R / F_G = 5.78 \times 10^{-5} / [(1000)(1)] = 5.78 \times 10^{-8}$$

Thus, radiation pressure has no appreciable effect on the probe.

The Poynting-Robertson effect (Section 7-5) will impede the probe's (assumed) circular motion, causing it to slowly spiral in toward the Sun. The time it takes for the probe to fall into the Sun is:

$$t = (7 \times 10^5) \rho r d^2 \text{ yrs}$$
$$= (7 \times 10^5)(1000)(1) d^2 \text{ years} = 7 \times 10^8 d^2 \text{ yrs}$$

where d is in AU.
If the probe started at an orbit of 1 AU, it would take nearly a billion years to fall into the Sun -- thus this effect is also insignificant.

7-2. At the origin of the solar system, the Sun's tidal forces and Roche limit might have played an important role. To avoid tidal disruption, what is the minimum density a protoplanet must have at the distance d (in AU) from the Sun? Comment on your result.

The Roche limit is given by (Equation 3-9):

$$d = 2.44 (\rho_M / \rho_m)^{1/3} R$$

where R is the radius of the primary.
Assume the Sun's density and radius were the same during the origin of the solar system as they are today (which is not really true).

$$d = 2.44 [(1.4 \times 10^3 \text{ kg/m}^3) / \rho_m]^{1/3} (7.0 \times 10^8 \text{ m})$$
$$= 1.9 \times 10^{10} (\rho_m)^{-1/3}$$

where d is in m and ρ_m in kg / m^3.

For d in AU,
$$d_{[AU]} = 1.3 \times 10^{-1} (\rho_m)^{-1/3}$$
$$(\rho_m)^{1/3} = 2.1 \times 10^{-3} (d_{[AU]})^{-3}$$
At a distance of 0.4 AU (Mercury's distance from the Sun),
$$\rho = (2.1 \times 10^{-3}) (0.4)^{-3} \text{ kg/m}^3$$
$$= 3.3 \times 10^{-2} \text{ kg/m}^3$$

The minimum density needed for a protoplanet to form would be less at greater distances from the protosun.

The terrestrial planets have densities of a few thousand kg/m^3, so protoplanets would have had to be 10^{-5} the current density in order to be tidally disrupted. Thus, the protoplanets were not tidally disrupted by the Sun.

7-3. According to theoretical models, 4 billion years ago Jupiter may have had a luminosity as high as 10^{-2} L$_\odot$ and a surface temperature of 1000 K. What was Jupiter's radius then if it radiated as a blackbody?

For a blackbody, the luminosity is (see Section 2-1C):
$$L = 4 \pi R^2 \sigma T^4$$
$$(10^{-2}) (3.9 \times 10^{26} \text{ W}) = 4 \pi R^2 (5.67 \times 10^{-8} \text{ W/m}^2 \text{ K}^4) (10^3 \text{ K})^4$$
$$R^2 = [(3.9 \times 10^{24}) / (7.1 \times 10^5)] \text{ m}^2 = 5.5 \times 10^{18} \text{ m}^2$$
$$R = 2.3 \times 10^9 \text{ m} = 3.4 \text{ R}_\odot = 33 \text{ R}_J$$

This proto-Jupiter would be 33 times the current size of Jupiter -- so the four Galilean satellites (Io, Europa, Ganymede, and Callisto) would have been inside the proto-planet. They could not have formed until the proto-Jupiter had contracted further.

7-4. (a) Estimate the present rate at which the Earth is accreting planetary dust.

Assume that the density of interplanetary particles in the Earth's neighborhood is 10^{-8} particles/m^3 (see Section 7-5). Due to the Earth's orbital motion, the Earth sweeps up a volume per unit time of:
$$V/t = \text{(cross sectional area of Earth)} \times \text{(velocity)}$$
$$= (\pi R_\oplus^2) v_\oplus$$
$$= \pi (6.4 \times 10^6 \text{ m})^2 (3.0 \times 10^4 \text{ m/s})$$
$$= 3.9 \times 10^{18} \text{ m}^3/\text{s}$$

The Earth sweeps up
$$n = (3.9 \times 10^{18} \text{ m}^3/\text{s}) / (10^{-8} \text{ particles/m}^3)$$
$$\approx 3.9 \times 10^{10} \text{ particles/s} \approx 1.2 \times 10^{18} \text{ particles/yr}$$

Assuming each particle has a density ≈ 5000 kg/m^3 and radius 50 μm, or a mass:
$$m = (4/3) \pi r^3 \rho = (4/3) \pi (5 \times 10^{-5} \text{ m})^3 (5 \times 10^3 \text{ kg/m}^3)$$
$$= 2.6 \times 10^{-9} \text{ kg}$$

then the total mass swept up is:

$$M = (1.2 \times 10^{18} \text{ particles / yr}) (2.6 \times 10^{-9} \text{ kg}) = 3.1 \times 10^9 \text{ kg / yr}$$

(This is $\approx$ 3.5 million tons / yr.)

(b) Calculate how long it will take the Earth at this rate to increase its mass by 50%.
At this rate, to accrete 50% of the Earth's mass would take:

$$t = (0.5 \times 6 \times 10^{24} \text{ kg}) / (3.1 \times 10^9 \text{ kg / yr}) = 10^{15} \text{ yr}$$

-- a very long time indeed!

Compare this result with that of Problem 7-10, where the mass influx of larger meteoroids (m ≈ 1 g) is calculated ($\approx 1.8 \times 10^5$ kg / yr). The smaller particles contribute more to the infall of material than do the larger visible meteors.

7-5. Calculate how fast the Sun would rotate if all the angular momentum of the planets were added to it. The Sun's angular momentum is ≈ 1% of the total angular momentum of the solar system (Section 7-6A). The angular momentum of a sphere is proportional to its angular velocity of rotation (assuming the mass and radius remain constant):

$$L \propto M V R$$

So, if the Sun were to contain nearly all the solar system's angular momentum, it would rotate ≈ 100 times faster than it currently does, or in a period of ≈ 1/4 day.

7-6. Compare the gravitational impact parameters for the Earth (surface) and for Jupiter (top of atmosphere) for bodies falling in at their respective escape velocities.

The gravitational impact parameter, S, is given by (Figure 7-30; Section 7-6C):

$$S = [R^2 + (2 G M R / v_0^2)]^{1/2}$$

If the infall speed, v_0, is equal to the escape speed, v_{esc}, from the surface (Equation 2-8):

$$v_0 = v_{esc} = (2 G M / R)^{1/2}$$
$$v_0^2 = v_{esc}^2 = 2 G M / R$$

so

$$S = \{R^2 + [(2 G M R) / (2 G M / R)]\}^{1/2}$$
$$= (R^2 + R^2)^{1/2} = \sqrt{2} R$$

The impact parameter would be $\sqrt{2}$ times the radius of the planet.

For the Earth:

$$S_\oplus = \sqrt{2} R_\oplus = \sqrt{2} (6.4 \times 10^3 \text{ km}) = 9 \times 10^3 \text{ km}$$

For Jupiter:

$$S_J = \sqrt{2} R_J = \sqrt{2} (6.9 \times 10^4 \text{ km}) = 1 \times 10^5 \text{ km}$$

For objects infalling at the escape speed of the planet's surface (or atmosphere), Jupiter's impact cross-sectional area is

$$(1 \times 10^5 \text{ km} / 9 \times 10^3 \text{ km})^2 \approx 1.2 \times 10^2 \text{ that of the Earth.}$$

7-7. Compare the escape velocities from Titan, Dione, and Hyperion. Estimate the lifetime of methane on each.

This problem is similar to Problems 2-13 and 5-9 which consider carbon monoxide in the atmosphere of the terrestrial planets and 6-10 which considers the lifetime of methane in the atmospheres of the Jovian planets.

The escape speed is given by (Equation 2-8):

$$v_{esc} = (2GM/R)^{1/2}$$

For Titan (Appendix Table 3-6), $R = 2.56 \times 10^6$ m, and

$$M = (2.36 \times 10^{-4} M_{Sat})(5.69 \times 10^{26} \text{ kg}) = 1.34 \times 10^{23} \text{ kg}$$

so $v_{esc} = [2(6.67 \times 10^{-11} \text{ N m}^2/\text{kg}^2)(1.34 \times 10^{23} \text{ kg})/(2.56 \times 10^6 \text{ m})]^{1/2}$

$= (7.0 \times 10^6 \text{ m}^2/\text{s}^2)^{1/2}$

$= 2.6 \times 10^3 \text{ m/s} = 2.6 \text{ km/s}$

For Dione, $R = 5.6 \times 10^5$ m, and

$$M = (1.85 \times 10^{-6} M_{Sat})(5.69 \times 10^{26} \text{ kg}) = 1.05 \times 10^{21} \text{ kg}$$

so $v_{esc} = [2(6.67 \times 10^{-11} \text{ N m}^2/\text{kg}^2)(1.05 \times 10^{21} \text{ kg})/(5.6 \times 10^5 \text{ m})]^{1/2}$

$= (2.5 \times 10^5 \text{ m}^2/\text{s}^2)^{1/2}$

$= 5.0 \times 10^2 \text{ m/s} = 0.5 \text{ km/s}$

For Hyperion, $R = 1.4 \times 10^5$ m (the geometric mean of the 3 axial radii), and

$$M = (3 \times 10^{-8} M_{Sat})(5.69 \times 10^{26} \text{ kg}) = 1.7 \times 10^{19} \text{ kg}$$

so $v_{esc} = [2(6.67 \times 10^{-11} \text{ N m}^2/\text{kg}^2)(1.7 \times 10^{19} \text{ kg})/(1.5 \times 10^5 \text{ m})]^{1/2}$

$= (6.1 \times 10^3 \text{ m}^2/\text{s}^2)^{1/2}$

$= 7.8 \times 10^1 \text{ m/s} = 0.078 \text{ km/s}$

The escape speed of Hyperion is very low, while that of Titan is the largest.

Methane (CH_4) has a mass,

$$m = (12 + 4 \times 1) m_H = 16 m_H = 16(1.67 \times 10^{-27} \text{ kg}) = 2.67 \times 10^{-26} \text{ kg}$$

The mean speed (Equation 2-7) at $T \approx 100$ K (Titan's atmospheric temperature) is

$v_{rms} = (3kT/m)^{1/2}$

$= [3(1.38 \times 10^{-23} \text{ J/K})(100 \text{ K})/(2.67 \times 10^{-26} \text{ kg})]^{1/2}$

$= (1.6 \times 10^5 \text{ m}^2/\text{s}^2)^{1/2}$

$= 3.9 \times 10^2 \text{ m/s} = 0.39 \text{ km/s}$

For a gas to be retained by a body (Section 2-1D):

$$v_{rms} < (1/10) v_{esc}$$

Comparing the above values, we see that methane will readily escape from Dione and Hyperion (since $v_{esc} \approx v_{rms}$ for Dione and $v_{esc} \ll v_{rms}$ for Hyperion), but will be retained for a while (but not for an infinite time) by Titan ($v_{rms} \approx (1/7) v_{esc}$).

7-8. Ceres has an infrared flux (at 10 μm) that is four times its visible flux. What is the albedo of Ceres? What is its radius?

The ratio of the infrared flux (reradiated energy from energy absorbed) to visual flux (energy reflected), where A is the albedo, is (see Section 7-2):

$$F_{IR} / F_{vis} = (1 - A) / A$$

So, $4 = (1 - A) / A$
$4A = 1 - A$
$A = 1/5 = 0.2$

The albedo is 20%. This is a relatively high value for an asteroid (the albedo is usually in the range 2 - 35%), indicating that Ceres is probably an S-type asteroid.

To determine the size of the asteroid we need to know either the visual or infrared flux from the asteroid, which the problem does not give. Let us take the asteroid to be in the asteroid belt (D = 2.8 AU), and to be at opposition (so d = 1.8 AU). Assume that the asteroid's flux as seen from the Earth is 1.4×10^{-11} W / m^2 (this is 10^{-14} the flux of the Sun at the Earth's orbit). From Section 7-2,

$$F_{vis} = (L_\odot / 4\pi D^2)(\pi R^2)(A / 4\pi d^2)$$

$$R^2 = F_{vis}(4\pi D^2)(4\pi d^2)/(\pi A L_\odot) = F_{vis}(16\pi) D^2 d^2 / (A L_\odot)$$

$$= [(1.4 \times 10^{-11} \text{ W/m}^2)(16\pi)(2.8 \times 1.5 \times 10^{11} \text{ m})^2 \times$$
$$(1.8 \times 1.5 \times 10^{11} \text{ m})^2] / [(0.2)(3.9 \times 10^{26} \text{ W})]$$

$$= 1.16 \times 10^{11} \text{ m}^2$$

$$R = 3.4 \times 10^5 \text{ m} = 340 \text{ km}$$

This is smaller than (but of the right order of magnitude) the accepted value of 470 km, implying that the assumed flux was somewhat incorrect.

7-9. Assume the asteroid Herculina has a moon of radius 100 km, orbiting 1000 km from Herculina. Make both bodies rocky.

(a) Calculate the moon's orbital period.

We need to know the diameter of Herculina which the text does not give. Assume it has a radius of 110 km (from 2nd edition of the text). If the moon has a radius of 100 km, it is approximately the same size as Herculina. Since it is a rocky body, its density is ≈ 3000 kg/m^3 (see Section 7-4), so its mass is:

$$M = V <\rho> = [(4/3)\pi r^3]<\rho>$$

$$= (4/3)\pi(110 \times 10^3 \text{ m})^3 (3000 \text{ kg/m}^3)$$

$$= 1.7 \times 10^{19} \text{ kg}$$

and its moon has a mass

$$M = (4/3)\pi(100 \times 10^3 \text{ m})^3 (3000 \text{ kg/m}^3)$$

$$= 1.3 \times 10^{19} \text{ kg}$$

Using Kepler's Third Law (Equation 1-24):

$$P^2 = 4\pi^2 a^3 / [(G(m_1 + m_2)]$$
$$= 4\pi^2 (10^6 \text{ m})^3 /$$
$$[(6.67 \times 10^{-11} \text{ N m}^2/\text{kg}^2)(1.7 \times 10^{19} \text{ kg} + 1.3 \times 10^{19} \text{ kg})]$$
$$= 2.0 \times 10^{10} \text{ s}^2$$
$$P = 1.4 \times 10^5 \text{ s} = 39 \text{ hr}$$

(b) Compare the tidal force of the Sun pulling the bodies apart with the mutual gravitational forces holding them together. (*Hint:* Assume both bodies are of the same density.)

The differential tidal force on Herculina and its moon from the Sun (Equation 3-8), assuming they are R = 3 AU from the Sun and dR = 1000 km apart, is:
$$dF = -(2 G M_\odot m / R^3) dR$$
$$= -[2(6.67 \times 10^{-11} \text{ N m}^2/\text{kg}^2)(2.0 \times 10^{30} \text{ kg})(1.3 \times 10^{19} \text{ kg}) /$$
$$(3 \times 1.5 \times 10^{11} \text{ m})^3](1.0 \times 10^6 \text{ m})$$
$$= -3.8 \times 10^{10} \text{ N}$$

while the mutual gravitational force holding the moon and Herculina together is (Equation 1-15):
$$F = G M m / r^2$$
$$= (6.67 \times 10^{-11} \text{ N m}^2/\text{kg}^2)(1.7 \times 10^{19} \text{ kg})(1.3 \times 10^{19} \text{ kg}) / (1.0 \times 10^6 \text{ m})^2$$
$$= 1.5 \times 10^{16} \text{ N}$$

So $|dF|/F = |-3.8 \times 10^{10} \text{ N}| / (1.5 \times 10^{16} \text{ N}) = 2.5 \times 10^{-6}$

Herculina and its moon are thus stable against tidal disruption by the Sun.

7-10. On a clear night, if you are patient and away from city lights, you can see about 10 meteors per hour. The meteoroid that creates a meteor has a mass of about 1 g and burns about 100 km above the Earth's surface. *Estimate* the amount of mass of meteoroids that enters the Earth's atmosphere per year. If this influx has been constant, how much mass has been added to the Earth since it was formed?

The observer can see only those meteors which burn up above the local horizon. Assume that the observer sees meteors higher than an altitude of 20° above his horizon. Ignoring, for simplicity, the curvature of the Earth and the atmosphere, the observer sees meteors that burn up within a horizontal distance of
$$\tan(20°) = 100 \text{ km} / x$$
$$x = 100 \text{ km} / (0.36) = 280 \text{ km}$$

She thus sees meteors that fall above a region of area on the Earth's surface of
$$A = \pi x^2 = \pi (280 \text{ km})^2 = 2.5 \times 10^5 \text{ km}^2$$

The surface area of the Earth is
$$S = 4\pi R_\oplus^2 = 4\pi (6.4 \times 10^3 \text{ km})^2 = 5.1 \times 10^8 \text{ km}^2$$

The observer thus sees a fraction of the meteors that fall into the atmosphere:
$$A/S = (2.5 \times 10^5 \text{ km}^2) / (5.1 \times 10^8 \text{ km}^2) = 5 \times 10^{-4}$$

The total mass of meteor entering the Earth's atmosphere is

$$M = (10 \text{ meteors / hr}) (1 \text{ g / meteor}) / (5 \times 10^{-4})$$
$$= 2 \times 10^4 \text{ g / hr} = 20 \text{ kg / hr} \approx 1.8 \times 10^5 \text{ kg / yr}$$

In 4.5 billion years, a mass of $M_{tot} = 8 \times 10^{14}$ kg $\approx 10^{-11} M_\oplus$ has been added to the Earth by meteoroid infall.

Compare this with the result of Problem 7-4, where the infall rate of small interplanetary material was found to be 3.1×10^9 kg / yr. The smaller particles contribute more to the infall of material than do the larger visible meteors.

7-11. Comet Ikeya-Seki (the great sun-grazing comet of 1965) had an elliptical orbit with a period of about 700 years. As it rounded the Sun, the comet had a perihelion distance of 0.008 AU.
(a) How far away from the Sun (in AU) will the comet be at aphelion?
Using Kepler's Third Law (Equation 1-24), the semi-major axis is:
$$a^3 = G(m_1 + m_2) P^2 / (4 \pi^2)$$
$$= (6.67 \times 10^{-11} \text{ N m}^2 / \text{kg}^2) (2.0 \times 10^{30} \text{ kg}) \times$$
$$[700 \text{ yr} \times (3.16 \times 10^7 \text{ s / yr})] / (4 \pi^2)$$
$$= 1.65 \times 10^{39} \text{ m}^3$$
$$a = 1.18 \times 10^{13} \text{ m} = 1.18 \times 10^{10} \text{ km} = 7.9 \times 10^1 \text{ AU} = 79 \text{ AU}$$
At aphelion, Comet Ikeya-Seki will be
$$r_a = 2a - r_p = (2 \times 79 \text{ AU}) - (0.008 \text{ AU}) = 158 \text{ AU from the Sun}$$

(b) If its perihelion velocity was 500 km/s, what will its orbital velocity be at aphelion?
The perihelion speed is given by (Equation 1-17a):
$$v_p = (2 \pi a / P) [(1 + e) / (1 - e)]^{1/2}$$
$$500 \text{ km / s} = [2 \pi (1.18 \times 10^{10} \text{ km}) / (2.21 \times 10^{10} \text{ s})] \times [(1 + e) / (1 - e)]^{1/2}$$
$$[(1 + e) / (1 - e)]^{1/2} = (500 \text{ km / s}) / (3.35 \text{ km / s}) = 1.49 \times 10^2$$
so $[(1 - e) / (1 + e)]^{1/2} = 6.7 \times 10^{-3}$
The aphelion speed is (Equation 1-17b):
$$v_a = (2 \pi a / P) [(1 - e) / (1 + e)]^{1/2}$$
$$= (3.35 \text{ km / s}) (6.7 \times 10^{-3})$$
$$= 2.2 \times 10^{-2} \text{ km / s} = 22 \text{ m / s}$$

[Although not needed to solve this problem, we can solve for the orbital eccentricity
$$[(1 - e) / (1 + e)]^{1/2} = 6.7 \times 10^{-3}$$
$$(1 - e) / (1 + e) = (6.7 \times 10^{-3})^2 = 4.49 \times 10^{-5}$$
$$1 - e = 4.49 \times 10^{-5} + 4.49 \times 10^{-5} e$$
$$(1 + 4.49 \times 10^{-5}) e = 1 - 4.49 \times 10^{-5}$$
$$e = 0.999910$$

(c) Make some reasonable assumption about the average density of the comet's nucleus. Did the comet pass close enough to the Sun to be ripped apart by tidal gravitational forces? (*Hint:* Work out the Roche limit for the Sun.)

Assuming that the comet is predominantly ice, its density is $\rho_c \approx 1000$ kg / m^3. A typical comet nucleus radius is ≈ 5 km.

The Roche limit for the Sun ($\rho_\odot = 1400$ kg / m^3) is given by (Equation 3-9):

$$d \approx 2.44 \, (\rho_\odot / \rho_c)^{1/3} \, R_\odot$$
$$= 2.44 \, [(1400 \text{ kg / m}^3) / (1000 \text{ kg / m}^3)]^{1/3} \, R_\odot$$
$$\approx 2.7 \, R_\odot$$
$$= 2.7 \, (7.0 \times 10^5 \text{ km}) / (1.5 \times 10^8 \text{ km / AU})$$
$$\approx 0.013 \text{ AU}$$

Comet Ikeya-Seki's perihelion distance of 0.008 AU was comparable to the Roche limit -- we would expect that there was a good chance that its nucleus was ripped apart by the tidal gravitational force of the Sun. In fact, the comet nucleus did survive this close passage -- but other comets, such as Comet West, were disrupted.

7-12. The crushing strength (pressure for material deformation) for iron meteorites is about 4×10^8 Pa. At what size would an iron asteroid have this central pressure? What would happen to it if it were this size or larger? (*Hint:* Use the equation of hydrostatic equilibrium.)

We follow the derivation in Section 4-3A (see also Problem 4-11). The equation of hydrostatic equilibrium (Equation 4-1) is:

$$dP / dr = -\rho(r) \, (G M / r^2)$$

If we assume that the density is constant, $\langle \rho \rangle$, throughout the volume, integrating the equation of hydrostatic equilibrium gives

$$P_c = (2/3) \, \pi \, G \, \langle \rho \rangle^2 \, R^2$$
$$R = [P_c \, (3 / 2\pi G)]^{1/2} / \langle \rho \rangle$$

For an iron meteorite, $\langle \rho \rangle \approx 7500$ kg / m^3, so for a central pressure 4×10^8 N/m^2,

$$R = [(4 \times 10^8 \text{ N / m}^2) \, \{3 / (2\pi \times 6.67 \times 10^{-11} \text{ N m}^2 / \text{kg}^2)\}]^{1/2}$$
$$/ \, (7.5 \times 10^3 \text{ kg / m}^3)$$

$$R = 2.3 \times 10^5 \text{ m} = 230 \text{ km}$$

If a pure iron meteoroid were larger than 230 km, the central pressure would be too great - the meteoroid would collapse under its own weight.

7-13. Assume that the particles in Saturn's rings have circular orbits ranging from 87×10^3 to 137×10^3 km from the center of Saturn.
(a) Calculate the orbital velocities at the inner and outer edges of the rings, assuming they are made of individual particles obeying Kepler's laws.

This part of the problem is similar to Problem 6-1, with slightly different orbital radii.

Using Kepler's Third Law (Equation 1-24):
$$P^2 = 4\pi^2 a^3 / [G(M+m)]$$
For Saturn, $M_{Sat} = 5.69 \times 10^{26}$ kg $\gg m_{particle}$.
The inner ring particles have a period
$$P_{inner}^2 = 4\pi^2 (8.7 \times 10^7 \text{ m})^3 / [(6.67 \times 10^{-11} \text{ N m}^2/\text{kg}^2)(5.69 \times 10^{26} \text{ kg})]$$
$$= 6.9 \times 10^8 \text{ s}^2$$
$$P_{inner} = 2.6 \times 10^4 \text{ s} = 7.3 \text{ hr}$$
and the outer ring particles have a period
$$P_{outer}^2 = 4\pi^2 (1.37 \times 10^8 \text{ m})^3 / [(6.67 \times 10^{-11} \text{ N m}^2/\text{kg}^2)(5.69 \times 10^{26} \text{ kg})]$$
$$= 2.7 \times 10^9 \text{ s}^2$$
$$P_{outer} = 5.2 \times 10^4 \text{ s} = 14.4 \text{ hr}$$
The inner particles orbit the planet faster than Saturn rotates (10.3 hr), while the outer particles orbit slower than Saturn rotates.

(b) What resolution (what accuracy in wavelength measurement) would a spectrograph have to have in order to distinguish between the inner and outer edge speeds of the rings?

The orbital velocity is
$$v = 2\pi a / P$$
$$v_{inner} = 2\pi (8.7 \times 10^7 \text{ m}) / (2.6 \times 10^4 \text{ s})$$
$$= 2.1 \times 10^4 \text{ m/s} = 21 \text{ km/s}$$
$$v_{outer} = 2\pi (1.37 \times 10^8 \text{ m}) / (5.2 \times 10^4 \text{ s})$$
$$= 1.7 \times 10^4 \text{ m/s} = 17 \text{ km/s}$$
The difference between the orbital velocity of the outer and inner ring particles is
$$\Delta v = v_{inner} - v_{outer} \approx 4 \text{ km/s}$$

To distinguish between these two velocities, the spectrograph would need a wavelength resolution $\Delta\lambda$ given by (Equation 3-4):
$$\Delta\lambda / \lambda = v/c = (4 \text{ km/s})/(3.0 \times 10^5 \text{ km/s}) \approx 1.3 \times 10^{-5}$$
For a wavelength of ≈ 500 nm (in the visual part of the spectrum),
$$\Delta\lambda = (1.3 \times 10^{-5})(500 \text{ nm}) = 6.5 \times 10^{-3} \text{ nm}$$

7-14. Compare the expected brightness of Halley's Comet at 3 AU from the Earth to that at 1 AU.

The distance of Comet Halley from the Sun is not given, so let's solve this problem for two conditions. First, assume that Comet Halley is 2 AU from the Sun, and that we view Halley successively at conjunction (distance 2 AU + 1 AU = 3 AU) and opposition (distance 2 AU - 1 AU = 1 AU). Second, assume that Comet Halley is at opposition at the times of the observations, so the distance from the Sun is 4 AU and 2 AU, respectively.

The observed brightness of a comet depends on its distance from the Sun and from the observer on the Earth, Equation 7-1,

$$B \propto R^{-n} r^{-2}$$

where n varies depending on the distance from the Sun. Far from the Sun, n = 2, while near the Sun where fluorescence occurs n ranges from 2 to 6. For Halley's Comet, n ≈ 5 for distances closer than 6 AU (Section 7-3).

In the first case,

$$B_{3\,AU} / B_{1\,AU} = (2\,AU / 2\,AU)^{-5} (3\,AU / 1\,AU)^{-2} = (1)^{-5} (3)^{-2}$$
$$= 1/9 = 0.11$$

Comet Halley is 9 times brighter when seen at 1 AU distance than at 3 AU distance, when the comet is 2 AU from the Sun.

In the second case,

$$B_{3\,AU} / B_{1\,AU} = (4\,AU / 2\,AU)^{-5} (3\,AU / 1\,AU)^{-2} = (2)^{-5} (3)^{-2}$$
$$= (1/32)(1/9)$$
$$= 1/288 = 0.0035$$

Comet Halley is 288 times brighter when viewed at opposition when it is at 1 AU distance (when 2 AU from the Sun) than when viewed at 3 AU distance (when 4 AU from the Sun). Note the strong dependence on the Sun-comet distance.

It is interesting to note that Comet Halley brightened unexpectedly, perhaps due to outgassing or collision with a meteoroid, in the early 1990s when it was several AU from the Sun.

7-15. Assume that a typical semimajor axis is 50,000 AU for a cometary nuclei in the Oort Cloud. What is the typical orbital period?

This problem is similar to Problem 2-14. Apply Kepler's third law (Equation 1-24),

$$P^2 = a^3$$

where the orbital period, P, is in years and the semimajor axis, a, in AU.

$$P = a^{3/2} = (50{,}000)^{3/2} = 1.1 \times 10^7 \text{ yr}$$

A comet in the Oort Cloud will have an orbital period on the order of 11 million years.

7-16. Compare the characteristics for the ring systems of the four Jovian planets.

The four Jovian ring systems are markedly different from each other, though gravitational interaction with shepherding moons is responsible for the appearance of several features in the ring systems. All systems are thin and consist of small particles up to meters in size, varying from rocky to icy in composition.

Jupiter: discovered by Voyager 1
  thin (< 30 km)
  narrow - main ring extends from 1.72 to 1.81 planetary radii
  outer, faint ring (gossamer ring)

  particles small, ≈ 3μm in size
  most likely rocky composition

Saturn: known since invention of telescope, ≈ 1610
thin (few km)
wide - system extends from 1.1 to 2.8 planetary radii
composed out thousands of ringlets, widths as small as 2 km
structural and compositions difference in different ring areas
water ice or rocky coated with ice
size typically 1 m, ranging from cm to tens of meters

Uranus: discovered 1977 by stellar occultation, imaged by Voyager 2
discrete ringlets, widely separated
ringlets 5 - 100 km wide
particles about 1 m size
dark (albedo 3%), implying black ice composition

Neptune: implied from ground-based occultation studies, imaged by Voyager 2
five individual rings, some tenuous
widths up to 2000 km
rings are clumpy and not complete (brighter segments)

7-17. (a) The Voyager mission found that Saturn's moon Titan has a substantial nitrogen atmosphere. Calculate the ratio of the root mean square speed to escape velocity of nitrogen in Titan's atmosphere. Make a statement about the implications of your result.
[See also Problem 7-7, which deals with methane in Titan's atmosphere.]
The mean speed (Equation 2-7) is

$$v_{rms} = (3\,k\,T\,/\,m)^{1/2}$$

and the escape speed is given by (Equation 2-8) is

$$v_{esc} = (2\,G\,M\,/\,R)^{1/2}$$

so

$$v_{rms}\,/\,v_{esc} = (3\,k\,T\,R\,/\,2\,G\,M\,m)^{1/2}$$

For a gas to be retained by a body (Section 2-1D):

$$v_{rms}\,/\,v_{esc} < 1/10$$

Nitrogen molecules (N_2) have a mass 28 x 1.67 x 10^{-27} kg (nitrogen atoms have atomic mass 14).

From Section 7-1c and 7-1e and Appendix tables 3-6 and 3-7:

$M_{Titan} = 1.37 \times 10^{23}$ kg, $R_{Titan} = 2.575 \times 10^6$ m, $T_{Titan} = 94$ K

$M_{Tritan} = 2.13 \times 10^{22}$ kg, $R_{Tritan} = 1.36 \times 10^6$ m, $T_{Tritan} = 37$ K

So, for Titan

$$v_{rms}\,/\,v_{esc} = \{\,[(3)\,(1.38 \times 10^{-23}\,\text{J/K})\,(94\,\text{K})\,(2.575 \times 10^6\,\text{m})]$$
$$/\,[(2)\,(6.67 \times 10^{-11}\,\text{N m}^2/\text{kg}^2)\,(1.37 \times 10^{23}\,\text{kg})$$
$$(28)\,(1.67 \times 10^{-27}\,\text{kg})]\,\}^{1/2}$$

$$v_{rms} / v_{esc} = 0.11$$

This is close to the requirement of $v_{rms} / v_{esc} < 1/10$ for the retention of an atmospheric gas.

(b) Do the same for Neptune's moon Triton, which was found by Voyager to have a less substantial nitrogen atmosphere.

For Triton

$$v_{rms} / v_{esc} = \{ [(3)(1.38 \times 10^{-23} \text{ J/K})(37 \text{ K})(1.36 \times 10^6 \text{ m})]$$
$$/ [(2)(6.67 \times 10^{-11} \text{ N m}^2/\text{kg}^2)(2.13 \times 10^{22} \text{ kg})$$
$$(28)(1.67 \times 10^{-27} \text{ kg})] \}^{1/2}$$
$$= 0.13$$

The ratio of mean to escape velocity is similar for Titan and Triton. In both cases v_{rms} / v_{esc} is a just little larger than 1/10, so the nitrogen atmospheres would be bound for a long (though, not infinite) time. There may be a recent source of nitrogen (outgassing) for both moons.

7-18. The Orionid meteor shower is believed to be produced by dust in the orbit of Halley's Comet. What is the orbital speed of meteors during the Orionid meteor shower?

The orbital speed for an object of mass m at distance r from the Sun, with an orbital semimajor axis a is (Equation 1-28)

$$v^2 = G(M_\odot + m)[(2/r) - (1/a)] .$$

For a dust particle ($m \ll M_\odot$) from Comet Halley, the orbital semimajor axis can be calculated from Kepler's third law (Equation 1-24), knowing the orbital period of Halley to be ≈ 76 years,

$$a = P^{2/3} = (76)^{2/3} = 18 \text{ AU} .$$

So the speed of a dust particle from Comet Halley when 1 AU from the Sun is

$$v^2 = (6.67 \times 10^{-11} \text{ N m}^2/\text{kg}^2)(1.99 \times 10^{30} \text{ kg})$$
$$[2/(1.496 \times 10^{11} \text{ m}) - 1/(18 \times 1.496 \times 10^{11} \text{ m})]$$
$$= 1.73 \times 10^9 \text{ m}^2/\text{s}^2$$
$$v = 4.2 \times 10^4 \text{ m/s} = 42 \text{ km/s}$$

So dust particles from Comet Halley will have an orbital speed of 42 km/s when they encounter the Earth in the Orionid meteor shower. (Note that this is close to the escape speed from the Solar System at this distance from the Sun, which is to be expected since the orbital eccentricity is close to 1 and the semi-major axis is large.) The velocity relative to the Earth will be depend on the angle at which the Earth encounters the particles.

7-19. A newly discovered faint asteroid and Mars are observed at opposition. Mars is observed to be 100 million times brighter than the asteroid. Assume Mars and the asteroid are 1.5 and 3.0 AU from the Sun, respectively.

(a) Estimate the radius of the asteroid if it has an albedo of 16%, similar to some S-type asteroids. The visible flux for Mars or the asteroid, where all observed light is reflected sunlight, is given by (Section 7-2)

$$F_{vis} = (L_\odot / 4\pi D^2)(\pi R^2)(A / 4\pi d^2)$$

so the ratio of the observed brightness of Mars to that of the asteroid is

$$F_{Mars} / F_{asteroid} = (D_{Mars} / D_{asteroid})^{-2} (d_{Mars} / d_{asteroid})^{-2}$$
$$(A_{Mars} / A_{asteroid})(R_{Mars} / R_{asteroid})^2$$

where D is the Sun-object distance, d is the Earth-object distance, A is the albedo, and R is the radius of the object.

Rearranging,

$$(R_{Mars} / R_{asteroid}) = (F_{Mars} / F_{asteroid})^{1/2} (D_{Mars} / D_{asteroid})$$
$$(d_{Mars} / d_{asteroid})(A_{Mars} / A_{asteroid})^{1/2}$$
$$= (1 \times 10^8)^{1/2} (1.5 \text{ AU} / 3.0 \text{ AU})(0.5 \text{ AU} / 2.0 \text{ AU})(0.16 / 0.16)^{1/2}$$
$$= 1.25 \times 10^3$$

where we have used the albedo of Mars from Appendix Table A3-3, and noted that at opposition the distance from the earth to the object is the Sun-object distance minus 1 AU.

Given the radius of Mars = 3.4×10^3 km, the radius of the asteroid is

$$3.4 \times 10^3 \text{ km} / 1.25 \times 10^3 = 2.7 \text{ km}.$$

(b) Estimate the radius of the asteroid if it has an albedo of 4%, typical of C-type asteroids. Following the steps as in part (a), with the only difference being the albedo,

$$(R_{Mars} / R_{asteroid}) = (1 \times 10^8)^{1/2} (1.5 \text{ AU} / 3.0 \text{ AU})$$
$$(0.5 \text{ AU} / 2.0 \text{ AU})(0.16 / 0.04)^{1/2}$$
$$= 6.25 \times 10^2$$

so the radius of the asteroid is

$$3.4 \times 10^3 \text{ km} / 6.25 \times 10^2 = 5.4 \text{ km}$$

Thus, for a reasonable range in albedo, we can estimate the diameter of the asteroid as being on the order of between ≈ 5 and ≈ 10 km -- comparable in size to the Martian satellite Deimos.

7-20. From the information given in this chapter, calculate the mass and density of Ida in the Ida/Dactyl binary asteroid system.

From Section 7-2, we know that Ida and Dactyl have diameters (or longest dimensions) of 56 and 1.5 km, respectively. Dactyl orbits Ida in a ≈ 100 km diameter orbit in ≈ 24

hours = 8.64 x 10⁴ s. The density of Ida is 2500 kg/m³. Assume that Ida is much more massive than Dactyl.

Using Kepler's Third Law (Equation 1-24):

$$P^2 = 4\pi^2 a^3 / [G(M_I + m_D)] \approx 4\pi^2 a^3 / (GM_I)$$

$$M_I = 4\pi^2 a^3 / (GP^2)$$

$$= 4\pi^2 (10^5 \text{ m})^3 / (6.67 \times 10^{-11} \text{ N m}^2/\text{kg}^2)(8.64 \times 10^4 \text{ s})^2$$

$$= 7.9 \times 10^{16} \text{ kg}$$

To compute the density of Ida, it is necessary to estimate its volume. The only dimension given is its length, 56 km. Asteroids are most likely to be elongated rather than spherical (see image of Gaspra, Figure 7-18). Assume the asteroid is more or less "boxy" with the other two dimensions being roughly one half and one third the longest dimension, say 28 and 19 km, respectively. The volume of the asteroid is then

$$V_I = (5.6 \times 10^4 \text{ m})(2.8 \times 10^4 \text{ m})(1.9 \times 10^4 \text{ m}) = 3.0 \times 10^{13} \text{ m}^3$$

The density is then

$$<\rho> = M_I / V_I \approx (7.9 \times 10^{16} \text{ kg}) / (3.0 \times 10^{13} \text{ m}^3)$$

$$= 2.6 \times 10^3 \text{ kg/m}^3$$

This is close to the value of 2500 kg/m³ given in the text. Of course, this answer was derived using unknown, estimated quantities!

7-21. In general, the dust tail and the plasma tail of a comet point in different directions. Explain.

The dust particles and ions are affected differently by the solar wind and the magnetic field produced by the charged particles, and by radiation pressure. The dust particles are unaffected by magnetic fields. As these particles blow away from the comet, they continue to follow a Keplerian orbit fairly similar to the comet's original orbit. Hence the dust tail curves on the comet's orbital path, away from the Sun. Radiation pressure from the Sun can alter the orbits of the dust particles (see Section 7-5). The ions, on the other hand, are affected by the magnetic fields produced by the solar wind's ions. Hence the cometary ions follow the path of the solar wind, so the ion tail points directly away from the Sun -- the solar wind's path.

Chapter 8: Electromagnetic Radiation and Matter

8-1. (a) Show that a beam of light obliquely incident upon and passing through a plane-parallel piece of glass is simply displaced without changing direction when it emerges from the glass.

Consider a plane parallel glass plate with a light beam incident at an angle θ.

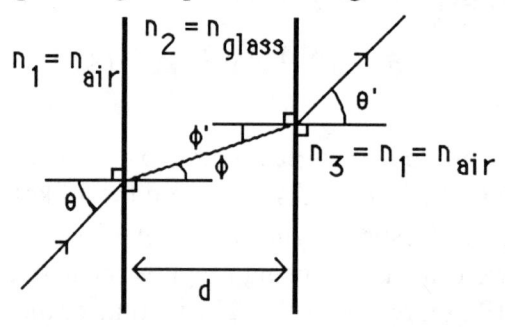

At the first surface, Snell's Law (Equation 8-6; Figure 8-4) tells us

$$n_1 \sin(\theta) = n_2 \sin(\phi)$$

(note, if $n_2 > n_1$ then $\sin(\phi) < \sin(\theta)$ and $\phi < \theta$),
and at the second surface

$$n_2 \sin(\phi') = n_3 \sin(\theta') \;.$$

If the glass has air on both sides, then $n_1 = n_3$, and if the two surfaces are parallel then $\phi = \phi'$. So, combining the two equations,

$$n_1 \sin(\theta) = n_2 \sin(f) = n_1 \sin(\theta') \;, \text{ and } \theta = \theta' \;.$$

Note that the result is generalizable to any number of layers, as long as the surfaces are parallel, and to any medium where $n_1 = n_3$.

(b) If the glass has a thickness d and index of refraction n, what is the linear displacement of the beam as a function of n and θ?

Consider the drawing below; we wish to find the distance t in terms of d, n, and θ.

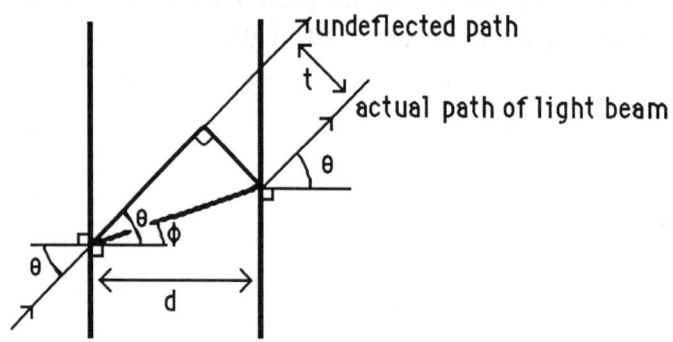

Consider the right triangle shown in bold. The hypotenuse of this triangle has length $d/\cos(\phi)$ -- this is the distance the beam travels in the glass. The distance t can be found from

$$\sin(\theta-\phi) = t / [d/\cos(\phi)]$$

or $t = d \sin(\theta-\phi) / \cos(\phi)$

Use the geometric identity:

$$\sin(x-y) = \sin(x)\cos(y) - \cos(x)\sin(y)$$

to simplify:

$$t = d [\sin(\theta) \cos(\phi) - \cos(\theta) \sin(\phi)] / \cos(\phi)$$

$$= d [\sin(\theta) - \sin(\phi) \cos(\theta) / \cos(\phi)]$$

Use Snell's Law: $\sin(\theta) = n \sin(\phi)$, where we assume $n_{air} \approx 1$. Thus,

$$t = d \{\sin(\theta) - [\sin(\theta)/n] \times [\cos(\theta)/\cos(\phi)]\}$$

Now, since

$$\cos(\theta) / \cos(\phi) = [\cos^2(\theta) / \cos^2(\phi)]^{1/2}$$

$$= \{[1 - \sin^2(\theta)] / [1 - \sin^2(\phi)]\}^{1/2}$$

$$= \{[1 - \sin^2(\theta)] / [1 - \sin^2(\theta)/n^2]\}^{1/2}$$

$$= n \times \{[1 - \sin^2(\theta)] / [n^2 - \sin^2(\theta)]\}^{1/2}$$

So

$$t = d \{\sin(\theta) - \sin(\theta) \times \{[1 - \sin^2(\theta)] / [n^2 - \sin^2(\theta)]\}^{1/2}\}$$

$$= d \sin(\theta) \{1 - \{[1 - \sin^2(\theta)] / [n^2 - \sin^2(\theta)]\}^{1/2}\}$$

8-2. What aperture is required to give 1-arcsec resolution for a wavelength of

(a) 500 nm (visible)

The theoretical resolution in radians of a telescope is given by (Equation 8-10):

$$\theta_{radian} = \lambda / d$$

[Note: Actually, the resolution would have an additional multiplicative factor of 1.22 due to the diffraction pattern of a circular aperture, but since this is not covered until Section 9-1, we will not include it here.]

In arcsec, since 1 radian = 206,265"

$$\theta arcsec = 206,265 \, \lambda / d$$

$$d = 206,265 \, \lambda / \theta arcsec$$

For 1" resolution at 500 nm = 500×10^{-9} m,

$$d = (2.06 \times 10^5) \times (5.00 \times 10^{-7} \text{ m}) / 1 = 1.0 \times 10^{-1} \text{ m} = 10 \text{ cm}$$

-- a very modest size telescope.

(b) 21 cm (radio).

For 1" resolution at 21 cm,

$$d = (2.06 \times 10^5) \times (2.1 \times 10^{-1} \text{ m}) / 1 = 4.3 \times 10^4 \text{ m} = 43 \text{ km}$$

Generalize from your results?

To achieve high resolution at radio wavelengths, we need to use interferometers (aperture synthesis). However, since the atmospheric "seeing" limits visible wavelength seeing to of order 1", Earth based radio telescopes can achieve better resolution than can optical telescopes -- we need only separate the antennae by more than ≈ 50 km.

8-3. (a) At what wavelengths will the following spectral lines be observed:
(i) a line emitted at 500 nm by a star moving toward us at 100 km/s.
To compute the Doppler shift of a line moving toward us at 100 km / s, the non-relativisitic formula is sufficient (Equation 8-11):

$$\lambda = \lambda_o (1 + v/c) \quad \text{here } v < 0$$
$$= 500 \text{ nm } [1 - (1.0 \times 10^2 \text{ km/s}) / (3.0 \times 10^5 \text{ km/s})]$$
$$= 499.8 \text{ nm} \quad \text{(a blue shift)}$$

(ii) the Ca II line (undisplaced wavelength of 397.0 nm) emitted by a galaxy receding at 60,000 km/s.
For a galaxy receding at 60,000 km / s, we must use the relativistic Doppler shift (Equation 8-14a):

$$\lambda = \lambda_o [(1 + v/c) / (1 - v/c)]^{1/2}$$
$$= 397.0 \text{ nm } \{[1 + (6.0 \times 10^4 \text{ km/s}) / (3.0 \times 10^5 \text{ km/s})] /$$
$$[1 - (6.0 \times 10^4 \text{ km/s}) / (3.0 \times 10^5 \text{ km/s})]\}^{1/2}$$
$$= 397.0 \text{ nm } (1.2 / 0.8)^{1/2}$$
$$= 486.2 \text{ nm} \quad \text{(a red shift)}$$

(b) A cloud of neutral hydrogen (H I) emits the 21-cm radio line (at rest frequency 1420.4 MHz) while moving away at 200 km/s. At what frequency will we observe this line?
The speed of the HI cloud is low enough that the non-relativisitic formula can be used (Equation 8-12):

$$\nu = \nu_o / (1 + v/c)$$
$$= 1420.4 \text{ MHz } / [1 + (2.0 \times 10^2 \text{ km/s}) / (3.0 \times 10^5 \text{ km/s})]$$
$$= 1419.5 \text{ MHz} \quad \text{(a red shift)}$$

8-4. A simple form of the binomial theorem states that

$$(1 + x)^n = 1 + nx + [n(n-1)x^2/2] + n(n-1)(n-2)x^3/6 + \cdots$$

when $x^2 < 1$. Starting with the relativistic expression, use this theorem to derive the classical equation for the Doppler shift in wavelength for the case in which $v \ll c$.
Starting with the relativistic form of the Doppler shift (Equation 8-14a):

$$\lambda = \lambda_o [(1 + v/c) / (1 - v/c)]^{1/2}$$
$$= \lambda_o [(1 + v/c)^{1/2} / (1 - v/c)^{1/2}] .$$

Using the binomial theorem (see also Appendix A9-5), for $v \ll c$:

$$(1 + v/c)^{1/2} \approx 1 + (1/2)(v/c) + \{(1/2)[(1/2) - 1](v/c)^2/2\}$$
$$+ (1/2)[(1/2) - 1][(1/2) - 2](v/c)^3/6 + \cdots$$
$$= 1 + (1/2)(v/c) - (1/8)(v/c)^2 + (1/16)(v/c)^6 + \cdots$$
$$\approx 1 + (1/2)(v/c)$$

where the higher order terms are ignored since $(v/c) \ll 1$.
Similarly,

$$(1 - v/c)^{-1/2} \approx 1 + (-1/2)(-v/c) = 1 + (1/2)(v/c)$$

Substituting,

$$\lambda \approx \lambda_0 \{[1 + (1/2)(v/c)] \times [1 + (1/2)(v/c)]\}$$
$$= \lambda_0 [1 + (v/c) + (1/4)(v/c)^2]$$
$$\approx \lambda_0 [1 + (v/c)] \quad \text{since } (v/c)^2 \ll 1$$

This is the classical Doppler shift, which can be rewritten as:

$$\Delta\lambda/\lambda_0 \equiv (\lambda - \lambda_0)/\lambda_0 = v/c.$$

8-5. (a) What is the energy of one photon of wavelength $\lambda = 300$ nm? Express your answer both in joules and in electron volts.

The energy of a photon is given by (Equation 8-15):

$$E = h\nu = hc/\lambda$$
$$= (6.625 \times 10^{-34} \text{ J s}) \times (3.00 \times 10^8 \text{ m/s}) / (300.0 \times 10^{-9} \text{ m})$$
$$= 6.63 \times 10^{-19} \text{ J}$$
$$= 4.14 \text{ eV}$$

(b) An atom in the second excited state ($n = 3$) of hydrogen is just barely ionized when a photon strikes the atom. What is the wavelength of the photon if all of its energy is transferred to the atom?

In the Bohr model of the hydrogen atom, the wavelength of a photon emitted/absorbed from a transition from energy level a to b is given by (Equation 8-25):

$$1/\lambda_{ab} = \nu_{ab}/c = R(1/n_b^2 - 1/n_a^2)$$
$$1/\lambda_{3\infty} = (1.097 \times 10^7 / \text{m})(1/\infty^2 - 1/3^2) = -1.219 \times 10^6 / \text{m}$$
$$\lambda_{3\infty} = -8.206 \times 10^{-7} \text{ m}$$
$$= -820.6 \text{ nm} \quad \text{(absorption of photon)}$$

8-6. The emission line of He II at 468.6 nm corresponds to what electronic transition?

Modify the Bohr model of the hydrogen atom for an H⁻ like atom, by including an additional factor of Z^2 in the energy levels (Equation 8-27):

$$1/\lambda_{ab} = RZ^2(1/n_b^2 - 1/n_a^2)$$
$$1/468.6 \times 10^{-9} \text{ m} = (1.097 \times 10^7 / \text{m})(2^2)(1/n_b^2 - 1/n_a^2)$$
$$(1/n_b^2 - 1/n_a^2) = 4.864 \times 10^{-2} = 0.04864$$

We have one equation with two unknowns, which in principle has an infinite number of solutions. But we have the additional requirement that n_a and n_b are integers. A trial and error solution can be found by trying an integer value of n_a and solving for n_b, where n_b must be less than n_a since an emission line is seen. This method gives:

$n_a = 1$ no transitions possible

$n_a = 2$ › $n_b = 1.83$ no good
$n_a = 3$ › $n_b = 2.50$ no good
$n_a = 4$ › $n_b = 2.9996 \approx 3$ -- this is it!

The emission line of **HeII** at 468.6 nm corresponds to a transition from the 4th to the 3rd energy level.

8-7. By applying the Boltzmann equation to the neutral hydrogen atom (neglect ionization), derive an expression for the population of the nth energy level relative to that of the ground state, at temperature T. Now, assuming that the multiplicity of each level is unity ($g_n = 1$), construct an appropriate graph showing your results for T = 6000 K. Do the same for 15,000 K. What are the major differences between them?

Starting with the Boltzmann equation (Equation 8-29):
$$N_B / N_A = (g_B / g_A) \exp[(E_A - E_B)/(kT)]$$
defining
$\quad A \equiv$ ground state (n = 1)
$\quad B \equiv$ the nth level of excitation
$\quad g_A = g_B \equiv 1$
gives
$$N_n / N_1 = \exp[(E_1 - E_n)/(kT)]$$
For convenience, take the energy of the ground level = 0 eV.
$$N_n / N_1 = \exp\{-E_n / [(8.61 \times 10^{-5} \text{ eV/K})(6000K)]\}$$
$$= \exp(-E_n / 0.517 \text{ eV})$$

The energy levels are given by (Equation 8-24):
$$E_n = 13.6 \text{ eV}(1 - 1/n^2)$$
so
$$N_n / N_1 = \exp[-13.6 \text{ eV}(1 - 1/n^2)/0.517 \text{ eV}] = \exp[-26.3(1 - 1/n^2)]$$
Evaluating for several different energy levels in neutral hydrogen at T = 6,000 K:

n	E_n (eV)	N_n / N_1	log (N_n / N_1)
1	0	1.00	0.0
2	10.19	2.7 x 10^{-9}	-8.57
3	12.07	7.1 x 10^{-11}	-10.15
4	12.73	2.0 x 10^{-11}	-10.70
5	13.04	1.1 x 10^{-11}	-10.96
10	13.46	4.9 x 10^{-12}	-11.31
100	13.599	3.8 x 10^{-12}	-11.42

These values are best plotted on a semi-log graph (of log $[N_n / N_1]$ vs. n).

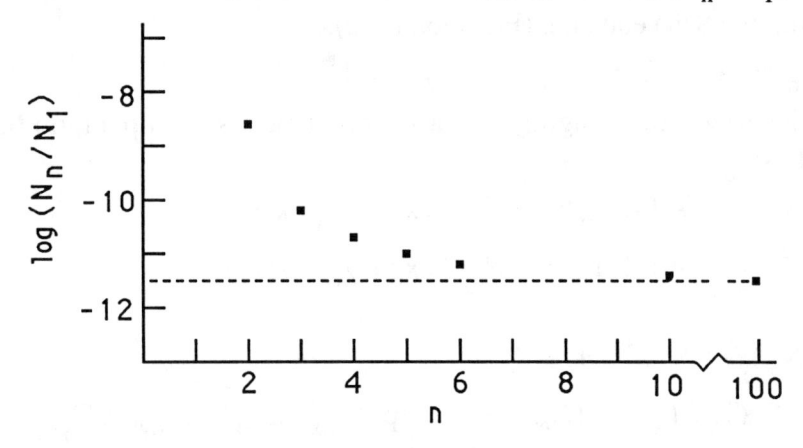

Similarly, at T = 15,000 K,

$$N_n / N_1 = \exp\{-E_n / [(8.61 \times 10^{-5} \text{ eV / K})(15{,}000\text{K})]\}$$
$$= \exp(-E_n / 1.292 \text{ eV})$$
$$= \exp[13.6 \text{ eV}(1 - 1/n^2) / 1.292 \text{ eV}] = \exp[-10.5(1 - 1/n^2)]$$

Evaluating for several different energy levels in neutral hydrogen at T = 15,000 K:

n	E_n (eV)	N_n / N_1	log (N_n / N_1)
1	0	1.00	0.0
2	10.19	3.8 x 10^{-4}	-3.42
3	12.07	8.8 x 10^{-5}	-4.05
4	12.73	5.3 x 10^{-5}	-4.28
5	13.04	4.2 x 10^{-5}	-4.38
10	13.46	3.1 x 10^{-5}	-4.51
100	13.599	2.8 x 10^{-5}	-4.56

The population of the upper levels increases at the higher temperatures relative to the population at lower temperatures, although at all temperatures the upper levels are less populated than the lower levels.

Note that we can generalize the relative populations of a level at two temperatures, Ta and Tb,

$$[N_n / N_1]_{Tb} = ([N_n / N_1]_{Ta})^{Ta/Tb}$$

8-8. To understand the relative importance of the different parameters in the Saha equation, perform the following experiment. Assume that T = 5000K, $N_e = 10^{15}/\text{cm}^3$, and $\chi = 12$ eV. By what factor does the ionization ratio (N_+ / N_0) change when we separately
  (a) double the temperature.

Which is more important during the temperature change, the exponential term or the $T^{3/2}$ term?
Beginning with the Saha equation (Equation 8-30):
$$N_+/N_o = A[(kT)^{3/2}/N_e]\exp(-\chi_o/kT)$$
to consider the effects of changing the various parameters, set up a ratio between the ionization ratios:
$$(N_+/N_o)_x = A[(kT_x)^{3/2}/N_{ex}]\exp(-\chi_{ox}/kT_x)$$
$$(N_+/N_o)_y = A[(kT_y)^{3/2}/N_{ey}]\exp(-\chi_{oy}/kT_y)$$
such that:
$$(N_+/N_o)_y/(N_+/N_o)_x =$$
$$(T_y/T_x)^{3/2}(N_{ex}/N_{ey})\exp[(\chi_{ox}/kT_x) - (\chi_{oy}/kT_y)]$$

Initially, $T = 5000$K, $N_e = 10^{15}$ cm^{-3}, $\chi_o = 12$ eV. Evaluating the exponential:
$$\exp(-\chi_{ox}/kT_x) = \exp\{(-12\text{ eV})/[(6.61 \times 10^{-5}\text{ eV/K})(5000\text{K})]\}$$
$$= \exp(-2.79 \times 10^1) \approx 7.64 \times 10^{-13}$$

Doubling the temperature causes the exponent to be reduced by the factor:
$$e^{(z - z/2)} = e^{z/2} = (e^{-z})^{-(1/2)},$$
where we have used: $\exp(a \times b) = [\exp(a)]^b$.
We see that the exponential is reduced to the square root of its previous value.
$$(N_+/N_o)_y/(N_+/N_o)_x = (2)^{3/2}(1)(7.64 \times 10^{-13})^{-1/2}$$
$$= 3.2 \times 10^6$$
So doubling the temperature increases the ionization ratio by 3.2 million times.

Note that the exponential term is much more important than the $T^{3/2}$ term in determining the ionization potential.

(b) double the electron density.
Doubling the electron density only influences the one factor:
$$(N_+/N_o)_y/(N_+/N_o)_x = (N_{ex}/N_{ey}) = 1/2$$
so the ionization ratio is half the previous value (the increased density results in additional recombinations).

(c) double the ionization potential.
Doubling the ionization potential only influences the exponential. Following arguments similar to those when we doubled the temperature (except now the numerator in the exponent doubles), doubling the ionization potential causes the exponent to be increased by a factor:
$$e^{(z - 2z)} = e^{-z} = (e^{-z})^{+1}.$$
So,

$$(N_+/N_o)_y / (N_+/N_o)_x = \exp[(\chi_{ox}/kT_x) - (\chi_{oy}/kT_y)]$$
$$= 7.64 \times 10^{-13} = 0.764 \times 10^{-12}$$

Doubling the ionization potential decreases the ionization ratio by a factor of about one trillionth.

The ionization ratio is very sensitive to changes in temperature or ionization potential, and relatively insensitive to changes in electron density.

8-9. Let N_2 be the number of second-level (first excited state) hydrogen atoms and N_1 be the number in the ground state. Using Figure 8-13, find the excitation ratio (N_2/N_1) and the excited fraction (N_2/N) for each of the following stars:

In producing Figure 8-13, the excitation ratio (N_2/N_1) of the number of neutral hydrogen atoms in the second energy level compared to the number in the first level was calculated from the Boltzmann equation; and the excited fraction (N_2/N) of the number of neutral hydrogen atoms in the second energy level compared to the total number of (ionized and un-ionized) atoms from combining the Boltzmann and Saha equations.

Using Figure 8-13 to estimate these ratios:
(a) Sirius, T = 10,000 K
$$\log(N_2/N_1) \approx -4.8 \qquad N_2/N_1 \approx 2 \times 10^{-5}$$
$$\log(N_2/N) \approx -5.2 \qquad N_2/N \approx 6 \times 10^{-6}$$

(b) Rigel (T = 15,000K):
$$\log(N_2/N_1) \approx -3.0 \qquad N_2/N_1 \approx 1 \times 10^{-3}$$
$$\log(N_2/N) \approx -6.2 \qquad N_2/N \approx 6 \times 10^{-7}$$

(c) the Sun (T = 5,800K):
$$\log(N_2/N_1) \approx -9 \qquad N_2/N_1 \approx 1 \times 10^{-9}$$
$$\log(N_2/N) \approx -9 \qquad N_2/N \approx 1 \times 10^{-9}$$

Which star will exhibit the strongest Balmer absorption lines? Explain your reasoning in arriving at this answer.

In order to see strong Balmer absorption, we need a large number of atoms to be neutral and in the n = 2 energy level (N_2/N large). Sirius will show the strongest Balmer absorption lines. Rigel is too hot -- even though many of those atoms which are neutral are in the n = 2 level, there are fewer atoms in the neutral state so the Balmer lines are weak. The Sun is too cool -- even though most atoms are neutral, the neutral atoms are not highly excited so the Balmer lines are weak.

8-10. (a) What is the speed of an electron with just sufficient energy to ionize by collision a sodium atom in the ground state?

From Table 8-3, we find that sodium has an ionization potential of 5.1 eV = 8.17 x 10^{-19} J. For an electron to just ionize this atom, its kinetic energy must be equal to the ionization potential.

$$(1/2)\, m_e\, v_e^2 = \chi_o$$
$$(1/2)\,(9.11 \times 10^{-31}\text{ kg})\, v_e^2 = 8.17 \times 10^{-19}\text{ J}$$
$$v_e^2 = 1.79 \times 10^{12}\text{ m}^2/\text{s}^2$$
$$v_e = 1.34 \times 10^6\text{ m/s} = 1{,}340\text{ km/s}$$

(b) What is the speed of a proton ionizing this atom?

Applying the same method for a collision by a proton gives:

$$(1/2)\, m_p\, v_p^2 = \chi_o$$
$$(1/2)\,(1.67 \times 10^{-27}\text{ kg})\, v_p^2 = 8.17 \times 10^{-19}\text{ J}$$
$$v_p^2 = 9.78 \times 10^8\text{ m}^2/\text{s}^2$$
$$v_p = 3.13 \times 10^4\text{ m/s} = 31.3\text{ km/s}$$

(c) What is the corresponding gas temperature?

Assuming perfect equilibrium and equating the ionization potential to the kinetic energy:

$$\chi_o = \langle (1/2)\, m\, v^2 \rangle = (3/2)\, k\, T$$
$$8.17 \times 10^{-19}\text{ J} = (3/2)\,(1.38 \times 10^{-23}\text{ J/K})\, T$$
$$T = 3.95 \times 10^4\text{ K} = 39{,}500\text{ K}$$

(d) At this temperature, what is the fractional thermal Doppler broadening ($\Delta\lambda/\lambda_o$) of a sodium spectral line?

The fractional thermal Doppler broadening is (Equation 8-13):

$$\Delta\lambda/\lambda_o = v/c$$

We need to calculate the speed for sodium atoms, $M_{Na} = 23\, m_H$:

$$(1/2)\,(23 \times 1.67 \times 10^{-27}\text{ kg})\, v_{Na}^2 = 8.17 \times 10^{-19}\text{ J}$$
$$v_{Na}^2 = 4.25 \times 10^7\text{ m}^2/\text{s}^2$$
$$v_{Na} = 6.52 \times 10^3\text{ m/s} = 6.52\text{ km/s}$$

Therefore,

$$\Delta\lambda/\lambda_o = (6.52 \times 10^3\text{ m/s})/(3.00 \times 10^8\text{ m/s})$$
$$= 2.2 \times 10^{-5}$$

Lines in the visible (500 nm) would have a thermal Doppler width of

$$\Delta\lambda = (2.2 \times 10^{-5})(500\text{ nm}) = 0.01\text{ nm}.$$

8-11. (a) How much more energy is emitted by a star at 20,000 K than one at 500 K?
The total energy per unit area emitted by a star at temperature T is (Equation 8-40):

$F = \sigma T^4$

Assuming that the two stars have the same radius, a star at 20,000 K emits

$(20,000 \text{ K} / 5,000 \text{ K})^4 = (4)^4 = 256$

times more energy than a star at 5,000K.

(b) What is the predominant color of each star in part (a)? Use the Wien displacement law, and express your answer in wavelengths.
The Wien displacement law gives the wavelength of maximum emission for a star at temperature T (Equation 8-39):

$\lambda_{max} T = 2.90 \times 10^{-3} \text{ m K} = 2.90 \times 10^6 \text{ nm K}$

For a T = 20,000K star:

$\lambda_{max} = (2.90 \times 10^6 \text{ nm K}) / (20,000 \text{K}) = 1.45 \times 10^2 \text{ nm} = 145 \text{ nm}$

this is in the ultraviolet, so the star would appear blue.
For a T = 5,000K star:

$\lambda_{max} = (2.90 \times 10^6 \text{ nm K}) / (5,000 \text{K}) = 5.80 \times 10^2 \text{ nm} = 580 \text{ nm}$

this is in the yellow part of the visible spectrum.

8-12. Deduce an approximate expression for Planck's radiation law (Equation 8-35a)
(a) at high frequencies ($h\nu/kT \gg 1$), the Wien distribution.
The Planck Law (Equation 8-35a) is

$I_\nu \Delta\nu = (2 h \nu^3 / c^2) [1 / \exp(h \nu / k T) - 1] \Delta\nu$

If $h \nu / k T \gg 1$, then $\exp(h \nu / k T) \gg 1$, so $\exp(h \nu / k T) - 1 \approx \exp(h \nu / k T)$.
This gives:

$I_\nu \Delta\nu = (2 h \nu^3 / c^2) [\exp(-h \nu / k T)] \Delta\nu$

which is the Wien Distribution for high frequencies (or low temperatures) (Equation 8-37a).

(b) at low frequencies ($h\nu/kT \ll 1$), the Rayleigh-Jeans approximation.
You may use the approximation $\exp(h\nu/kT) \approx 1 + h\nu/kT$ when $h\nu/kT \ll 1$. To what wavelength does $h\nu/kT = 1$ correspond?
If $h \nu / k T \ll 1$, then $\exp(h \nu / k T) \approx 1 + (h \nu / k T)$,

so $\exp(h \nu / k T) - 1 \approx (h \nu / k T)$.

This gives:

$I_\nu \Delta\nu = (2 h \nu^3 / c^2) [1 / (h \nu / k T)] \Delta\nu$

$= (2 h \nu^3 / c^2) (k T / h \nu) \Delta\nu$

$= 2 k T (\nu^2 / c^2) \Delta\nu$

which is the Rayleigh-Jeans approximation for low frequencies (Equation 8-38a).

If we set $h\nu/kT = 1$ and solve for λ, we get

$$\nu = kT/h = [(1.38 \times 10^{-23} \text{ J/K})/(6.625 \times 10^{-34} \text{ J s})] T$$
$$= (2.08 \times 10^{10} \text{ s}^{-1} \text{ K}^{-1}) T$$
$$\lambda = c/\nu = (1.44 \times 10^{-2} \text{ m K})/T$$

For a given temperature, we can estimate the wavelength ranges for which the above approximations will apply.

8-13. Estimate the line width from thermal Doppler broadening for the Ca II K line at 393.3 nm for a star at 3000 K; a star at 6000 K; and one at 12,000 K. Comment on the degree of importance of temperature on the Ca II K line broadening.

From Section 8-5B, the most probable speed of a Maxwellian distribution of particles is related to the temperature by

$$\langle v \rangle = (2kT/m)^{1/2}$$

For Calcium, $m = 40 \, m_H$,

$$\langle v \rangle = \{[2(1.38 \times 10^{-23} \text{ J/K}) T / [(40)(1.67 \times 10^{-27} \text{ kg})]\}^{1/2}$$
$$= 2.03 \times 10^1 \, T^{1/2}$$

At $T = 3{,}000$K: $\langle v \rangle = (2.03 \times 10^1) \, 3000^{1/2} = 1.11 \times 10^3 \text{ m/s} \approx 1.1 \text{ km/s}$
At $T = 6{,}000$K: $\langle v \rangle = (2.03 \times 10^1) \, 6000^{1/2} = 1.57 \times 10^3 \text{ m/s} \approx 1.6 \text{ km/s}$
At $T = 12{,}000$K: $\langle v \rangle = (2.03 \times 10^1) \, 12{,}000^{1/2} = 2.22 \times 10^3 \text{ m/s} \approx 2.2 \text{ km/s}$

The spectral line is broadened to twice the Doppler shift (Equation 8-13):

$$\Delta\lambda = 2(v/c)\lambda_o$$

At $T = 3{,}000$K: $\Delta\lambda = 2[(1.1 \text{ km/s})/(3.0 \times 10^5 \text{ km/s})](393.3 \text{ nm}) = 2.0 \times 10^{-3}$ nm
At $T = 6{,}000$K: $\Delta\lambda = 2[(1.6 \text{ km/s})/(3.0 \times 10^5 \text{ km/s})](393.3 \text{ nm}) = 4.2 \times 10^{-3}$ nm
At $T = 12{,}000$K: $\Delta\lambda = 2[(2.2 \text{ km/s})/(3.0 \times 10^5 \text{ km/s})](393.3 \text{ nm}) = 5.8 \times 10^{-3}$ nm

The line width increases linearly with the speed of the gas, which increases as the square root of the temperature. Hence the line width goes as the square root of the stellar temperature.

8-14. A star behind a nebula is 25% as bright at 500 nm as it would be if it were not behind the nebula. What is the optical depth of the nebula at 500 nm? (Assume that the nebula does not contribute to the light observed at 500 nm.)

In the transfer equation (Equation 8-42) the second term will be zero since the source function is zero ($S_\nu = 0$) -- the nebula does not contribute to the luminosity. Since the star is directly behind the nebula, $\mu \equiv \cos(\theta) = \cos(0°) = 1$. Letting the optical depth of the nebula be $\tau = \tau_2 - \tau_1$, the transfer equation becomes

$$I_\nu = I_\nu(0)\, e^{-\tau}$$

so $I_\nu / I_\nu(0) = 0.25 = e^{-\tau}$

$$\tau = -\ln(0.25)$$
$$= 1.39 \approx 1.4$$

The optical depth of the nebula is $\tau = 1.4$.

8-15. Given that the J = 1 to 0 rotational transition of carbon monoxide (CO) corresponds to a photon frequency of 115.27 GHz, estimate the distance between the carbon and oxygen atoms in a carbon monoxide molecule.

The molecular rotational energy levels are given by (Section 8-3C):

$$E_J = (h/2\pi)^2\, J(J+1) / 2\mu r^2$$

where J is the energy level, μ the reduced mass and r the atomic separation.

A transition from energy level J+1 to level J is given by

$$\Delta E = E_{J+1} - E_J = [(h/2\pi)^2 / 2\mu r^2]\, [(J+1)(J+2) - (J)(J+1)]$$
$$= [(h/2\pi)^2 / 2\mu r^2]\, [(J^2 + 3J + 2) - (J^2 + J)]$$
$$= [(h/2\pi)^2 / 2\mu r^2]\, [2J + 2]$$
$$= [(h/2\pi)^2 / \mu r^2]\, [J + 1]$$

The frequency corresponding to this transition is

$$\nu_{J+1 \to J} = \Delta E / h$$

so

$$\nu_{J+1 \to J} = [h / (4\pi^2 \mu r^2)]\, [J + 1]$$
$$= [(6.623 \times 10^{-34}\, J\, s) / (4\pi^2 \mu r^2)]\, [J + 1]$$
$$= [1.68 \times 10^{-35}\, J\, s / (\mu r^2)]\, [J + 1]$$

Rearranging the equation,

$$r^2 = (1.68 \times 10^{-35}\, J\, s)(J + 1) / (\mu)(\nu_{J+1 \to J}).$$

The reduced mass, μ, for carbon monoxide is

$$\mu = [mM / (m + M)] = [(12 \times 16)/(12 + 16)]\,(1.67 \times 10^{-27}\, kg)$$
$$= 1.15 \times 10^{-26}\, kg.$$

so

$$r^2 = (1.68 \times 10^{-35}\, kg\, m^2\, s^{-2}\, s)(0+1) / (1.15 \times 10^{-26}\, kg)(115.27 \times 10^9 /s)$$
$$= 1.27 \times 10^{-20}\, m^2$$
$$r = 1.13 \times 10^{-10}\, m = 1.13\, \text{Å} = 0.113\, nm$$

The separation between the carbon and oxygen atoms in a carbon monoxide molecule is

≈ 1/9 nm. (The Ångstrom unit, Å, is useful since atomic dimensions are on the order of an Å.)

8-16. Consider a pure hydrogen gas at a temperature of 10,080 K. What is the ratio of the populations of the ground state (n = 1) to the first excited state (n = 2)? Note that the energy difference is 10.2 eV between these two states. At what temperature would both levels have equal populations?
Use the Boltzmann equation (Equation 8-29):

$$N_B / N_A = (g_B / g_A) \exp[(E_A - E_B) / (kT)]$$

where $E_B > E_A$.

Note: Because the text gives no specific information about the values of the statistical weights, g, the instructor should provide them (or the students will assume that they have value 1).

For neutral hydrogen, $g_{n=1} = 2$, $g_{n=2} = 8$. We'll want the Boltzmann constant, k, in units of eV/K, so, from Appendix Table A7-2,

$$k = 1.38 \times 10^{-23} \text{ J/K} = (1.38 \times 10^{-23} \text{ J/K}) / (1.60 \times 10^{-19} \text{ J/eV})$$
$$= 8.61 \times 10^{-5} \text{ eV/K}.$$

So,

$$N_{n=2} / N_{n=1} = (8/2) \exp[(-10.2 \text{ eV}) / (8.61 \times 10^{-5} \text{ eV/K})(10080 \text{ K})]$$
$$= 4 \exp(-11.75) = 3.1 \times 10^{-5}$$
$$N_{n=1} / N_{n=2} = 3.2 \times 10^{4}$$

Both levels will have equal populations at

$$N_{n=2} / N_{n=1} = 1.0 = 4 \exp[(-10.2 \text{ eV}) / (8.61 \times 10^{-5} \text{ eV/K})(T)]$$
$$\ln(0.25) = (-10.2 \text{ eV}) / (8.61 \times 10^{-5} \text{ eV/K})(T)$$
$$T = [(-10.2 \text{ eV}) / (8.61 \times 10^{-5} \text{ eV/K})] / \ln(0.25)$$
$$= 8.5 \times 10^{4} \text{ K} = 85{,}000 \text{ K}$$

This is hotter than a typical stellar surface, but cooler than in the interior of a star.

8-17. Consider a pure hydrogen gas at a temperature of 10,080 K and a gas pressure of 10^{-5} atm. Calculate the ratio H^+/H. (*Hints:* $N_e = H^+$ and $N = N_e + H^+ + H$. The ionization potential is 13.6 eV.)
Use the Saha equation (Equation 8-30):

$$H^+ / H \equiv N_+ / N_0 = A [(kT)^{3/2} / N_e] \exp(-\chi_0 / kT).$$

Note: The constant A is not defined in the text, so the instructor must provide some information to the students.

$$A = (2/h^3)(2\pi m)^{3/2} (g_+ / g_0) \equiv A' (g_+ / g_0).$$

Taking $g_+ / g_0 = 1$,

$$A' \approx [2 / (6.62 \times 10^{-34} \text{ J s})^3][2\pi (9.11 \times 10^{-31} \text{ kg})]^{3/2}$$
$$\approx 9.4 \times 10^{55} \text{ kg}^{3/2} \text{ J}^{-3} \text{ s}^{-3}$$

Using the Boltzmann constant, k, in units of eV/K (see Problem 8-16)
$$k = 8.61 \times 10^{-5} \text{ eV/K}$$
the Saha equation becomes
$$H^+/H = A\,[(kT)^{3/2}/N_e]\exp(-\chi_o/kT)$$
$$= (9.4 \times 10^{55} \text{ kg}^{3/2} \text{ J}^{-3} \text{ s}^{-3})\{[(1.38 \times 10^{-23} \text{ J/K})(10080 \text{ K})]^{3/2}/N_e\}$$
$$\times \exp[-13.6 \text{ eV}/(8.61 \times 10^{-5} \text{ eV/K})(10080 \text{ K})]$$
$$= \{[4.88 \times 10^{27} \text{ kg}^{3/2} \text{ J}^{-3/2} \text{ s}^{-3})(\exp(-15.7)]\}/N_e$$
$$= 7.41 \times 10^{20} \text{ kg}^{3/2} \text{ J}^{-3/2} \text{ s}^{-3}/N_e$$
$$= 7.41 \times 10^{20} \text{ m}^{-3}/N_e$$

For pure hydrogen, $N_e = N_+ \equiv H^+$.
$$N = N_e + H^+ + H \equiv N_e + N_+ + N_o = 2N_e + N_o$$
$$\approx 2N_e \quad \text{if } N_+ \gg N_o$$
The ideal (or perfect) gas law (see Section 13-1A):
$$P = NkT \approx 2N_e kT$$
$$N_e = P/2kT$$
For $P \approx 10^{-5}$ atm = 1 N/m²,
$$N_e = 1 \text{ N/m}^2/[2(1.38 \times 10^{-23} \text{ J/K})(10080 \text{ K})]$$
$$= 3.6 \times 10^{18} \text{ m}^{-3}$$

Thus,
$$H^+/H = N_+/N_o$$
$$\approx 7.41 \times 10^{20} \text{ m}^{-3}/N_e = 7.41 \times 10^{20} \text{ m}^{-3}/3.6 \times 10^{18} \text{ m}^{-3}$$
$$= 2 \times 10^2 = 200$$
This justifies our assumption that $N_+ \gg N_o$.

[Note: Equations 13-11 or 13-12 could be used if the constants in these equations were known.]

8-18. A pure helium gas has a temperature of 15,000 K and a gas pressure of 10^{-5} atm. Calculate the ratio He^+/He. The ionization potential for He is 24.5 eV.
Follow the same method as in Problem 8-17, making similar assumptions.
For helium, take $g_+/g_o = 2$, so
$$A \equiv A'\,(g_+/g_o) \approx 9.4 \times 10^{55} \text{ kg}^{3/2} \text{ J}^{-3} \text{ s}^{-3}\,(g_+/g_o)$$

$$He^+ / He = A' (g_+ / g_o) [(kT)^{3/2} / N_e] \exp(-\chi_o / kT)$$
$$= (9.4 \times 10^{55} \text{ kg}^{3/2} \text{ J}^{-3} \text{ s}^{-3}) (2) \{[(1.38 \times 10^{-23} \text{ J/K})(15000 \text{K})]^{3/2} / N_e\}$$
$$\times \exp[-24.5 \text{ eV} / (8.61 \times 10^{-5} \text{ eV/K})(15000 \text{ K})]$$
$$= \{[1.77 \times 10^{28} \text{ kg}^{3/2} \text{ J}^{-3/2} \text{ s}^{-3})(\exp(-19.0)]\} / N_e$$
$$= 7.41 \times 10^{20} \text{ kg}^{3/2} \text{ J}^{-3/2} \text{ s}^{-3} / N_e$$
$$= 1.0 \times 10^{20} \text{ m}^{-3} / N_e$$

For $P \approx 10^{-5}$ atm = 1 N/m^2,
$$N_e = 1 \text{ N/m}^2 / [2 (1.38 \times 10^{-23} \text{ J/K})(15000 \text{ K})]$$
$$= 2.4 \times 10^{18} \text{ m}^{-3}$$

then
$$He^+ / He \approx 1.0 \times 10^{20} \text{ m}^{-3} / 2.4 \times 10^{18} \text{ m}^{-3}$$
$$= 4.1 \times 10^1 = 40$$

Helium is more difficult to ionize (greater ionization potential) than hydrogen, so a smaller fraction of atoms are ionized (even at this higher temperature) than for hydrogen in Problem 8-17.

8-19. The flux from the Sun above the Earth's atmosphere is about 1370 W/m^2. This quantity is called the solar constant S and equals πf (Sun). Use the angular radius of the Sun as seen from the Earth and find πF, the surface flux of the Sun.

The Sun's angular size in the sky (diameter $\approx 0.5°$; an apparent diameter of 32 arcsec would give a better approximation), coupled with the distance to the Sun (d = 1.5 × 10^{11} m) can be used to find the radius of the Sun, $R_\odot$.

$$\tan(\theta) = 2 R_\odot / d$$
$$R_\odot = (1.5 \times 10^{11} \text{ m})(\tan(0.5°))/2 = 6.5 \times 10^8 \text{ m} \approx 7 \times 10^8 \text{ m}$$

The solar constant or flux, πf, at 1 AU and the flux, πF, at the solar surface are related by
$$\pi f d^2 = \pi F R_\odot^2$$
$$\pi F = \pi f d^2 / R_\odot^2$$
$$= (1370 \text{ W/m}^2)(1.5 \times 10^{11} \text{ m})^2 / (6.5 \times 10^8 \text{ m})^2$$
$$= 7.3 \times 10^7 \text{ W/m} \quad \text{-- at the surface of the Sun}$$

Chapter 9: Telescopes and Detectors

9-1. Compare the resolving power and the light-gathering power of the human eye with those of a

(a) 10-cm telescope

Assume that a dark adapted human eye has a pupil diameter of 0.5 cm (Section 9-1). The minimum resolvable angle (or resolution) depends on the inverse of the diameter of the lens (the resolving power is the inverse of the resolution and thus depends on the diameter of the lens), and the light-gathering power depends on the area (see Section 9-1):

$$RP_1 / RP_2 = D_1 / D_2$$

$$LGP_1 / LGP_2 = (D_1 / D_2)^2.$$

The resolution is

$$1.22 \, \theta_{min} = 1.22 \times 206{,}265 \, \lambda/d \text{ arcsec}$$

or, for the eye (d = 0.5 cm = 5×10^6 nm) at visual wavelengths (λ = 500 nm) is

$$1.22 \, \theta_{min} = 1.22 \times 206{,}265 \, (5.0 \times 10^2 \text{ nm} / 5 \times 10^6 \text{ nm}) \text{ arcsec}$$

$$= 26 \text{ arcsec}$$

For a 10-cm telescope

$$RP_{10\text{-cm}} / RP_{eye} = 10 \text{ cm} / 0.5 \text{ cm} = 20$$

$$LGP_{10\text{-cm}} / LGP_{eye} = (10 \text{ cm} / 0.5 \text{ cm})^2 = 400$$

(b) 4-m telescope

For a 4-m (= 400-cm) telescope

$$RP_{4\text{-m}} / RP_{eye} = 400 \text{ cm} / 0.5 \text{ cm} = 800$$

$$LGP_{4\text{-m}} / LGP_{eye} = (400 \text{ cm} / 0.5 \text{ cm})^2 = 640{,}000 = 6.4 \times 10^5$$

9-2. For what astronomical purposes would one use a telescope having a long focal length and a large f-ratio?

A longer focal length telescope will produce a larger image and thus have a larger plate scale ($s \propto f$), a smaller image area, and larger f-ratio (see Section 9-1). In other words, the image will be spread out (magnified) more, the field of view will be less, and the image will be fainter. This characteristic can be useful for producing detailed images (on film or a CCD camera, perhaps) of a small region of the sky. The bright planets are particularly well suited for observations with long focal length telescopes.

9-3. The maximum possible antenna separation of the VLA is 40 km. What is the resolving power when operating at 1.5 cm? Why is it that the Earth's atmosphere does not limit the resolving power?

The minimum resolvable angle (in arcsec) is given by (Section 9-1):

$$\theta_{min} = 206{,}265 \, \lambda / d$$

where λ is the wavelength and d is the diameter of the objective (or, in this case, the largest separation of the interferometer elements). Since the system is an interferometer -- producing a double slit interference pattern, the 1.22 factor is not used here. λ and d must be expressed in the same units.

For the VLA operating at $\lambda = 1.5$ cm,

$$\theta_{min} = 206{,}265\ (1.5\ \text{cm} / 40\ \text{km})$$
$$= 206{,}265\ (1.5\ \text{cm} / (4.0 \times 10^6\ \text{cm}))$$
$$= 7.7 \times 10^{-2}\ \text{arcsec} = 0.077" \approx 0.08"$$

The Earth's atmosphere does not limit the resolving power of the VLA since the wavelength of radiation which it sees is much longer than is the case for optical telescopes. Effects of the turbulence of the air are insignificant at these radio wavelengths.

9-4. The transmission of the Earth's atmosphere at 3 μm is about 10%. Astronomers define the optical depth τ at some wavelength λ such that (Section 8-7)

$$I_\tau = I_o \exp(-\tau)$$

where I_o is the original intensity in a light beam and I_τ the intensity after passing through a material with optical depth τ. What is the optical depth through the Earth's atmosphere at the zenith at 3 μm?

At 3 μm, the transmission of the atmosphere is 10%, or

$$I(\tau) / I_o = e^{-\tau} = 0.1.$$

Thus,

$$\ln(e^{-\tau}) = \ln(0.1)$$
$$-\tau = -2.30$$
$$\tau = 2.30.$$

The optical depth of the Earth's atmosphere at 3 μm is 2.30. This value is quite large as compared with optical wavelengths -- the atmosphere is less transparent (or more opaque) at 3 μm.

9-5. Calculate the scale height for water vapor in the Earth's atmosphere. Use this information to state whether telescopes atop Mauna Kea would be better for infrared astronomy than ones at sea level.

This problem is similar to Problem 4-17.

The scale height is given by (see Section 4-5B for a discussion of scale height):

$$H = (kT) / (\mu m_H g)$$

where μ is the molecule weight of the atom or molecule being considered (=18 for water), m_H is the mass of the hydrogen atom, g is the gravitational acceleration, k is Boltzmann's constant, and T is the temperature.

For water, assuming a temperature of ≈ 300K,

$$H = [(1.38 \times 10^{-23}\ \text{J/K})(300\text{K})] / [(18)(1.67 \times 10^{-27}\ \text{kg})(9.8\ \text{m}/\text{s}^2)]$$
$$= 1.4 \times 10^4\ \text{m} = 14\ \text{km}$$

Mauna Kea is at an altitude of ≈ 5 km, or 1/3 of a scale height. The water vapor content is therefore reduced to about $e^{-1/3} \approx 70\%$ of the amount at sea level. (Mauna Kea is also above the cloud level.) This results in less absorption (lower opacity) of infrared radiation by the water absorption bands.

9-6. The shortest wavelength at which VLBI is routinely done is 1.3 cm.
(a) What is the minimum resolvable angle of an interferometer operating at 1.3 cm if the baseline is about one Earth diameter?
The minimum resolvable angle (considering the two antennae as a double slit, so the 1.22 factor does not apply) is:

$$\theta_{min} = 206{,}265 \, \lambda / d$$

where λ is the wavelength and d is the largest separation of the interferometer elements (Section 9-1).

For VLBI at 1.3 cm operating with a baseline of the Earth's diameter ($\approx 2 \times 6400$ km $\approx 1.3 \times 10^9$ cm),

$$\theta_{min} \approx 206265 \, [1.3 \text{ cm} / (1.3 \times 10^9 \text{ cm})]$$
$$\approx 2.1 \times 10^{-4} \text{ arcsec} = 0.21 \text{ milli-arcsec}$$

(b) VLBI experiments are now being attempted at a wavelength of 3 mm. What is the resolving power at this wavelength for an Earth-diameter baseline?
At wavelengths of 3 mm, VLBI with Earth diameter baselines would have a minimum resolvable angle of

$$\theta_{min} \approx 206265 \, [0.3 \text{ cm} / (1.3 \times 10^9 \text{ cm})]$$
$$\approx 4.8 \times 10^{-5} \text{ arcsec} \approx 0.05 \text{ milli-arcsec}$$

9-7. One of the disadvantages of an interferometer is that it is insensitive to objects with angular sizes much larger than the minimum resolvable angle of the shortest baseline (essentially because different portions of the source destructively interfere with each other). The flux from a source of angular diameter θ'' measured by an interferometer of baseline L (in kilometers) is approximately

$$F_{meas} \approx F_{true} \exp[-0.3(L \theta / \lambda)^2]$$

where λ is measured in centimeters. For a wavelength of 6 cm, what is the largest angular size of a source that can be observed with an interferometer of minimum baseline 1000 km if the lowest detectable flux is 0.1 F_{true}? What would be the properties of sources that one would *not* want to observe using an interferometer?

$$F_{meas} / F_{true} = 0.1 = \exp[-0.3(L \theta / \lambda)^2]$$

For L = 1000 km, and λ = 6 cm,

$$0.1 = \exp[-0.3(1000 \times \theta / 6)^2] = \exp(-8.33 \times 10^3 \, \theta^2)$$
$$\ln(0.1) = -2.30 = \ln[\exp(-8.33 \times 10^3 \, \theta^2)] = -8.33 \times 10^3 \, \theta^2$$
$$\theta^2 = -2.30 / (-8.33 \times 10^3) = 2.76 \times 10^{-4}$$

$\theta = 1.66 \times 10^{-2}$ arcsec $= 0.017" \approx 0.02"$

With this system one would not want to observe extended objects larger than $\approx 0.02"$.

9-8. (a) What is the theoretical resolution of the *Hubble Space Telescope* (objective diameter = 2.4 m) at
(i) $\lambda = 500$ nm (visible)

Because the Hubble Space Telescope (HST) is located above the Earth's atmosphere, its resolving power is limited by the optics not the atmosphere. With its high quality (corrected) optics it has nearly the theoretical resolving power. The space telescope has a diameter of 2.4 m.

At $\lambda = 500$ nm $= 5.0 \times 10^{-7}$ m, referring to Section 9-1 we see:

$1.22\, \theta_{min} = 1.22 \times 2.06 \times 10^5 \, \lambda / d$
$= 1.22 \times 2.06 \times 10^5 \, (5.0 \times 10^{-7}$ m $/ 2.4$ m$) = 0.052" \approx 0.05"$

(ii) $\lambda = 200$ nm (ultraviolet)

At the shorter ultraviolet wavelength $\lambda = 200$ nm $= 2.0 \times 10^{-7}$ m,

$1.22\, \theta_{min} = 1.22 \times 2.06 \times 10^5 \, (2.0 \times 10^{-7}$ m $/ 2.4$ m$) = 0.021" \approx 0.02"$

(iii) $\lambda = 2000$ nm (infrared)

At the longer infrared wavelength $\lambda = 2000$ nm $= 2.0 \times 10^{-6}$ m,

$1.22\, \theta_{min} = 1.22 \times 2.06 \times 10^5 \, (2.0 \times 10^{-6}$ m $/ 2.4$ m$) = 0.21" \approx 0.2"$

(b) Considering your answers to (a), why would it ever be advantageous to observe in the infrared with the space telescope?

Even though the HST does not have as good a resolving power in the infrared as it does in the optical or ultraviolet, it will still have many infrared applications. The molecules in the Earth's atmosphere produce many absorption bands in the infrared which make the atmosphere only partially transparent. The HST allows us to observe at infrared wavelengths which are not observable from the Earth's surface. The Hubble telescope can also be colder, thus decreasing its infrared emission.

9-9. What are the observational characteristics of an object that is better studied by the *Hubble Space Telescope* than by a very large ground-based optical telescope? What are the characteristics best suited for a very large ground-based optical telescope?

The corrected Hubble Space Telescope has the advantage of better resolving power and no atmospheric absorption. It is therefore most useful for observations requiring high resolving power, as for an object with much fine scale structure. Since the image of small angular size objects will not be spread out by "seeing" (as is true for ground based telescopes), faint objects can be recorded. The Space Telescope is also useful for observations at wavelengths where the atmosphere is opaque (such as at ultraviolet and infrared wavelengths).

A large ground-based optical telescope will not have as good resolution because of atmospheric limitations. It will, however, have a far greater light gathering power than the Space Telescope, thus being able to observe faint objects. For example, a 10-m telescope has $(10m / 2.4m)^2 \approx 17$ times the light gathering power of a 2.4-m telescope. The large ground-based telescope is then superior to the Space Telescope for objects that are faint but which do not require particularly good resolving power. In addition, the accessibility of ground-based telescopes allows for more easy interchange of detectors.

9-10. One of the main advantages of photographic plates or CCDs is their ability to integrate, that is, to collect light for a time much longer than the human eye can. The human eye effectively integrates for about 0.2 s (which is why we don't notice the individual frames of movie films). The flux of the faintest object that can be detected is proportional to $1 / t^{1/2}$, where t is the integration time. How much fainter an object can be detected using a 1-h integration on the Palomar 5-m telescope than can be seen with the naked eye?

The minimum detectable flux is proportional to the inverse of the integration time, or $F_{min} \propto 1 / (\sqrt{t})$. To compare the unaided eye to a one hour exposure with a CCD on the Palomar 5-m telescope, we have to consider both the increased light gathering power and integration time of the telescope system.

$$LGP_{Palomar} / LGP_{eye} = (500 \text{ cm} / 0.5 \text{ cm})^2 = 1.0 \times 10^6$$

$$F_{1hr} / F_{0.2s} = (0.2 / 3600)^{0.5} = 7.5 \times 10^{-3}$$

Combining these two factors gives that the 5-m telescope with a CCD camera can detect objects with a flux of 7.5×10^{-9} that of the human eye. This corresponds to objects which are about 20 magnitudes fainter (see Section 11-2 for a discussion of magnitudes).

An additional factor, which isn't presented in the problem, is that CCD detectors have higher quantum efficiency than the eye. That is, the CCD can convert nearly 100% of the incident photons into a detectable signal, while the human eye converts only about 1% of the incident light into a detectable signal. A good CCD integrating on the Palomar 5-m telescope for an hour can thus detect objects with a flux 7.5×10^{-11} that of the human eye, or ≈ 25 magnitudes fainter.

9-11. One of the major difficulties with gamma-ray astronomy is the poor resolving power, with angles less than 2° being unresolved even for the best instruments. Nevertheless, gamma-ray images of our Galaxy have been produced using satellite data. Discuss the difficulty in interpreting these images in light of the poor resolving power.

There are a number of difficulties with the $\approx 2°$ resolution of γ-ray telescopes. In a 2° field, there are many objects which could be contributing to the observed flux. It is thus difficult to tell which object is the source of the γ-rays. For example, in the galactic center there are a number of radio, infrared, and x-ray sources within a field smaller than 2°. If one observes γ-rays from the galactic center, one must decide which of the sources seen at other wavelengths are contributing to the total flux. It may even be possible that the γ-ray source may not be visible (or strong) at other wavelengths, making identification difficult. Also, it is not possible to accurately map extended structures, or to see detailed structure smaller than 2°. These limitations make it difficult to compare the γ-ray images with those obtained at other wavelengths.

9-12. Compare the theoretical resolution of a 5-m optical telescope at 500 nm with a 300-m radio telescope at 21 cm. Comment on why radio astronomers regularly use the technique of interferometry in order to map small angular scale structures.

The theoretical resolution (Section 9-1) is given by

$$\text{resolution} = 1.22\, \theta_{min} = 1.22\,(206265)\, \lambda/d \text{ arcsec}$$
$$= 2.52 \times 10^5\, \lambda/d \text{ arcsec}$$

For a 5-m telescope at 500 nm,

$$\text{resolution}_{5\text{-m}} = 2.52 \times 10^5\, (500 \times 10^{-9}\, m)/5\, m \text{ arcsec}$$
$$= 2.5 \times 10^{-2} \text{ arcsec} = 0.025 \text{ arcsec}$$

However, the atmospheric seeing limits, in practice, optical telescopes to ≈1 arcsec resolution.

For a 300-m telescope at 21 cm,

$$\text{resolution}_{300\text{-m}} = 2.52 \times 10^5\, (0.21\, m)/300\, m \text{ arcsec}$$
$$= 1.76 \times 10^2 \text{ arcsec} = 176 \text{ arcsec} \approx 3 \text{ arcmin}$$

this is ≈ 1/10 the angular diameter of the Moon.

Clearly, for radio astronomers to achieve the same angular resolution of optical astronomers (either the 0.025 arcsec theoretical resolution or the 0.3 - 1 arcsec practical resolution limit from atmospheric seeing), radio telescopes would need to be located several kilometers apart.

9-13. You are observing a faint star and would like 1% accuracy in your photometric measurement (e.g., signal to noise 100:1). After 5 minutes of integration time, you have a signal-to-noise ratio of 10:1 (10% accuracy).

(a) What is the total integration time required to achieve the desired signal-to-noise ratio of 100:1? Do you think it is worth striving for this accuracy of measurement?

The signal-to-noise ratio (Section 9-3D)

$$S/N = (QE \times f_p \times t)^{1/2}$$

depends on the quantum efficiency (QE) of the detector, the flux, f_p, of photons on the detector (which depends on the brightness of the object and the collecting area and focal length of the telescope) and integration time, t. QE and f_p are determined by the instrumentation and the object being observed, and presumedly cannot be changed for this observation.

So, effectively, the signal to noise ratio is:

$$S/N \propto t^{1/2}.$$

Therefore, to increase the S/N by a factor of 10, from 10:1 to 100:1, requires increasing the integration time by a factor of $10^2 = 100$. A 500 minute integration (8 1/3 hour) would be required to achieve the required accuracy. This is an entire night of summer time observing! The astronomer would have to decide whether one measurement with S/N 100:1 is more worthwhile to obtain than 10 measurements (of ten different astronomically interesting objects, perhaps) with S/N of 10:1.

(b) If you could increase the quantum efficiency of your detector by a factor of 10, what would be the total integration time required to achieve the desired signal to noise ratio of 100:1? Do you think it would now be worth striving for this accuracy of measurement?

If the detector efficiency is improved by a factor of ten, f_p increased by factor of 10, then the integration time has to be increased by only a factor of 10, to 50 minutes, in order to achieve the desired 100:1 S/N. This is more reasonable for the astronomer, since several objects could be observed in one observing night.

This is why astronomers want to build larger telescopes and improve detector efficiency. The development of high quantum efficiency CCD cameras (to nearly 100% efficiency) has made it possible for astronomers to study objects today which a decade or more ago were nearly impossible to study with the same telescopes. The next advance will necessarily be the design and construction of much larger telescopes, since little improvement can be made in detector efficiency.

9-14. A CCD detector is mounted at the focus of an f/14 reflecting telescope with a 40-cm diameter mirror. The CCD chip contains 500 by 500 pixels, with each pixel being 20 μm square.

(a) What is the angular size (in arcsec) of the sky which is imaged on each pixel?

The plate scale, s, for a telescope of focal length f is given by (Section 9-1)

$$s = 0.01745 \, f \quad [\text{per } °]$$

where the units of s are units of f per degree.

For s in units of f per arcsec,

$$s = 4.85 \times 10^{-6} \, f \quad [\text{per ''}].$$

The focal length is given in units of the telescope diameter, d,

$$f = d \times \text{f-ratio}$$

so

$$s = 4.85 \times 10^{-6} \times d \times \text{f-ratio per ''}$$

For the 40-cm telescope,

$$s = 4.85 \times 10^{-6} \times 0.40 \text{ m} \times 14 \text{ per ''} = 2.7 \times 10^{-5} \text{ m}/'' = 27 \text{ μm}/''$$

or, the angular size is

$$1/s = 1''/27 \text{ μm} = 0.037 ''/\text{μm}$$

For 20 μm pixels,

angular size per pixel = 20 μm × 0.037 ''/μm = 0.74'' ≈ 3/4 arcsec

(b) What is the angular field (in arcminutes) of the CCD chip?
For a 500 x 500 pixel detector, the field of view is
   # pixels x angular size/pixel = 500 x 0.74" = 370" = 6.17'.
The field of view of the detector is just over 6 arcmin square.

9-15. To appreciate how advances in telescope design and instrumentation have allowed astronomers to study previously undetectable faint objects, compare a present-day imaging system with that of two decades ago. Given identical exposure times, how much fainter an object can the Keck 10-m telescope equipped with a CCD detector of quantum efficiency 70% detect than the Mt. Palomar 5-m telescope equipped with a photographic plate (quantum efficiency 1%)?
Consider the signal received by a telescope-detector system (Section 9-3D)
   $S = QE \times f_p \times t$
where QE is the quantum efficiency of the detector and f_p is the flux received at the detector, which here will be a function of the telescope collecting area (square of the diameter), for an integration time t.

Comparing the Keck and Palomar telescopes,
   $S_{Keck} / S_{Palomar}$ = (0.70 / 0.01) (10 m / 5 m)^2
      = 280
The new Keck telescope equipped with the best of detectors can detect objects 280 times fainter than the Palomar telescope with detectors in common use over a decade ago.

9-16. The sodium D-lines in the solar spectrum have wavelengths of 589.59 and 589.00 nm.

(a) Given a grating with 10^4 lines / cm, what is the angular separation of the lines in *first* order?
The angles for constructive interference are given by (Section 9-4B)
   $\sin \theta = (n \lambda) / d$
where n is the order of the spectrum (here, n = 1) for wavelength, λ, and d is the groove spacing (here, 10^{-6} m). For λ = 589.59 nm,
   $\theta = \arcsin (n \lambda) / d$
      = arcsin (1 x 589.59 x 10^{-9} m) / 10^{-6} m = arcsin (0.58959)
      = 36.1279°
and for λ = 589.00 nm,
   $\theta = \arcsin (n \lambda) / d$
      = arcsin (1 x 589.00 x 10^{-9} m) / 10^{-6} m = arcsin (0.58900)
      = 36.0861°
The angular separation for these two lines in the first order spectrum is thus
   $\Delta \theta$ = 36.1279° - 36.0861° = 0.0418° = 150.5 arcsec

(b) Given a grating with 10^4 lines / cm, what is the angular separation of the lines in *second* order?
For the second order spectrum, for λ = 589.59 nm,
   θ = arcsin (2 x 589.59 x 10^{-9} m) / 10^{-6} m = arcsin (1.17918) -- no solution!

and for $\lambda = 589.00$ nm,

θ = arcsin (2 x 589.00 x 10^{-9} m) / 10^{-6} m = arcsin (1.17800) -- no solution!

There is no solution to either of these angles.

For both of these lines, using this grating there will not be a second order spectrum. Note that the angle for the first order spectrum is also fairly large ($\approx 36°$).

A grating with < 8000 lines per cm would be needed to see a second order spectrum.

9-17. Suppose two stars have an angular separation of θ arcsec on the sky. How far apart are these images at the focal plane of a telescope with focal length f?

The plate scale of a telescope is given by (Section 9-1)

$s_{[per\ °]} = 0.01745\ f$

where f is the focal length and s is the plate scale in units of f per degree.

If the separation on the sky is measured in arcsec rather than degrees (it usually is), then, since there are 3600 arcsec in one degree, the plate scale will be

$s_{[per\ "]} = (0.01745\ f) / (3600) = 4.847 \times 10^{-6}\ f$

If the stars are separated by an angle θ arcsec, they will be

$s\ \theta = 4.847 \times 10^{-6}\ f\ \theta$

apart at the focal plane of the telescope.

9-18. A telescope has a temperature of about 300 K. Why is this a problem at infrared wavelengths (≈ 10 μm)? Why is this *not* a problem at visual wavelengths?

Consider Wien's displacement law for a blackbody (Equation 8-39), which follows Planck's law

$\lambda_{max} = 2.896 \times 10^{-3}$ m K / T

If T ≈ 300 K, then

$\lambda_{max} = 2.896 \times 10^{-3}$ m K / 300 K ≈ 10^{-5} m = 10 μm

At a wavelength of 10 μm, the blackbody thermal emission of the telescope, dome, and Earth's atmosphere (which are all around T ≈ 300 K) is near maximum. Hence doing infrared astronomy, especially near 10 μm, is roughly equivalent to doing optical astronomy from a city with very bright lights on in the dome.

The thermal emission from the dome and telescope are not a problem for optical astronomy, since the Planck blackbody curve falls off rapidly at wavelengths shorter than 10 μm for T = 300 K (see Figure 8-14).

9-19. Recall [Section 4-5B] that the pressure distribution of an atmosphere can be described by the barometric equation, where the scale height H is the important factor. Analogously, you can relate the number density to height h above the Earth's surface:

$n(h) = n(0) \exp(-h / H)$

where n(0) is the density of a particular atmospheric constituent at sea level. Now the optical depth (Section 8-7) is proportional to the number of particles in a column through the atmosphere.

(a) Write down an expression for the column density N(h) above an altitude h.

The column density is the number of particles per area of the entire column, so we have to integrate the number density from the altitude h to the edge of the atmosphere.

111

$$N(h) = {}_h\!\int^\infty n(h)\, dh = {}_h\!\int^\infty n(0)\, e^{-h/H}\, dh$$

Since n(0) is a constant, it can be brought outside the integral. Using $\int e^{ax}\, dx = (1/a)e^{ax}$

$$\begin{aligned}
N(h) &= n(0)\, {}_h\!\int^\infty e^{-h/H}\, dh \\
&= -n(0)\, H\, [e^{-h/H}]\big|_h^\infty \\
&= -n(0)\, H\, [e^{-\infty} - e^{-h/H}] \\
&= n(0)\, H\, e^{-h/H} \\
&= H\, n(h)
\end{aligned}$$

The column density is simply the scale height times the density at that height.

(b) Looking at the zenith from sea level, the optical depth in the near infrared is about 3. Water vapor contributes most to this opacity; its scale height is about 2 km. Compare the optical depths at zenith for infrared telescopes at 2, 4, and 10 km. What do you conclude about the placement of an infrared telescope?

At sea level,
$$\begin{aligned}
N(0\text{ km}) &= H\, n(0)\, e^{-(0\text{ km})/(2\text{ km})} = H\, n(0)\, e^{-0} \\
&= H\, n(0)
\end{aligned}$$
-- corresponding to an optical depth of about 3

At a height of 2 km,
$$N(2\text{ km}) = H\, n(0)\, e^{-(2\text{ km})/(2\text{ km})} = H\, n(0)\, e^{-1}$$
$$= 0.37\, H\, n(0) = 0.37\, N(0\text{ km})$$
-- corresponding to an optical depth of about $(0.37)(3) \approx 1.1$

At a height of 4 km,
$$N(4\text{ km}) = H\, n(0)\, e^{-(4\text{ km})/(2\text{ km})} = H\, n(0)\, e^{-2}$$
$$= 0.14\, H\, n(0) = 0.14\, N(0\text{ km})$$
-- corresponding to an optical depth of about $(0.14)(3) \approx 0.4$

At a height of 10 km,
$$N(10\text{ km}) = H\, n(0)\, e^{-(10\text{ km})/(2\text{ km})} = H\, n(0)\, e^{-5}$$
$$= 0.0067\, H\, n(0)$$
-- corresponding to an optical depth of about $(0.0067)(3) \approx 0.02$

The infrared opacity from water vapor decreases significantly with altitude (with a scale height much comparable to the height of mountains). Hence, it is advantageous to locate an infrared telescope on a mountain rather than at sea level. The optimum location of an infrared telescope would be even higher than a mountain, such as in an airplane (or of course, in space!). This is why astronomers fly infrared telescopes in airplanes and balloons.

9-20. A main goal with the next generation of large telescopes is to optimize their performance in the infrared. Why? At a site with excellent seeing and / or adaptive optics, a 10-m telescope can operate at its diffraction limit. What is that limit at 2 μm? At 10 μm? How do these values compare to the best seeing at an excellent site?

Astronomers are interested in studying the sky at infrared wavelengths since many types of objects emit a significant fraction of their energy at these wavelengths. Thermal emission from blackbodies with temperatures of 10's to 100's of degrees Kelvin peaks at infrared wavelengths. Of particular interest are "cool" objects, such as stars in their early stages of formation and dust clouds surrounding star forming regions (see Chapter 15).

The theoretical diffraction limit is given in Section 9-1,

$$\theta_{min} = (1.22)(206,265 \, \lambda / d)$$

where θ_{min} is in arcsec and the wavelength, λ, and telescope diameter, d, are in the same units. At 2 μm, a 10-m telescope has a theoretical resolution of

$$\theta_{min} = (1.22)(206,265)(2 \times 10^{-6} \text{ m}) / (10 \text{ m}) = 0.05 \text{ arcsec},$$

and at 10 μm,

$$\theta_{min} = (1.22)(206,265)(10 \times 10^{-6} \text{ m}) / (10 \text{ m}) = 0.25 \text{ arcsec}.$$

The theoretical diffraction limit at 10 μm is comparable to the atmospheric seeing limit under the best conditions. Hence, ground-based infrared telescopes can, under optimum conditions, achieve diffraction limited seeing.

Chapter 10: The Sun: A Model Star

10-1. Show that the Sun is not at rest in the center of the Solar System by calculating the distance from the Sun's center to the center of mass of the Sun-Jupiter system.

The distance, d, of the Sun's center from the center of mass of the Sun - Jupiter system is given by (see Figure 1-14; Equation 1-21):

$$d\,M_\odot = (D - d)\,M_J$$

where D is the Sun - Jupiter distance, $M_\odot$ is the mass of the Sun, and M_J is the mass of Jupiter.

$$d\,(M_\odot + M_J) = D\,M_J$$
$$d = D\,(M_J / M_\odot + M_J) \approx D\,(M_J / M_\odot) \text{ since } M_\odot \gg M_J$$
$$d = 7.8 \times 10^8 \text{ km } (1.9 \times 10^{27} \text{ kg} / 2.0 \times 10^{30} \text{ kg})$$
$$= 7.4 \times 10^5 \text{ km} \approx 1.06\,R_\odot$$

Thus the center of the Sun-Jupiter system is just outside the surface of the Sun.

10-2. Find the thermal Doppler width of a spectral line at 500 nm formed in the Sun's photosphere (T ≈ 5400 K).

To find the thermal Doppler width of a spectral line in the Sun's photosphere, we must first find the speed of the atom (or molecule). From Section 8-5B, the most probable speed is

$$v_p = (2\,k\,T / m)^{1/2}.$$

Since the problem does not specify what atom is producing the line, let us assume the general case where the mass of the atom is $m = z\,m_H$, where z is the atomic mass of the atom and m_H is the mass of the hydrogen atom = 1.67×10^{-27} kg. Thus,

$$v^2 = 2\,k\,T / (z\,m_H)$$
$$= (1/z)\,[2 \times (1.38 \times 10^{-23} \text{ J / K})\,(5400 \text{ K}) / (1.67 \times 10^{-27} \text{ kg})]$$
$$= (1/z)\,(8.93 \times 10^7 \text{ m}^2/\text{s}^2)$$

so, $v = (9.45 \times 10^3 / \sqrt{z})$ m/s

The spectral line is broadened to twice the Doppler shift (Equation 8-13; see also Section 8-5B):

$$\Delta\lambda = 2\,\lambda_o\,(v / c)$$
$$= 2 \times 500 \text{ nm } [9.45 \times 10^3 \text{ m/s} / (\sqrt{z})] / (3.0 \times 10^8 \text{ m / s})$$
$$= (3.1 \times 10^{-2} \text{ nm}) / (\sqrt{z})$$

For hydrogen, z = 1, the line width would be 0.031 nm. For atoms of atomic mass z, the line width would be $0.031 / \sqrt{z}$ nm.

10-3. (a) Given that the photosphere is at a temperature of 6000 K, would you expect collisional or radiative excitations to be more important in exciting hydrogen atoms to the second (n = 2) level?

For radiative excitation from the n=1 to n=2 level in the Sun's photosphere, we need a significant number of Lyman-alpha photons (λ = 121.6 nm). Looking at the diagram of the Planck curve for a 6000 K blackbody (Figure 8-14), we see that there are very few Ly-α photons. Also, from the excitation curve (Figure 8-13), we see that N_2/N_1 is very small for T = 6000 K.

To estimate the importance of collisional excitations, we can compare the velocity required for collisional excitation to the average velocity of atoms at a temperature of 6000 K. For this excitation, the collision must impart 10.2 eV to an atom in the n=1 state. Consider an electron colliding with the H atom in the ground state, and compute the velocity required for an electron to impart 10.2 eV of energy.

$$(1/2) m_e v^2 = 10.2 \text{ eV} = 1.63 \times 10^{-18} \text{ J}$$

$$v^2 = 3.26 \times 10^{-18} \text{ J} / m_e$$

$$= (3.26 \times 10^{-18} \text{ kg m}^2/\text{s}^2) / (9.11 \times 10^{-31} \text{ kg})$$

$$= 3.58 \times 10^{12} \text{ m}^2/\text{s}^2$$

$$v = 1.9 \times 10^6 \text{ m/s}$$

The mean velocity of electrons at this temperature is given by (Equation 8-28):

$$(1/2) m_e v^2 = (3/2) k T$$

$$v^2 = 3 k T / m_e = 3 (1.38 \times 10^{-23} \text{ J/K})(6000 \text{ K}) / (9.11 \times 10^{-31} \text{ kg})$$

$$= 2.73 \times 10^{11} \text{ m}^2/\text{s}^2$$

$$v = 5.2 \times 10^5 \text{ m/s}$$

We could have taken a shortcut by noticing that the thermal energy (3/2) k T at 6000K (=1.24 $\times 10^{-19}$ J) is less than the excitation energy from the n=1 to n=2 level. This would hold for any particle in the photosphere -- the ratio of the required to average velocity of any particle (electron, proton, etc.) is independent of the particle mass.

Thus we see that the average velocity of electrons and atoms at 6000 K is a factor of four lower than the velocity required for collisional excitation of the hydrogen atom from the n=1 to n=2 level. However, the wings of the Boltzmann velocity distribution are wide enough that some electrons have the necessary velocity (energy) for some collisional excitations to occur.

(b) Would you expect the Lyman α line to appear in emission or absorption?

If collisional excitation is more important than radiative excitation, we would expect to see Lyman α in emission rather than absorption. However the emission would be weak since the population of the n = 2 level would be small.

10-4. Consider the following two lines of similar excitation potential: Fe I at 414.4 nm and Fe II at 417.3 nm. Explain in general terms (with reference to the Boltzmann and/or Saha equation) why the 414.4-

nm line is the stronger of the two in the photospheric spectrum and the weaker of the two in the chromospheric spectrum.

The lines correspond to comparable energy transitions in the neutral and singly ionized atoms. To understand why the Fe**I** line at 414.4 nm is stronger than the Fe**II** line at 417.3 nm in the Sun's photosphere but weaker in the chromosphere we need to investigate the level of ionization of the iron atom in these two solar layers. From the Saha equation (Equation 8-31; see Section 8-4B for definition of variables),

$$(N_{i+1} / N_i) = [A (k T)^{3/2} / N_e] \exp[-\chi_i / k T]$$

we see that there is a strong dependence of the state of ionization on the temperature. As the temperature increases from the photosphere to the chromosphere layers, the number of atoms in the higher excitation state (N_{i+1}) will increase. Applied to the problem at hand, at the lower temperatures of the photosphere more iron atoms will be un-ionized, so the neutral Fe**I** lines will be stronger; in the hotter chromosphere more iron atoms will be singly ionized (but it is not hot enough to doubly ionize iron), so the Fe**II** lines will be stronger.

10-5. From our description of chromospheric and coronal radio opacities, describe how you would determine the motion of a solar radio burst through the solar atmosphere.

The solar corona is optically thin (transparent) at short wavelengths ($\approx$1 cm), less transparent at longer wavelengths, becoming optically thick (opaque) at 10 - 20 cm. So when observing at short wavelengths we see the chromosphere. At longer wavelengths we see progressively higher in the solar atmosphere -- the lower corona, and then the upper corona at $\approx$10 cm. We can use this property to follow the motion of a solar radio burst by making time dependent observations over a broad range of wavelengths. When we observe a burst at $\approx$1 cm we are seeing activity in the chromosphere. As the event propagates to higher levels in the atmosphere, we will see enhanced emission at longer wavelengths. The different radio opacities of the solar atmosphere thus allow us to follow the path of a solar radio burst as it propagates from the lower chromospheric layers to the upper coronal layers.

10-6. How can one determine the temperature structure of a sunspot using only its continuous spectrum?

We can estimate the temperature structure of a sunspot by measuring the wavelength of the peak intensity in the continuous spectrum at different small angular diameter locations within the sunspot. We then would use Wien's Law (Equation 8-39):

$$T = 2.898 \times 10^{-3} \text{ m K} / \lambda_{max}$$

to determine the temperature at each location, and therefore the temperature structure of the sunspot.

10-7. How would you unambiguously assign a sunspot near the sunspot minimum to the old or new cycle?

At the start of a new sunspot cycle, the sense of magnetic polarity of the sunspot groups within a hemisphere reverses (i.e. whether the north magnetic pole within a region precedes or follows the south pole). Examining the sense of the polarity will distinguish between spots associated with an old or a new cycle.

Additionally, a new sunspot cycle starts with spots predominately at high solar latitudes (±35°). Near the end of a cycle the spots are generally located at lower latitudes (±8°) -- the Butterfly diagram (Figure 10-20). The latitude of an observed sunspot may thus also provide clues as to whether the spot is associated with the previous or current sunspot cycle.

10-8. Some prominences are said to have speeds greater than the escape speed from the Sun at the chromosphere. What is the critical speed?

The escape speed is given by (Equation 2-8):

$$v_{escape} = (2\,G\,M\,/\,R)^{1/2}.$$

For prominences in the Sun's chromosphere,

$$v_{escape} = [2 \times (6.67 \times 10^{-11}\ N\ m^2/kg^2) \times (1.99 \times 10^{30}\ kg)/(6.96 \times 10^8\ m)]^{1/2}$$

$$= 6.2 \times 10^5\ m/s = 620\ km/s$$

10-9. Using the data on the solar wind given in this chapter, compute the average rate of mass loss of our Sun ($M_\odot$ / year) from

(a) the solar wind

To estimate the Sun's mass loss rate from the solar wind we need to know the speed and particle density at some specified distance from the Sun. The solar wind is time variable, so let's take an "average" proton density and speed to calculate the mass loss (Section 10-5). We will assume also that most of the mass loss is via protons. At the Earth's orbit, a distance of 1 A.U., the proton density is typically $5 \times 10^6\ m^{-3}$ with a speed of 400 km / s.

To estimate the number of protons escaping from the Sun in a time interval Δt, consider a box of cross-sectional area A and length $L = v\,\Delta t$. In a time Δt the number of particles crossing an area A is given by: $N = v\,\Delta t\,A\,n$, where n is the particle density. The mass loss for protons is: $\Delta M = N\,m_H = v\,\Delta t\,A\,n\,m_H$. To find the total mass loss rate, take A to be the area of a 1 A.U. radius sphere.

$$A = 4\pi (1.5 \times 10^{11}\ m)^2 = 2.83 \times 10^{23}\ m^2$$

$$m_H = 1.67 \times 10^{-27}\ kg$$

$$v = 4.0 \times 10^5\ m/s$$

$$n = 5.0 \times 10^6\ m^{-3}$$

$$\Delta M / \Delta t = (4.0 \times 10^5\ m/s)(2.83 \times 10^{23}\ m^2)(5.0 \times 10^6\ m^{-3})(1.67 \times 10^{-27}\ kg)$$

$$= 9.5 \times 10^8\ kg/s$$

$$= 4.8 \times 10^{-22}\ M_\odot / s$$

$$= 1.5 \times 10^{-14}\ M_\odot / yr$$

(b) energy generation

The luminosity of the Sun is $L_\odot = 3.9 \times 10^{26}$ J/s. This energy is the result of mass loss in the proton-proton chain (and other fusion reactions) within the core of the Sun. Use the conversion of mass to energy ($E = mc^2$),

$$L = \Delta E / \Delta t = (\Delta m / \Delta t) c^2$$

$$\begin{aligned}(\Delta m / \Delta t) &= L / c^2 \\ &= (3.9 \times 10^{26} \text{ J/s}) / (3 \times 10^8 \text{ m/s})^2 \\ &= 4.3 \times 10^9 \text{ kg/s} \\ &= 6.9 \times 10^{-14} \text{ M}_\odot / \text{yr.}\end{aligned}$$

This mass loss in energy generation is of the same order of magnitude as the mass loss from the solar wind.

10-10. Calculate and compare the scale heights for hydrogen in the Sun's photosphere, chromosphere, and corona.

The scale height is given by (Section 4-5B):

$$H = kT / (\mu\, m_H\, g)$$

where μ is the molecular weight and g is the local acceleration of gravity.

For the Sun, $g_\odot = G M_\odot / R_\odot^2$.

Because the photosphere and chromosphere are relatively thin compared with the solar radius, we can use the same value of $g_\odot$ for all three layers.

$$\begin{aligned}g_\odot &= G M_\odot / R_\odot^2 \\ &= (6.67 \times 10^{-11} \text{ N m}^2/\text{kg}^2) \times (1.99 \times 10^{30} \text{ kg}) / (6.96 \times 10^8 \text{ m})^2 \\ &= 274 \text{ m/s}^2\end{aligned}$$

Assuming that the gas is predominately ionized hydrogen, $\mu = 0.5$,

$$\begin{aligned}H &= kT / (\mu\, m_H\, g) \\ &= \{(1.38 \times 10^{-23} \text{ J/K}) / [(0.5) \times (1.67 \times 10^{-27} \text{ kg}) \times (274 \text{ m/s}^2)]\}\, T \\ &= (6.0 \times 10^1 \text{ m K}^{-1})\, T\end{aligned}$$

In the photosphere, $T = 6000$ K, so

$$H_{photosphere} = 3.6 \times 10^5 \text{ m} = 360 \text{ km}$$

In the lower chromosphere, $T \approx 10{,}000$ K (dependent upon level within the layer), so

$$H_{chromosphere} = 6.0 \times 10^5 \text{ m} = 600 \text{ km}$$

In the corona, $T \approx 10^6$ K, so

$$H_{corona} = 6.0 \times 10^7 \text{ m} = 60{,}000 \text{ km}$$

10-11. Calculate the magnetic pressure exerted by a sunspot and compare it to the kinetic gas pressure in the photosphere. Conclusion? Note that the magnetic pressure is

$$P_{mag} = B^2 / 2\mu_o$$

where the constant $\mu_o = 4\pi \times 10^{-7}$ N/A^2. Recall that 1 T = 1 N/A·m.

[*NOTE:* This problem may be difficult for students since it uses physical concepts not covered at this point in the text. Refer to Section 13-1.]

For a sunspot of magnetic field 0.3 T = 0.3 N/A·m (Table 10-1),

$$P_{mag} = (0.3 \text{ T})^2 / [2 (4\pi \times 10^{-7} \text{ N/A}^2)] = 3.6 \times 10^4 \text{ N/m}^2$$

The kinetic pressure is given by the ideal gas law (Equation 13-1)

$$P_{kin} = nkT.$$

To estimate the density, n, use Figure 10-1 where the density of the photosphere is given to vary from 4×10^{-4} to 0.8×10^{-4} kg/m^3. Taking a value of 1×10^{-4} kg/m^3 and using the relationship (Equation 13-2) for the number density in terms of density, ρ, and reduced mass, μ,

$$n = \rho / \mu \, m_H .$$

If the photosphere is entirely ionized hydrogen, $\mu = 1/2$, so

$$n = (1 \times 10^{-4} \text{ kg/m}^3) / [(1/2)(1.67 \times 10^{-27} \text{ kg})]$$
$$= 1.2 \times 10^{23} \text{ m}^{-3}$$

and therefore, using a photospheric temperature of 5800 K,

$$P_{kin} = (1.2 \times 10^{23} \text{ m}^{-3})(1.38 \times 10^{-23} \text{ J/K})(5,800 \text{ K})$$
$$= 9.6 \times 10^3 \text{ N/m}^2$$

The kinetic pressure is thus a little (factor of 4) lower than the magnetic pressure. The magnetic field is thus important in determining the structure in sunspot regions.

10-12. A sunspot's umbra has a temperature of only 4000 K. At an equal optical depth of $\tau = 1$, the umbra is about 2600 K cooler than the photosphere. Calculate the umbral intensity contrast, I_u/I_p at 550 nm and 1.0 μm. Compare to the observed values of 0.1 at 550 nm and 0.23 at 1.0 μm. Comment?
The monochromatic intensity is given by (Equation 8-37b)

$$I_\lambda(T) = (2hc^2/\lambda^5) \exp(-hc/\lambda kT)$$

so comparing the umbral to photosphere intensity contrast

$$I_u/I_p = \exp(-hc/\lambda kT_u) / \exp(-hc/\lambda kT_p)$$
$$= \exp[(-hc/\lambda k)(1/T_u - 1/T_p)]$$
$$I_u/I_p = \exp[-\{(6.625 \times 10^{-34} \text{ J s})(3.0 \times 10^8 \text{ m/s})$$
$$/ \lambda (1.38 \times 10^{-23} \text{ J/K})\}(1/T_u - 1/T_p)]$$
$$= \exp[\{(-1.44 \times 10^{-2} \text{ m K})/\lambda\}(1/T_u - 1/T_p)]$$

For $\lambda = 550$ nm $= 5.5 \times 10^{-7}$ m,
$$I_u / I_p = \exp[\{(-1.44 \times 10^{-2} \text{ m K}) / (5.5 \times 10^{-7} \text{ m})\}(1/4000 \text{ K} - 1/6600 \text{ K})]$$
$$= 0.076 \quad \text{-- a bit less than the observed value of 0.1}$$

For $\lambda = 1$ μm,
$$I_u / I_p = \exp[\{(-1.44 \times 10^{-2} \text{ m K}) / (1 \times 10^{-6} \text{ m})\}(1/4000 \text{ K} - 1/6600 \text{ K})]$$
$$= 0.242 \quad \text{-- the same as the observed value of 0.23}$$

10-13. Estimate the angular size of a granule and supergranule as viewed from the Earth.

Granules and supergranules have physical sizes of 700 km (Section 10-2A) and 30,000 km (Section 10-2D). Using the small angle approximation,
$$\theta \approx \tan(\theta) = d / D \quad \text{[in radians]}$$
and in arc seconds
$$\theta \approx 206265 \, d / D \quad \text{[in arcsec]}$$
where d is the size of the solar feature and D is the distance to the Sun.

$$\theta_{granule} \approx 2.06 \times 10^5 \, (7.0 \times 10^2 \text{ km}) / (1.5 \times 10^8 \text{ km})$$
$$= 0.97 \text{ arcsec} \approx 1.0 \text{ arcsec}$$

$$\theta_{supergranule} \approx 2.06 \times 10^5 \, (3.0 \times 10^4 \text{ km}) / (1.5 \times 10^8 \text{ km})$$
$$= 41 \text{ arcsec}$$

For comparison, the Sun subtends about 1/2 degree in the sky.

10-14. Use the blackbody radiation law to estimate the wavelength of maximum intensity for the light from the Sun's photosphere.

Use the Wien's displacement law (Equation 8-39) to estimate the wavelength of maximum light for different temperature regions of the Sun:
$$\lambda_{max} = 2.898 \times 10^{-3} \text{ m K} / T$$

For the 5800 K photosphere,
$$\lambda_{max} = 2.898 \times 10^{-3} \text{ m K} / 5800 \text{ K}$$
$$= 5.0 \times 10^{-7} \text{ m} = 500 \text{ nm}$$

This is in the visible part ($\approx$ yellow-green) of the electromagnetic spectrum.

10-15. Using the Stefan-Boltzmann law with reasonable values for temperature and radius, estimate the energy output of the Sun's photosphere and the corona in watts. Compare your result for the photosphere and the corona to the value of the solar luminosity given in the text. Comment on the validity of using the blackbody approximation for the corona.

The luminosity or energy output for a layer on the Sun can be estimated from the Stefan-Boltzmann law (Equation 8-40 and Section 8-6C) for a blackbody:

$$L = 4\pi R^2 \sigma T^4$$
where R and T are the temperature of the region, respectively.

For the photosphere, with $R = R_\odot = 7 \times 10^8$ m and T = 5800 K,
$$L = 4\pi (7 \times 10^8 \text{ m})^2 (5.67 \times 10^{-8} \text{ W/m}^2 \text{ K}^4)(5800 \text{ K})^4$$
$$= 4.0 \times 10^{26} \text{ W}$$
This is comparable to the value given in Table 10-1, which is what we expect since the photospheric layer is the layer we see on the Sun.

For the corona, with $R \approx 2R_\odot = 1.4 \times 10^9$ m (the radius goes from ≈ 1 to $3 R_\odot$) and $T = 2 \times 10^6$ K,
$$L = 4\pi (1.4 \times 10^9 \text{ m})^2 (5.67 \times 10^{-8} \text{ W/m}^2 \text{ K}^4)(2 \times 10^6 \text{ K})^4$$
$$= 2.2 \times 10^{37} \text{ W}$$
The result for the corona is quite a bit (11 orders of magnitude!) greater than the Sun's luminosity. The reason for this is that the blackbody approximation is not valid for the corona, which is transparent. One must be careful in applying the blackbody approximations to objects or features which are not opaque blackbodies.

10-16. Consider a sphere of uniform density of radius R and absorption coefficient κ. We look into the sphere along a series of paths, passing different distances d from the center of the edge.

(a) Derive and expression for the optical depth τ as a function of d.

Refer to Figure 10-5 for the geometry being referred to. [Note that κ is called "absorption coefficient" in the problem but "opacity" in the text.]
The optical depth and absorption coefficient (opacity) are given by Equation 10-3:
$$d\tau = \kappa_\lambda \rho \, dx.$$
Integrating from x = 0 to x = d, assuming uniform density, ρ, and κ_λ,
$$\int_{\tau(x=0)}^{\tau(x=d)} d\tau = \int_0^d \kappa_\lambda \rho \, dx = \kappa_\lambda \rho \int_0^d dx$$
If $\tau_{x=0} = 0$,
$$\tau_d = \kappa_\lambda \rho \, d$$

(b) Calculate the rate of change of τ with d; that is, dτ/dd.
$$d\tau/dd = d(\kappa_\lambda \rho \, d)/dd = \kappa_\lambda \rho \, dd/dd = \kappa_\lambda \rho$$
The rate of change of the optical depth is a constant, assuming that both the density, ρ, and opacity, κ_λ, are constant.

(c) Use the result to discuss the observed sharpness of the solar limb.

Referring to the explanation of limb darkening in Section 10-2B and Figure 10-5, if the rate of change of the optical depth, τ, with depth, d, was constant, we would be able to see the same distance d into the interior areas of the Sun's surface (as in Figure 10-5A) -- so the intermediate regions would have some small darkening effect. Hence, the limbs would be gradual. The sharpness of the solar limb suggests that this constant rate of change is not the case, as shown in Figure 10-5B. The opacity is not constant, but rather increases with depth.

10-17. Consider the Sun's corona.

(a) For what reason can you not see it when you cover the Sun with your thumb?

We cannot see the Sun's corona when we cover the Sun's photosphere with our thumbs since we are located below the light scattering in the Earth's atmosphere. The scattered photospheric light in the sky is much brighter than the corona. Thus, although we have blocked light from the photosphere from directly reaching our eyes, the photospheric light scattered in the atmosphere is still visible -- as that nice bright shade of Carolina blue. The corona is visible during a solar eclipse because the Moon is located above the Earth's atmosphere and the scattering it produces. Hence, when the Moon blocks out the photosphere's disk of the Sun during an eclipse, there is little photospheric light being scattered by the earth's atmosphere into our eyes.

(b) The corona's temperature is about 2 million K. Why don't we observe the Sun as a blackbody emitter at this temperature?

A blackbody is an object that absorbs <u>all</u> the radiant energy that strikes it -- it would be opaque. The coronal density is so low that it is transparent to most radiant energy. Hence, the solar corona is <u>not</u> a blackbody. It is not even a good approximation. So we don't see the corona as a blackbody emitter.

(c) Why is the low density in the corona conducive to high levels of ionization?

According to the Saha equation (Equation 8-31)

$$N_{i+1} / N_i = A \left[(kT)^{3/2} / N_e \right] \exp(-\chi_i / kT)$$

the degree of ionization depends on the temperature and density of the gas. Note that N_{i+1} is inversely proportional to the electron density. The total density of the corona is proportional to the electron density (but not exactly proportional to it, due to multiple levels of ionization). Hence a low density, for a given temperature, is conducive to a higher degree of ionization. Note, however, that the high temperature of the corona is also a very important factor in the ionization state (see Problem 8-8). In a more intuitive way, the low density inhibits recombination once an atom is ionized. The free electron and ion have less chance of colliding and recombining.

Chapter 11: Stars: Distances and Magnitudes

11-1. Astronomers living on Mars would define their astronomical unit in terms of the orbit of Mars. If they defined parsec in the same manner as we do, how many Martian astronomical units would such a parsec contain? How many Earth astronomical units would equal a Martian parsec? How many Earth parsecs are there in a Martian parsec?

A Martian parsec would contain the same number of Martian astronomical units as our parsec contains our astronomical units because of the relation between a parsec and an AU does not depends on the size of the AU. (It depends on the number of arcsec in a radian.) So,
$$1 \, pc_{Mars} = 206{,}265 \, AU_{Mars}.$$

Since $1 \, AU_{Mars} = 1.524 \, AU_{Earth}$,
by simple scaling: $1 \, pc_{Mars} = 1.524 \, pc_{Earth}$
and
$$1 \, pc_{Mars} = 1.524 \times 206{,}265 \, AU_{earth} = 3.143 \times 10^5 \, AU_{Earth}.$$

Note that a Martian astronomer could determine distances by parallax with 1.5 times better accuracy than an Earth based astronomer (ignoring atmospheric effects); or a Martian astronomer could determine parallaxes for stars 1.5 times more distant than an Earth astronomer could, with comparable accuracy -- but it would take about twice as long to make these measurements since the Martian year is ≈ 2.1 times longer than the Earth year.

11-2. A variable star changes in brightness by a factor of 4. What is the change in magnitude?
A brightness ratio of 4 corresponds to a magnitude difference given by (Equation 11-5):
$$m - n = 2.5 \log (f_n / f_m)$$
$$= 2.5 \log (4) \approx 1.5.$$
The change is ≈ 1.5 magnitudes.

11-3. What is the combined apparent magnitude of a binary system consisting of two stars of apparent magnitudes 3.0 and 4.0?
Since the magnitude scale is logarithmic, you cannot just add the magnitudes of the two stars. Convert the magnitudes of the two stars to brightness (fluxes), add the brightnesses, and reconvert this to the magnitude scale. Using Equation 11-5:
$$m_1 - m_2 = 2.5 \log (f_2 / f_1)$$
$$\log (f_2 / f_1) = (3.0 - 4.0) / 2.5 = -0.4$$
$$f_2 / f_1 = 0.4$$
$$f_2 = 0.4 \, f_1$$
$$f_1 + f_2 = 1.4 \, f_1$$
Reconverting to the magnitude scale,
$$m_1 - m_{1+2} = 2.5 \log (f_{1+2} / f_1)$$

$$3.0 - m_{1+2} = 2.5 \log (1.4) = 0.37$$
$$m_{1+2} = 2.63$$

The combined magnitude of two stars of magnitude 3.0 and 4.0 is 2.63.

11-4. If a star has an apparent magnitude of -0.4 and a parallax of 0.3", what is
(a) the distance modulus?
The distance modulus is the difference between the apparent and absolute magnitude (Equation 11-8).
$$m - M = -5 - 5 \log (\pi"), \text{ where } \pi" \text{ is the stellar parallax in arc seconds.}$$
$$m - M = -5 - 5 \log (0.3) = -5 - 5 \times (-0.523) = -2.39 \approx -2.4$$

(b) the absolute magnitude?
$$M = m - (-2.4) = -0.4 + 2.4 = +2.0$$

11-5. What is the distance (in parsecs) of a star whose absolute magnitude is +6.0 and whose apparent magnitude is +16.0?
Using the distance modulus, where d is the distance in parsecs (Equation 11-6):
$$m - M = 5 \log (d) - 5$$
$$16 - 6 = 5 \log (d) - 5$$
$$\log (d) = 3$$
$$d = 10^3 \text{ pc} = 1{,}000 \text{ pc}$$

11-6. What are the absolute magnitudes of the following stars:
Write the distance modulus (Equations 11-7 and 11-8) as:
$$M = m + 5 - 5 \log (d_{pc})$$
$$M = m + 5 + 5 \log (\pi")$$

(a) m = 5.0, d = 100 pc
$$M = 5.0 + 5 - 5 \log (100) = 10.0 - 5 \times 2 = 0.0$$
(b) m = 10.0, d = 1 pc (Is there such a star?)
$$M = 10.0 + 5 - 5 \log (1) = 15.0 - 5 \times 0 = 15.0$$
There is no star this close to the Sun.
(c) m = 6.5, d = 250 pc
$$M = 6.5 + 5 - 5 \log (250) = 11.5 - 5 \times 2.4 = -0.5$$
(d) m = -3.0, d = 5 pc
$$M = -3.0 + 5 - 5 \log (5) = 2.0 - 5 \times 0.7 = -1.5$$
(e) m = -1.0, d = 500 pc
$$M = -1.0 + 5 - 5 \log (500) = 4.0 - 5 \times 2.7 = -9.5$$
(f) m = 6.5, parallax π = 0.004"
$$M = 6.5 + 5 + 5 \log (0.004) = 11.5 + 5 \times (-2.4) = -0.5$$

11-7. What would the expression for absolute magnitude be, in terms of apparent magnitude and distance, if absolute magnitude were defined as the magnitude a star would have at 100 pc?
If the absolute magnitude were defined at a standard distance of 100 pc rather than 10

pc, the absolute magnitude of stars would be 5 magnitudes more (fainter) since the increase in distance by a factor of 10 corresponds to a brightness decrease of 1/100 -- or 5 magnitudes.

So rather than the distance modulus being:

$$m - M = 5 \log (d) - 5, \text{ or } M = m + 5 - 5 \log (d_{pc})$$

the expressions would be:

$$m - M = 5 \log (d) - 10, \text{ or } M = m + 10 - 5 \log (d_{pc}).$$

11-8. The Sun has an apparent magnitude of -26.75.

(a) Calculate its absolute visual magnitude.

The Sun is at a distance of 1 A.U. = (1 / 206,265) pc = 4.85×10^{-6} pc, so using Equation 11-7:

$$M = m + 5 - 5 \log (d_{pc}) = -26.75 + 5 - 5 \log (4.85 \times 10^{-6})$$
$$= -21.75 - 5 \times (-5.314) = 4.82 \approx 4.8$$

The Sun, at a distance of 10 parsec, would as bright (faint) in the sky as the dimmer stars seen with the unaided eye.

(b) Calculate its magnitude at the distance of Alpha Centauri (1.3 pc).

At a distance of 1.3 pc, the Sun would have an apparent magnitude:

$$m = M - 5 + 5 \log (d_{pc}) = 4.8 - 5 + 5 \log (1.3) = -0.2 + 5 \times 0.114 = 0.4$$

The Sun would be one of the brighter stars in the sky from Alpha Centauri.

(c) The Palomar Sky Survey is complete to magnitudes as faint as +19. How far away (in parsecs) would a star identical to the Sun have to be in order to just barely be bright enough to be visible on Sky Survey photographs?

For the Sun to be barely visible on the Palomar Sky Survey, it would have an apparent magnitude, m = 19, so

$$m - M = 5 \log (d_{pc}) - 5$$

$$19 - 4.8 = 5 \log (d_{pc}) - 5$$

$$\log (d_{pc}) = 3.84$$

$$d = 6.9 \times 10^3 \text{ pc} = 6,900 \text{ pc} \approx 7 \text{ kpc}$$

A star identical to the Sun at a distance of 7 kpc would be barely visible on the Palomar Sky Survey. Those stars like the Sun further away than 7 kpc would not be visible.

11-9. Using the data from this chapter, determine the apparent magnitude difference between Sirius and the Sun, as seen from the Earth. How much more luminous is Sirius than the Sun?

From Problem 11-8, we know that $m_\odot$ = -26.75, and $M_\odot$ = 4.8.

For Sirius, m_{Sirius} = -1.47, and M_{Sirius} = 1.4 (Appendix Table A4-1).

The differential apparent magnitude is

$$m_\odot - m_{Sirius} = -26.75 - (-1.47) = -25.28$$

so, the Sun is ≈ 25 magnitudes brighter than Sirius as seen from the Earth.

This corresponds to a brightness ratio of (Equation 11-3):

$$f_\odot / f_{Sirius} \approx (100)^{25/5} = 100^5 = 10^{10} = 10 \text{ billion times brighter.}$$

The differential absolute magnitude is
$$M_\odot - M_{Sirius} = 4.8 - 1.4 = 3.4$$
Sirius is 3.4 absolute magnitude brighter than the Sun.

This corresponds to a luminosity difference of (Equation 11-3)
$$L_{Sirius} / L_\odot = 10^{0.4(4.8 - 1.4)} = 10^{1.36} = 22.9 \approx 23 \text{ times more luminous.}$$

11-10. A certain globular cluster has a total of 10^4 stars; 100 of them have $M_v = 0.0$, and the rest have $M_v = +5.0$. What is the integrated visual magnitude of the cluster?

There are 100 stars with absolute magnitude $M_v = 0.0$ and 9,900 with $M_v = +5.0$. Find the absolute magnitude of each group, convert to luminosity, add the luminosity from the two groups, then convert the luminosity of the ensemble to magnitudes.

For the 100 stars of $M = 0.0$ (Equation 11-5):
$$M_{1star} - M_{100stars} = 2.5 \log (L_{100stars} / L_{1star})$$
$$0.0 - M_{100stars} = 2.5 \log (100) = 2.5 \times 2 = 5.0$$
$$M_{100stars} = -5.0$$

For the 9,900 stars of $M = +5.0$:
$$M_{1star} - M_{9900stars} = 2.5 \log (L_{9900stars} / L_{1star})$$
$$5.0 - M_{9900stars} = 2.5 \log (9900) = 2.5 \times 4.0 = 10.0$$
$$M_{9900stars} = -5.0$$

Thus each group contributes an absolute magnitude of -5.0,
$$\text{or } L_{100starsofM=0} \approx L_{9900starsofM=+5} \,.$$

For the ensemble of all stars
$$M_{100stars} - M_{all} = 2.5 \log (L_{all} / L_{100stars})$$
$$-5.0 - M_{all} = 2.5 \log (2) = 0.75$$
$$M_{all} = -5.75.$$

The integrated absolute magnitude of the cluster is $M_v = -5.75$.

This problem could have been done simply by noting that 100 stars of $M = 0.0$ are 100 times (or 5 magnitudes) brighter than a 0.0 magnitude star; or $M = -5.0$. The 9,900 $\approx 10^4$ stars of $M = +5.0$ are $10^4 = (10^2)^2$ brighter (or 10 magnitudes) than a 5.0 magnitude star; or $M = -5.0$. Remembering that a factor of 2 in luminosity corresponds to 0.75 magnitude, we see that the sum of both groups is:
$$-5.0 - 0.75 = -5.75 \text{ magnitude.}$$

As an aside, note that if the globular cluster was at a distance of 10 kpc, its apparent magnitude would be (Equation 11-6):

$$m - M = 5 \log (d_{pc}) - 5$$
$$m = -5.75 + 5 \log (10,000) - 5 = -5.75 + 20 - 5 = 9.25.$$

11-11. The V magnitudes of two stars are both observed to be 7.5, but their blue magnitudes are $B_1 = 7.2$ and $B_2 = 8.7$.

(a) What is the color index of each star?

The color index is defined (Equation 11-9):

$$CI = B - V = m_B - m_V = M_B - M_V.$$
$$CI_1 = 7.2 - 7.5 = -0.3 \quad \text{-- the star is blue}$$
$$CI_2 = 8.7 - 7.5 = 1.2 \quad \text{-- the star is red}$$

(b) Which star is bluer and by what factor is it brighter at blue wavelengths that the other star?

Star 1 is the bluer star since the blue magnitude is less (brighter) than the visual magnitude. Star 1 is 1.5 magnitudes brighter in the blue than star 2, so star 1 is (Equation 11-3):

$$f_1 / f_2 = 100^{1.5/5} = 2.512^{1.5} = 4.0 \text{ times brighter in the B band than star 2.}$$

11-12. What is the color index of a star at a distance of 150 pc with $m_V = 7.55$ and $M_B = 2.00$?

The color index is defined (Equation 11-9):

$$CI = B - V = m_B - m_V = M_B - M_V.$$

Finding the absolute visual magnitude (Equation 11-6) from the information given:

$$m_V - M_V = 5 \log (d_{pc}) - 5$$
$$7.55 - M_V = 5 \log (150) - 5 = 5.88$$
$$M_V = 1.67$$

Now compute the color index,

$$CI = M_B - M_V = 2.00 - 1.67 = 0.33$$

-- The star is a bit redder than Vega, but would be more yellow than red.

11-13. What is the absolute bolometric magnitude of a star with a luminosity of 10^{33} W?

The luminosity and absolute bolometric magnitude are related by (Equation 11-20b):

$$\log (L_* / L_\odot) = 1.9 - 0.4 M_{bol(*)}$$
$$\log (10^{33} \text{ W} / 3.9 \times 10^{26} \text{ W}) = 1.9 - 0.4 M_{bol(*)}$$
$$\log (2.56 \times 10^6) = 6.409 = 1.9 - 0.4 M_{bol(*)}$$
$$M_{bol(*)} = -11.3$$

11-14. Given the expressions for the luminosity of a star (Equation 11-19) and its bolometric magnitude in terms of that of the Sun (Equation 11-20a), find an expression for the bolometric magnitude of the star as a function of its temperature and radius. The effective temperature of the Sun is 5780 K.

The luminosity of a star is (Equation 11-19)
$$L_* = 4\pi R_*^2 \sigma T_{eff*}^4 \text{ J/s},$$
and the bolometric magnitude in terms of that of the Sun is (Equation 11-20a)
$$M_{bol\odot} - M_{bol*} = 2.5 \log (L_*/L_\odot).$$
Substitute Equation 11-19 (for both the Sun and a star) into Equation 11-20a:
$$\begin{aligned}M_{bol\odot} - M_{bol*} &= 2.5 \log [(4\pi R_*^2 \sigma T_{eff*}^4)/(4\pi R_\odot^2 \sigma T_{eff\odot}^4)] \\ &= 2.5 \log [(R_*^2 T_{eff*}^4)/(R_\odot^2 T_{eff\odot}^4)] \\ &= 5.0 \log [(R_* T_*^2)/(R_\odot T_\odot^2)] \\ &= 5.0 \log [(R_*/R_\odot) \times (T_*/T_\odot)^2]\end{aligned}$$

Putting in values for the Sun, $T_{eff} = 5780$ K, $R = 6.96 \times 10^8$ m, $M_{bol} = +4.7$,
$$4.7 - M_{bol*} = 5.0 \log(R_* T_*^2) - 5.0 \log[(6.96 \times 10^8 \text{ m})(5780 \text{ K})^2]$$
$$M_{bol*} = 86.5 - 5.0 \log[R_{*(m)} T_{*(K)}^2]$$
where the star's temperature is in Kelvin and radius in meters.

11-15. The bolometric correction for a star is -0.4, and its apparent visual magnitude is +3.5. Find its apparent bolometric magnitude.

The bolometric magnitude can be found using the bolometric correction (Equation 11-21):
$$BC = m_{bol} - m_V = M_{bol} - M_V$$
$$m_{bol} = BC + m_V = -0.4 + 3.5 = +3.1$$

11-16. Two astronomers, located 100 km apart along a north-south line, simultaneously observe an asteroid near the zenith. Comparison of their observations indicates that the asteroid had a parallax of 5 arcsec. Estimate the distance to the asteroid in kilometers. How many times more distant is it than the Moon?

Using the small angle approximation,
$$\theta \approx \tan(\theta) = d/D \text{ [in radians]}$$
or
$$\theta \approx \tan(\theta) = 206265\, d/D \text{ [in arcsec]}$$

where θ is the observed parallax for an object at distance D seen between two sites of separation d.

The distance to the asteroid is
$$\begin{aligned}D &= 2.06 \times 10^{5}{}''\, d/\theta \\ &= 2.06 \times 10^{5}{}'' (1.0 \times 10^2 \text{ km})/5'' \\ &= 4.1 \times 10^6 \text{ km}\end{aligned}$$
The asteroid is about 4 million km distant.

The Moon's orbital semi-major axis is 3.84×10^5 km, so the asteroid is $1.07 \times 10^1 = 10.7 \approx 11$ times more distant than the Moon.

11-17. Observational astronomers often use a rule of thumb that a 1% change in brightness roughly corresponds to a change of 0.01 magnitude. Justify this approximation and comment on its validity.
The relationship between brightness ratios and magnitude differences is (Equation 11-5):
$$m - n = 2.5 \log (f_n / f_m).$$
For a 1% difference in brightness, $f_n / f_m = 1 \pm 0.01 = 1.01$ or 0.99.
For $f_n / f_m = 1.01$, $m - n = 2.5 \log (1.01) = 0.0108$
For $f_n / f_m = 0.99$, $m - n = 2.5 \log (0.99) = -0.0109$
The approximation that a 1% difference in brightness ratio corresponds to a 0.01 magnitude difference is good to better than 10%. This approximation becomes progressively less accurate for larger brightness ratios, particularly for brightness ratios greater than 20 or 30%.

11-18. What is the largest distance that a star of absolute magnitude -6 could be detected by the Palomar 5-m telescope? By the *Hubble Space Telescope*? [Use the limiting magnitudes given in Section 11-2.]
The limiting magnitudes for the Palomar 5-m telescope and Hubble Space telescope are +23.5 and +25, respectively. To determine the distance of a star of absolute magnitude -6 which can be seen with these instruments use Equation 11-6:
$$m - M = 5 \log(d_{pc}) - 5$$
$$d_{pc} = 10^{(m - M + 5) / 5}$$
For Palomar
$$d_{pc} = 10^{(23.5 - (-6) + 5) / 5} = 10^{6.9} = 7.9 \times 10^6 \text{ pc} = 7.9 \text{ Mpc}$$
For Hubble
$$d_{pc} = 10^{(25 - (-6) + 5) / 5} = 10^{7.2} = 1.6 \times 10^7 \text{ pc} = 16 \text{ Mpc}$$

11-19. A variable star is observed to change its B - V color index from 0.5 to 0.7.
(a) Assuming the star radiated as a blackbody, what would be the temperatures corresponding to the two color indices?
Stars don't radiate as perfect blackbodies. If they did, the relationship between the B-V color index of a star and its blackbody temperature would be (Equation 11-10b)
$$T = 7090 / [(B - V)] + 0.71]$$

A color index of B - V = 0.5 corresponds to a temperature of
$$T = 7090 / [(0.5)] + 0.71] = 5860 \text{ K}$$
and a color index of B - V = 0.7 corresponds to a temperature of
$$T = 7090 / [(0.7)] + 0.71] = 5030 \text{ K}$$

(b) Assuming the star is like the Sun, what would be the temperatures corresponding to the two color indices?
For real stars, which aren't perfect blackbodies, the relationship in part (a) is approximated by Equation 11-11b
$$T = 8540 / [(B - V)] + 0.865]$$

For this more realistic model, a color index of B - V = 0.5 corresponds to a temperature of
$$T = 8540 / [(0.5)] + 0.865] = 6260 \text{ K}$$
and a color index of B - V = 0.7 corresponds to a temperature of
$$T = 8540 / [(0.7)] + 0.865] = 5460 \text{ K}$$
The star thus changes temperature from 6260 K to 5460 K, comparable to the Sun's effective temperature ≈ 5800 K.

The "perfect blackbody" and "realistic model" results for the stellar temperatures corresponding to color indices of 0.5 and 0.7 differ by 7 to 8% in this example. It is clear that one cannot naively use the perfect blackbody relationship, but must understand better the real stellar spectra.

11-20. (a) Assuming an extinction coefficient of 0.2, how much fainter (in magnitudes) would a star appear when at an altitude (angular height above the horizon) of 30° compared to when it is at an altitude of 90°?

The observed magnitude of a star at the bottom of the atmosphere, $m(\lambda)$, compared to that outside the Earth's atmosphere, $m_o(\lambda)$, is given by (Section 11-4B):
$$m(\lambda) - m_o(\lambda) = k_o(\lambda) \sec(z)$$
where $k_o(\lambda)$ is the extinction coefficient and z is the zenith angle.
[Note: the zenith angle = 90° - altitude; sec(z) = 1/ cos(z).]

For the zenith, the zenith angle ≡ 0°, so
$$m_{zenith}(\lambda) - m_o(\lambda) = (0.2) \sec(0°) = 0.2 / \cos(0°) = 0.2$$
so at the zenith a star is 0.2 magnitudes fainter than it would be outside the atmosphere. For an altitude 30° (zenith angle = 90° - altitude = 60°),
$$m_{alt30°}(\lambda) - m_o(\lambda) = (0.2) \sec(60°) = 0.2 / \cos(60°) = 0.4$$
so a star at altitude 30° is 0.4 magnitude fainter than it would be outside the atmosphere. A star at altitude 30° is thus 0.4 - 0.2 = 0.2 magnitudes fainter than it would be at the zenith.

Note that we can generalize the magnitude difference for a star at any two zenith angles by:
$$m_{z1}(\lambda) - m_{z2}(\lambda) = k_o(\lambda) [\sec(z_1) - \sec(z_2)]$$
in which case we would have
$$m_{alt30°}(\lambda) - m_{zenith}(\lambda) = 0.2 [\sec(60°) - \sec(0°)] = (0.2) [2 - 1] = 0.2$$

(b) At what zenith angle would the star appear 1 magnitude fainter than when at the zenith?
For a star to be 1 magnitude fainter than when at the zenith, we can use the last equation we derived in part (a),

$$m_{z1}(\lambda) - m_{z2}(\lambda) = k_o(\lambda)[\sec(z_1) - \sec(z_2)]$$

$$\sec(z_1) = [m_{z1}(\lambda) - m_{z2}(\lambda)] / k_o(\lambda) + \sec(z_2)$$
$$= [1.0]/0.2 + \sec(0°)$$
$$= 5.0 + 1.0 = 6.0$$

$$\cos(z_1) = 1/6$$

$$z_1 = 80.4$$

or the altitude = $90° - z_1 = 9.6°$.

A star at zenith angle 80.4° (altitude 9.6°) will appear 1 magnitude fainter than it would at the zenith.

11-21. Derive an expression for the bolometric correction for a star whose continuous emission is exactly like that of a blackbody at temperature T.

From Equation 11-21, the bolometric correction for a star is defined as

$$BC = M_{bol} - M_V.$$

Now Equation 11-20a can be used to give an expression for M_{bol}.

$$M_{bol} = -2.5 \log(L_*/L_\odot) + M_{bol\odot}$$

where $M_{bol\odot} = 4.72$.

Thus,

$$BC = -2.5 \log(L_*/L_\odot) + 4.72$$
$$= -2.5 \log(L_*) + 2.5 \log(L_\odot) + 4.72$$

The luminosity of the Sun, $L_\odot$, is 3.90×10^{26} W (Appendix Table A7-1), so

$$2.5 \log(L_\odot) = 66.48$$

$$BC = -2.5 \log(L_*) + 66.48 + 4.72 = -2.5 \log(L_*) + 71.20$$

If the star is a blackbody of temperature T, then its luminosity is given by Equation 11-19:

$$L_* = 4\pi R_*^2 \sigma T_{eff*}^4 \text{ J/s}$$

so

$$BC = -2.5 \log(4\pi R_*^2 \sigma T_{eff*}^4) + 71.20$$
$$= -2.5 \log(4\pi \sigma) - 2.5 \log(R_*^2) - 2.5 \log(T_{eff*}^4) + 71.20$$
$$= -2.5 \log[(4\pi)(5.67 \times 10^{-8})] - 5.0 \log(R_*) - 10.0 \log(T_{eff*}) + 71.20$$
$$= 86.57 - 5 \log(R_*) - 10 \log(T_{eff*})$$

where the star's radius, R, is in meters and its surface temperature, T, in Kelvin.

11-22. Most photometers today can measure magnitudes with a precision of 0.01 magn with little difficulty. Given this error in measuring the apparent magnitude of a star, what error results in its distance (assuming we know its absolute magnitude exactly)?

The apparent magnitude and distance are related directly by Equation 11-6:

$$m - M = 5 \log d - 5.$$
Solving for the distance,
$$\log d = (m - M + 5) / 5$$
$$d = 10^{(m - M + 5)/5}$$

If we make an error of +0.01 in the apparent magnitude, the calculated distance is
$$d' = 10^{(m+0.01 - M + 5)/5} = 10^{(m - M + 5)/5} \times 10^{(+0.01)/5}$$
$$= d \times 10^{0.002} = 1.0046 \, d$$

If m is in error by ±0.01 magnitude, the calculated value of the distance would be in error by about half a percent. This assumes we know the absolute magnitude with perfect accuracy -- which we don't!

An alternate solution to this problem:
From Problem 11-17, we found that a difference of 0.01 magnitude corresponds to a 1% error in the flux. The flux is inversely proportional to the square of the distance (the inverse square law); or the distance is inversely proportional to the square root of the flux. When taking the square root of a quantity close to unity, the percent difference from 1 is approximately halved -- see the binomial expansion, Appendix A9-6, where the first two terms in the expansion are given by:
$$(1 \pm x)^{1/2} \approx 1 \pm (1/2)x + \dots .$$
So the error in the distance is roughly one half the percent error in the flux. Thus, the error in distance will be roughly $1/2$% for a 1% error in flux or a 0.01 magnitude error.

11-23. Consider a star with $M_V = 4.4$ and a surface temperature the same as that of the Sun. If we do not have to consider atmospheric and interstellar extinction, what flux density would we measure at UBV?
From Section 11-4C, the UBV flux densities corresponding to an $M_{bol} = 0.0$ star are:
$$F_U = 4.34 \times 10^{-12} \text{ W/m}^2 \text{ Å}$$
$$F_B = 6.60 \times 10^{-12} \text{ W/m}^2 \text{ Å}$$
$$F_V = 3.54 \times 10^{-12} \text{ W/m}^2 \text{ Å}$$

For a star with the same temperature as the Sun, the bolometric correction will be the same as for the Sun, so BC = -0.07. From Equation 11-21, $BC = M_{bol} - M_V$, so for this star,
$$M_{bol} = M_V + BC = 4.4 - 0.07 = 4.33$$

From Equation 11-3, the flux density ratios are related to the magnitude difference by
$$f_n / f_m = 100^{(m-n)/5}$$
so at UBV we would measure the flux densities:
$$F_U = (4.34 \times 10^{-12} \text{ W/m}^2 \text{ Å})(100^{(0-4.33)/5}) = 8.04 \times 10^{-14} \text{ W/m}^2 \text{ Å}$$
$$F_B = (6.60 \times 10^{-12} \text{ W/m}^2 \text{ Å})(100^{(0-4.33)/5}) = 1.22 \times 10^{-13} \text{ W/m}^2 \text{ Å}$$
$$F_V = (3.54 \times 10^{-12} \text{ W/m}^2 \text{ Å})(100^{(0-4.33)/5}) = 6.55 \times 10^{-14} \text{ W/m}^2 \text{ Å}$$

Chapter 12: Stars: Binary Systems

12-1. From the information given in Section 12-2A, show that Kepler's harmonic law holds for an apparent orbit of a visual binary.

For an elliptical orbit Kepler's Harmonic Law (Equation 12-1; see also Equation 1-24) states that:

$$(M_1 + M_2) P^2 = A^3$$

where M_1 and M_2 are the masses in solar masses ($M_\odot$), P is the orbital period in years, and A is the semi-major axis in A.U. When the inclination is zero, i = 0, we see the true orbit, so the harmonic law holds. When i ≠ 0, we must correct the value of A that we observe by a trigonometric factor to find the true value of A.

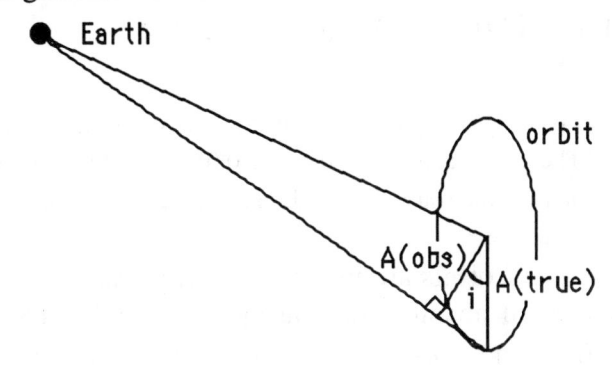

The observed semi-major axis is related to the true semi-major axis,

$A_{obs} = A_{true} \cos(i)$, or $A_{true} = A_{obs} / \cos(i)$.

The harmonic law thus becomes

$$(M_1 + M_2) P^2 = [A_{obs} / \cos(i)]^3.$$

Because for a given stellar system, i and cos(i) are constant, we still have

$P^2 \propto A_{obs}^3$, although the proportionality constant contains an extra $\cos^3(i)$ factor.

12-2. Demonstrate the correctness of Equation 12-2. Use a diagram if necessary.

For a visual binary at distance d (in parsecs) or parallax π'' (in arcsec), the observed angular semi-major axis a" (in arcsec), and the true orbital semi-major axis a (in AU) are related as shown in the figure. [The figure represents a 'skinny triangle' (a" small) -- it has been enlarged for clarity.]

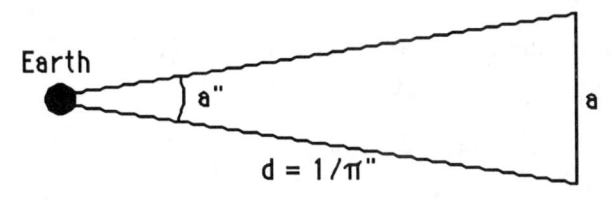

Since stars are far away from the Earth, we can use the small angle formula

$a''_{radians} \approx \tan(a''_{radians}) = a_{pc} / d_{pc} = a_{pc} \pi''$,

but using 1 radian = 206,265 arcsec, and 1 pc = 206,265 AU, we get

$a''_{arcsec} / 206,265 = (a_{AU} / 206,265) / d_{pc} = (a_{AU} / 206,265) \pi''$

so
$$a"_{arcsec} = a_{AU} / d_{pc} = a_{AU}\, \pi"$$
or since our units are now as we want, we get:
$$a_{AU} = a" / \pi" = a"\, d_{pc}.$$

12-3. What is the sum of the stellar masses in a visual binary of period 40 years, maximum separation 5.0", and parallax 0.3"? Assume an orbital inclination of zero and a circular orbit.

Since the orbit is circular and in the plane of the sky, the maximum observed separation is the semi-major axis. Using Kepler's Harmonic Law (Equations 12-1 and 12-3), in terms of solar masses, years and AU's:
$$(M_1 + M_2) P^2 = a^3 = (a"/\pi")^3$$
$$(M_1 + M_2) = (5.0"/0.3")^3 / (40)^2 = 2.9\, M_\odot$$

12-4. Find the distance in parsecs to a visual binary that consists of stars of absolute bolometric magnitudes of +5.0 and +2.0. The mean angular separation is 0.05", and the observed orbital period is ten years. The stars obey the mass-luminosity relation, Equations 12-5a, b, and c. What assumptions have you made to arrive at your answer?

Assume that the stars' orbits are in the plane of the sky. Using Figure 12-3, we can infer that the stars of absolute bolometric magnitudes +5.0 and +2.0 are of mass $\log(M/M_\odot) = -0.05$ and +0.2, respectively, corresponding to masses of 0.9 and 1.6 solar masses. Using Kepler's Harmonic Law (Equations 12-1 and 12-3),
$$(M_1 + M_2) P^2 = a^3 = (a"/\pi")^3 = (a"\, d_{pc})^3$$
$$(0.9 + 1.6)(10)^2 = (0.05 \times d_{pc})^3$$
$$d_{pc} = [(250)^{1/3}] / 0.05$$
$$d \approx 125\ pc$$
(or $\pi = 1/d_{pc} \approx 0.008"$)

12-5. Show that binary systems with small orbits have high orbital speeds.

Assume circular orbits. For circular orbits, the orbital speed is (Equation 12-7):
$$v = 2\pi r / P.$$
From Kepler's Harmonic Law (Equation 12-1),
$$P = a^{3/2} / (M_1 + M_2)^{1/2}.$$
Substituting, where, since the orbit is circular, $r = (1/2) a$,
$$v = 2\pi r (M_1 + M_2)^{1/2} / a^{3/2}$$
$$= 2\pi (a/2)(M_1 + M_2)^{1/2} / a^{3/2}$$
$$= \pi (M_1 + M_2)^{1/2} / a^{1/2}$$

Thus the smaller the orbit, the greater the velocity of stars in a binary system. This general result also applies to elliptical orbits, although the algebra is more complex.

12-6. The velocity curves of a double-line spectroscopic binary are observed to be sinusoidal, with amplitudes of 20 and 60 km/s and a period of 1.5 years.

(a) What is the orbital eccentricity?
The velocity curves are observed to be sinusoidal, so the orbit is circular -- eccentricity, e = 0.

(b) Which star is more massive, and what is the ratio of the stellar masses?
From center of mass arguments, the ratio of the stellar masses is related to the ratio of the radii of the orbits and also the orbital velocities (Equation 12-4). See the figure below:

$$M_1 / M_2 = R_2 / R_1 = v_2 / v_1 = (60 \text{ km/s}) / (20 \text{ km/s}) = 3$$

The star with the smaller velocity (20 km/s) is the most massive star, being three times more massive than the other star.

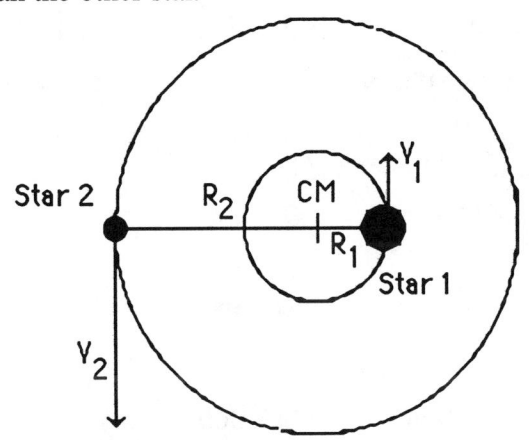

(c) If the orbital inclination is 90°, find the relative semimajor axis (in astronomical units) and the individual stellar masses (in solar masses).
If the orbital inclination is 90°, the orbital plane is along the line-of-sight to the binary system, so the observed velocities are the orbital velocities. For circular orbits (Equation 12-7),

$$v = 2\pi R / P$$
$$R = P v / 2\pi$$

so $a = R_1 + R_2 = P(v_1 + v_2) / 2\pi$

$= (1.5 \text{ yr}) \times (80 \text{ km/s}) \times (3.16 \times 10^7 \text{ s/yr}) \times (1 \text{ AU} / 1.5 \times 10^8 \text{ km}) / 2\pi$

$= 4.02 \text{ AU}$

Using Kepler's Harmonic Law (Equation 12-1)

$(M_1 + M_2) = (3M_2 + M_2) = 4M_2$

$= a^3 / P^2 = (4.02)^3 / (1.5)^2 = 28.9 \, M_\odot$

so $M_2 = 7.2 \, M_\odot$, $M_1 = 21.7 \, M_\odot$

12-7. An eclipsing binary has an orbital period of $2^d 22^h$, the duration of each eclipse is 18^h, and totality lasts 4^h.

(a) Find the stellar radii in terms of the circular orbital radius a.

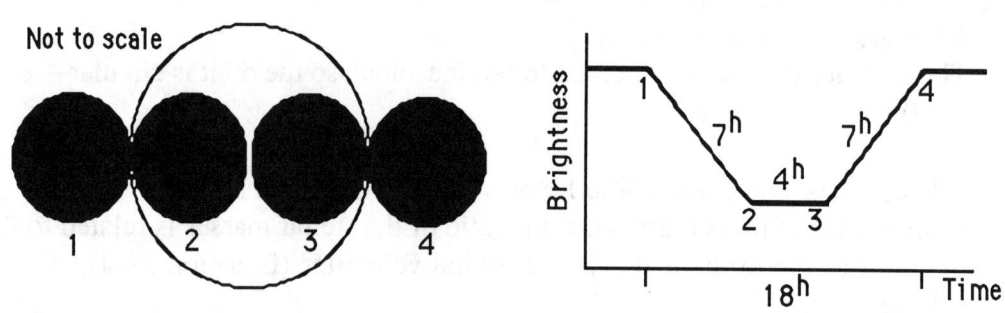

The stellar radii relative to the orbital radius are given by (Equation 12-11):
$$R_s / a = \pi (t_2 - t_1) / P$$
and $R_l / a = \pi (t_4 - t_2) / P$
where R_s and R_l are the radii of the small and larger star, respectively, and the times of contact t are shown in the figure. a is the semi-major axis of the orbit, and P is the orbital period ($2^d 22^h$ = 70 hours).
$$R_s / a = \pi (7 \text{ hr}) / (70 \text{ hr}) = 0.31$$
$$R_l / a = \pi (11 \text{ hr}) / (70 \text{ hr}) = 0.49$$

(b) If spectroscopic data indicate a relative orbital speed of 200 km/s, what are the actual stellar radii (in kilometers and solar radii)?
If the orbits are circular (equation 12-10)
$$a = v P / 2\pi = (200 \text{ km/s})(70 \text{ hr} \times 3600 \text{ s/hr}) / 2\pi$$
$$= 8.0 \times 10^6 \text{ km} = 11.5 \, R_\odot$$
Therefore, from (a)
$$R_s = 0.31 \times 11.5 \, R_\odot = 3.6 \, R_\odot = 2.5 \times 10^6 \text{ km}$$
$$R_l = 0.49 \times 11.5 \, R_\odot = 5.6 \, R_\odot = 3.9 \times 10^6 \text{ km}$$

12-8. The surface temperature of one component of an eclipsing binary is 15,000 K, and that of the other is 5000 K. The cooler star is a giant with a radius four times that of the hotter star.
(a) What is the ratio of the stellar luminosities?
The stellar luminosity (Equation 11-19),
$$L \propto R^2 T^4$$
Where
$$R_{cool} = 4 R_{hot}$$
$$T_{hot} = 3 T_{cool}$$
so $L_{hotter} / L_{cooler} = (R_{hot} / R_{cool})^2 \times (T_{hot} / T_{cool})^4$
$$= (1/4)^2 \times (3)^4 = 5.06$$

(b) Which star is eclipsed at primary minimum?

136

Since the area covered is the same for the primary and secondary eclipses (assuming circular orbits), the star with the higher surface temperature (highest luminosity per unit area) is eclipsed at primary minimum.

(c) Is primary minimum a total or an annular eclipse?
Since the higher surface temperature star is smaller than the other star, the primary eclipse is total (if i ≈ 90°; otherwise partial eclipse occurs, the degree of eclipse depending on i). The secondary eclipse is annular, when the smaller hotter star covers part of the larger cooler star.

(d) Primary minimum is how many times deeper than secondary minimum (in energy units)?
Outside of eclipse, the total brightness is 1.00 + 5.06 = 6.06 units.
At primary minimum, only the cooler fainter star is seen. The brightness is 1.00 unit.
At secondary minimum, we see the entire hot star and 15/16 of the larger cooler star (1/16 of the cool star is eclipsed). The total brightness is 5.06 + (15/16) x 1.00 = 6.00 units.
The primary minimum is 5.06 [5 1/16] units (out of 6.06 at maximum light), and the secondary minimum is 0.06 [1/16] units. The primary minimum is thus 5.06 / 0.06 = 81 times deeper than the secondary minimum.

12-9. The star Sirius A has a surface temperature of 10,000 K, a radius 1.8 $R_\odot$, and M_{bol} = 1.4; the radius of its white dwarf companion, Sirius B, is 0.01 $R_\odot$ and M_{bol} = 11.5.

(a) What is the ratio of their luminosities?
The ratio of luminosities is (Equation 11-3):

$$L_{SiriusA} / L_{SiriusB} = L_1 / L_2 = 100^{0.4(m_2 - m_1)}$$
$$= 100^{0.4(11.5 - 1.4)} = 1.10 \times 10^4$$

(b) What is the ratio of their effective temperatures?
The effective temperature, T, is given by (Equation 11-19):

$$T \propto (L/R^2)^{1/4}$$

$$T_{SiriusA} / T_{SiriusB} = [(L_{SiriusA} / L_{SiriusB}) / (R_{SiriusA} / R_{SiriusB})^2]^{1/4}$$
$$= [(1.10 \times 10^4) / (1.8 / 0.01)^2]^{1/4}$$
$$= 0.76 \quad \text{Sirius B is hotter than Sirius A!!}$$

(c) If they orbit at i = 90°, which star is eclipsed at primary minimum?
Since Sirius B (the white dwarf) is the hotter star, it would be eclipsed at primary minimum.

(d) If your photometer can measure magnitudes to an accuracy of ≈ 0.001, would you be able to detect the hypothetical primary eclipse? [Hint: Use $\log_{10}(1 + x) = x / 2.3$ for x ≪ 1.]

The total brightness of the system outside of eclipse is $1 + 1.10 \times 10^4$ units, while at primary minimum the brightness is diminished by 1 unit (to 1.10×10^4 units). Using Equation 11-5:

$$m_{primary} - m_{no\ eclipse} = 2.5 \log (L_{no\ eclipse} / L_{primary})$$
$$= 2.5 \log [(1.10 \times 10^4 + 1) / (1.10 \times 10^4)]$$
$$= 2.5 \log [1 + 1 / (1.10 \times 10^4)]$$
$$= 2.5 \log [1 + (9.09 \times 10^{-5})]$$
$$= 2.5 [(9.09 \times 10^{-5}) / 2.3] \quad \text{using the given approximation}$$
$$= 9.9 \times 10^{-5}$$
$$\approx 1.0 \times 10^{-4} = 0.0001$$

Thus, if our photometer can measure magnitudes only to an accuracy of ≈ 0.001, the instrument is not sensitive enough (by a factor of 10) to detect the primary eclipse.

12-10. Derive an expression giving the stellar angular diameter in milliarcseconds when the actual stellar diameter (in solar radii) and distance from us (in parsecs) are known.

Let $D \equiv$ the diameter of a star, $d \equiv$ the distance to the star.

D (in AU) is given by: $D_{AU} = \theta'' / \pi''$ (see Equation 12-2 or Problem 12-2).
So
$$\theta_{arcsec} = D_{AU}\, \pi_{arcsec} = D_{AU} / d_{pc}$$
or
$$\theta_{milli-arcsec} = (D_{AU} / d_{pc}) \times 10^3$$

The solar radius is 6.96×10^5 km / 1.5×10^8 km = (1 / 215) AU, so
$$D_{AU} = D_{R\odot} / 215$$
Substituting,
$$\theta_{milli-arcsec} = (D_{R\odot} / d_{pc}) \times 10^3 / 215 = 4.65 \times (D_{R\odot} / d_{pc})$$
At a distance of 1 pc, the Sun would have an angular diameter of 4.65 milli-arcsec.

12-11. Estimate the orbital separation and velocity for a binary star system composed of solar mass stars that have an orbital period of 12 hours (assume circular orbits).

From Kepler's third law (Equation 12-1; see also Equation 1-24):
$$(M_1 + M_2)\, P^2 = a^3$$
for the period, P, in years, the semimajor axis, a, in AU, and masses, M, in solar masses. The orbital separation is thus
$$a^3 = (1 + 1)\,(0.5\ \text{day} / 365.26\ \text{day/year})^2 = 3.75 \times 10^{-6}$$
$$a = 1.55 \times 10^{-2}\ \text{AU} = 2.3 \times 10^6\ \text{km}$$

The orbital velocities are given by Equation 12-7:
$$v_1 = 2\pi r_1 / P \qquad v_2 = 2\pi r_2 / P$$
where $r_1 = r_2 = a/2$, and thus $v_1 = v_2$, for $M_1 = M_2$ from Equation 12-4. So
$$v_1 = v_2 = 2\pi (a/2) / P$$

$$= 2\pi\,(2.3 \times 10^6 \text{ km}/2)/(12 \text{ hours} \times 3600 \text{ s/hour})$$
$$= 1.7 \times 10^2 \text{ km/s}$$

12-12. (a) Use the mass-luminosity relation to compute the luminosity range of stars from the observed mass range of $0.085\, M_\odot$ to $100 M_\odot$.

The mass-luminosity relationship is given by Equation 12-5a:

$$L/L_\odot \propto (M/M_\odot)^\alpha$$

where (Equations 12-5b and 12-5c):

$$L/L_\odot = (M/M_\odot)^{4.0} \qquad \text{for stars with } M > 0.43\, M_\odot$$
$$L/L_\odot = 0.23\,(M/M_\odot)^{2.3} \qquad \text{for stars with } M < 0.43\, M_\odot$$

Stars of mass $0.085\, M_\odot$ thus have a luminosity

$$L/L_\odot = 0.23\,(0.085)^{2.3} = 7.9 \times 10^{-4}\, L_\odot$$

and stars of mass $100\, M_\odot$ thus have a luminosity

$$L/L_\odot = (100)^4 = 1.0 \times 10^8\, L_\odot$$

Stars of mass range $0.085\, M_\odot$ to $100 M_\odot$ thus correspond to a luminosity range of $\approx 8 \times 10^{-4}\, L_\odot$ to $10^8\, L_\odot$. These values correspond, roughly, to Figure 12-3.

(b) What is the mass of a star that is 0.1 the luminosity of the Sun? What is the mass of a star that is 1000 times the luminosity of the Sun?

Using $L/L_\odot = (M/M_\odot)^{4.0}$, a star $0.1\, L_\odot$ would have a mass

$$M = (L/L_\odot)^{0.25}\, M_\odot = (0.1)^{0.25}\, M_\odot = 0.56\, M_\odot$$

This is greater than $0.43\, M_\odot$, so we have used the correct index α.

A star of $1000\, L_\odot$ would have a mass

$$M = (1000)^{0.25}\, M_\odot = 5.6\, M_\odot$$

12-13. (a) Refer to the light curve for the eclipsing binary star WY Cancri (Figure 12-8). What qualitative information can be deduced from a visual inspection of the light curve?

From the relative depths of the primary and secondary minima, one can determine the relative luminosities (see for example, Problem 12-8).

From the duration of the primary and secondary eclipses, one can determine the stellar radii in terms of the orbital radii (see, for example, Problem 12-7); and with the relative luminosities we can then determine the relative temperatures.

From the observation that the secondary minimum is located symmetrically between the primary minima, we can deduce that either the orbit is nearly circular or that the major axis of the system's orbit is pointed toward us; an accurate determination of the times of primary and secondary minima could help us determine the orbital eccentricity and orientation to our line of sight.

From the observation that the magnitude is not constant between the minima, we can deduce that either the stars are distorted, there is gas between the stars which we see at different angles, or that there are different number of starspots at different stellar longitudes on one or both of the stars.

From the observation that the minima are not flat-bottomed, we can conclude that the eclipses aren't total and thus the inclination of the orbit to our line of sight its not identically zero; an accurate model fit to the shape of the eclipses can determine the inclination.

(b) If WY Cancri has a period of 19.9 hours, estimate the durations of primary and secondary eclipse (R_l/a and R_s/a).

Measuring Figure 12-8, the entire primary eclipse extends over a phase of
$$(7 \text{ mm} / 66 \text{ mm}) \times 360° = 38°.$$
The secondary eclipse extends over the same phase.

For a 19.9 hour period, this phase range would correspond to a time interval of
$$(38° / 360°) \times 19.9^{hr} = 2.1^{hr}.$$
The primary and secondary eclipses each last 2.1 hours, easily observable in an evening of observing.

To determine the radii of the stars relative to the orbital semimajor axis, we measure the times of contacts (see Figure 12-9) and assume central eclipse. The light curve actually looks like a partial eclipse, but for the sake of this problem let us try to determine the possible duration of minimum, assuming it is total. [Note that these times are not easy to measure.]

The time from first to second contact is
$$(3 \text{ mm} / 66 \text{ mm}) \times 360° = 16°$$
which corresponds to (Equation 12-11)
$$R_s / a = \pi (16°) / 360° = 0.14$$
while the time from first to third contact is
$$(4 \text{ mm} / 66 \text{ mm}) \times 360° = 22°$$
which corresponds to (Equation 12-11)
$$R_L / a = \pi (22°) / 360° = 0.19$$

12-14. Consider a binary system with an orbital period of ten years. The stars have radial velocities of 10 and 20 km / s. Find the individual masses of the stars if the orbital inclination is
(a) 90°

From Kepler's third law (Equation 12-1; see also Equation 1-24):
$$(M_1 + M_2) P^2 = a^3 = (r_1 + r_2)^3$$
for the period, P, in years, the semimajor axis, a, in AU, and masses, M, in solar masses. The ratio of the stellar masses is given by (Section 12-3A)
$$M_1 / M_2 = r_2 / r_1 = v_2 / v_1 = [v'_2 / \sin(i)] / [v'_1 / \sin(i)] = v'_2 / v'_1$$

where r is the distance of the star from the center of mass of the system, v is the orbital speed, v' is the observed speed, and i is the orbital inclination.

Note that sin(i) is the same for both stars in the system. Hence, the mass ratio is determined by the (inverse of the) ratio of the radial velocities, and does not require knowing the orbital inclination. Note also that the more massive star has the smaller velocity, since it is closer to the center of mass of the system and thus has a smaller orbit.

$$M_1 / M_2 = (20 \text{ km/s}) / (10 \text{ km/s}) = 2$$

Using Equation 12-7 for circular orbits, $r = VP/2\pi$, to find the semi-major axis of the system, where here, $\sin(i) = \sin(90°) = 1$,

$$a = r_1 + r_2 = (v_1 + v_2) P / 2\pi = [(v'_1 + v'_2) / \sin(i)] P / 2\pi$$
$$= (10 \times 10^3 \text{ m/s} + 20 \times 10^3 \text{ km/s}) (10 \text{ yr} \times 3.16 \times 10^7 \text{ s/yr}) / 2\pi$$
$$= 1.5 \times 10^{12} \text{ m} = 10 \text{ AU} \qquad \text{(how convenient!)}$$

Applying Kepler's third law

$$(M_1 + M_2) = a^3 / P^2 = 10^3 / 10^2 \, M_\odot = 10 \, M_\odot$$
$$(2 M_2 + M_2) = 3 M_2 = 10 \, M_\odot$$

So, $M_2 = 3.33 \, M_\odot \qquad M_1 = 6.67 \, M_\odot$

Note that an inclination of 90° corresponds to the smallest masses for the stars, given the observed radial velocities.

(b) 45°

For inclination 45°, the observed radial velocity is less than the true orbital velocity. The orbital semi-major axis is

$$a = r_1 + r_2 = (v_1 + v_2) P / 2\pi = [(v'_1 + v'_2) / \sin(45°)] P / 2\pi$$
$$= [(30 \times 10^3 \text{ m/s}) / 0.7071] (10 \text{ yr} \times 3.16 \times 10^7 \text{ s/yr}) / 2\pi$$
$$= 2.12 \times 10^{12} \text{ m} = 14.14 \text{ AU}$$

Applying Kepler's third law,

$$(M_1 + M_2) = a^3 / P^2 = (14.14)^3 / 10^2 \, M_\odot = 28.3 \, M_\odot$$
$$3 M_2 = 28.3 \, M_\odot$$

So, $M_2 \approx 9.4 \, M_\odot \qquad M_1 \approx 18.9 \, M_\odot$

12-15. The star Vega has a measured angular diameter of 3.24 mas and a flux of 2.84×10^{-8} W/m². Its distance is 8.1 pc. What are Vega's diameter and effective temperature?

The diameter of Vega can be found from knowing the distance and angular size. The diameter of the star in AU is given by Equation 12-2 (see also problem 12-2):

$$D = D'' / \pi'' = D'' \, d$$

where D and D" are the diameter and angular size (in arcsec) of the star, π is the parallax, and d is the distance in parsecs. Using conversions in Appendix Table A7-1:

$$D = (3.24 \times 10^{-3})(8.1) \text{ AU} = 2.62 \times 10^{-2} \text{ AU} = 3.94 \times 10^9 \text{ m} = 5.6 \, R_\odot$$

-- this is about right for an A0 star (see Appendix Table A4-3)

The effective temperature can be found by combining the Stefan-Boltzmann law (Equation 11-19)

$$L = 4 \pi R^2 \sigma T_{eff}^4$$

and the the relationship between flux and luminosity (Equation 8-17)

$$F = L / 4 \pi d^2$$

giving

$$F = 4 \pi R^2 \sigma T_{eff}^4 / 4 \pi d^2 = (R/d)^2 \sigma T_{eff}^4$$

$$T_{eff}^4 = F (d/R)^2 / \sigma$$

$$= (2.84 \times 10^{-8} \text{ W/m}^2) [(8.1 \text{ pc}) \times (3.086 \times 10^{16} \text{ m/pc}) / (3.94 \times 10^9 \text{ m} / 2)]^2$$
$$/ (5.67 \times 10^{-8} \text{ W/m}^2 \text{ K}^{-4})$$

$$= 8.06 \times 10^{15} \text{ K}^4$$

$$T_{eff} = 9.5 \times 10^3 \text{ K} = 9500 \text{ K} \qquad \text{-- this is about right for an A0 star}$$

12-16. Consider a visual binary where the stars are 1" and 2" from the center of mass. The distance is 10 pc, and the orbital period is ten years. If $i = 90°$, what are the masses? The luminosities?

Using Kepler's third law (Equation 12-3) and Equation 11-2 relating distance and parallax

$$(M_1 + M_2)_{M_\odot} = (a''/\pi'')^3 / P_{yr}^2 = (a'' d_{pc})^3 / P_{yr}^2$$

$$= [(1'' + 2'')(10 \text{ pc})]^3 / (10 \text{ yr})^2$$

$$M_1 + M_2 = 270 \, M_\odot$$

The ratio of the stellar masses is given by (Section 12-3A)

$$M_1 / M_2 = r_2 / r_1 = a_2'' / a_1''$$

$$= 1'' / 2'' = 0.5$$

$$M_1 = 0.5 \, M_2$$

Therefore,

$$0.5 \, M_2 + M_2 = 270 \, M_\odot$$

$$M_2 = 180 \, M_\odot \qquad \text{--- This is an unrealistically high mass (see Section 12-2B)}$$

$$M_1 = 0.5 \, M_2 = 90 \, M_\odot$$

These are high masses for normal stars (Star 1 is near the upper end of stellar masses, while Star 2 is unrealistically high; see Section 12-2B). But assuming the mass-luminosity relationship (Equation 12-5b) holds even for these high masses

$$L / L_\odot = (M / M_\odot)^{4.0}$$

$$L_1 = (90)^{4.0} L_\odot = 6.6 \times 10^7 L_\odot$$

$$L_2 = (180)^{4.0} L_\odot = 1.0 \times 10^9 L_\odot \quad \text{-- wow!}$$

Chapter 13: Stars: The Hertzsprung-Russell Diagram

13-1. The absorption spectra of four stars exhibit the following characteristics. What are the appropriate spectral types?

Using the descriptions in Table 13-1 and the relative line strengths in Figure 13-6, the following approximate spectral types can be assigned:

(a) The strongest features are titanium oxide bands.
   M stars (M5 - M7)

(b) The strongest lines are those of ionized helium.
   O stars (O5 - O7)

(c) The hydrogen Balmer lines are very strong, and some lines of ionized metals are present.
   A stars (B5 - A5)

(d) There are moderately strong hydrogen lines, and lines of neutral and ionized metals are seen, but the Ca II H and K lines are the strongest in the spectrum.
   G stars (near G2 -- like our Sun)

13-2. To which spectral types may the following stars be *approximately* assigned if their continuous spectra are of maximum intensity at wavelengths given.

[*Hint:* Plot λ_{max} from Wien's law as a function of spectral class for the main sequence.)

Using Figure 13-6, an approximate spectral type can be estimated from the temperature -- which can be determined from Wien's Law (Equation 13-6):

$$\lambda_{max} \, T = 2.90 \times 10^6 \text{ nm K}$$

	λ_{max}	T	spectral type
(a)	50 nm	58,000K	O5 or earlier
(b)	300 nm	9,670K	A0
(c)	600 nm	4,830K	K0
(d)	900 nm	3,220K	M4
(e)	1.2 μm	2,420K	late M
(f)	1.5 μm	1,930K	later M (protostars?)

13-3. If the parallax of a main-sequence star is in error by 25%, how far and in what direction will this star be displaced from the main sequence in an H-R diagram?

A parallax error of 25% will produce an error in estimating distance of 25%. This distance error will not affect the estimate of the star's spectral class, but it will affect the absolute magnitude determination. Hence the displacement on the H-R diagram will be in the vertical direction. For a quick-and-dirty estimate, a 25% distance error will cause roughly a 50% error in luminosity (brightness $\propto$ distance^2), which would correspond to about 0.5 magnitudes. A more accurate estimate of the error in absolute magnitude is found by considering (Equation 11-8):

$$M = m + 5 + 5 \log (\pi")$$

or using natural logarithms

$$M = m + 5 + (5 / 2.30) \ln (\pi")$$

differentiating:

$$dM = (5 / 2.30) (d\pi/\pi) = 2.2 \, d\pi/\pi$$

If $d\pi/\pi = 0.25$, then

$$dM = 2.2 \times 0.25 = 0.55$$

A 25% error in parallax would produce a 0.55 magnitude error on the H-R diagram.

13-4. Which parameter in the Saha ionization-equilibrium equation is *most* important in explaining the spectral differences between

(a) giants and dwarfs of spectral type G

The Saha ionization equation is (Equation 13-9; see also Section 8-4B):

$$N_{i+1} / N_i = A \, [(kT)^{3/2} / N_e] \exp(-\chi_i / kT)$$

See Problem 8-8 for a discussion of the relative importance of T, N_e, and χ_i in determining the ionization ratio.

Giants and dwarfs of spectral type G have very nearly the same temperature, but there is a large difference in density between the G giants and dwarfs. Even though the temperature effect is generally much more sensitive than the electron density, the large difference in density results in the density effect being the most important in this instance (the effects of temperature and density nearly balance out). Because of this rough compensation, the G giants and dwarfs have essentially similar spectra. A major spectral difference occurs due to pressure broadening (due to electron density) of the lines in the dwarf stars as compared to the relatively narrow lines of the less dense giants.

(b) B and A dwarfs

A dwarfs have a temperature of $T \approx 10,000$ K, while B dwarfs have $T \approx 25,000$ K (Figure 13-6). This large temperature difference has profound effects on the spectra of these two classes of stars. In the A dwarfs the HI lines are very strong (hydrogen is predominately neutral), while in the B dwarfs much of the hydrogen is ionized so the HI lines are relatively weak. As we saw in problem 8-8, this 2.5 difference in temperature causes a significant change in ionization (due to the temperature dependence in the exponential). In the B0 dwarfs, the HeI lines are much stronger than in the A0 dwarfs. Again, the dominant factor is due to the temperature difference.

Also contributing to this effect is that the ionization potential, χ_i, for HeI is greater than that for HI, which partially compensates for the temperature differences between the B and A stars. This explains why the HeI lines are strongest in the hotter B stars, as are the HI lines strongest in the cooler A stars.

13-5. (a) What is the ratio of the surface gravities of a K0 V star and a K0 I star?

We can very roughly estimate the surface gravity difference between a K0V (dwarf) and a K0I (supergiant) star by estimating the ratio of their radii from their luminosity differences. The absolute magnitude difference of 10 magnitudes between these two luminosity classes of K0 stars corresponds to a luminosity ratio of 10,000 (Equation 11-3). Since the surface temperatures are comparable, this luminosity difference implies that the K0V star has 10,000 times the surface area of a K0I star -- or 100 times the radius (Equation 13-13). Assume that the masses are the same (which needn't be true). The surface gravity is given by (Section 1-4B; Equation 1-15):

$$g = GM/R^2$$

so the relative surface gravity is:

$$g_{K0V} / g_{K0I} \approx 10^{-4}$$

Doing the problem more realistically: since the masses are not the same, let's select representative stars from Appendix Table A4-3:

Spec K0V, $M_V = +5.9$, $R = 0.85 \, R_\odot$, $M = 0.8 \, M_\odot$

Spec K0I, $M_V = -4.5$, $R = 200 \, R_\odot$, $M = 13 \, M_\odot$

so, $g_{K0V} / g_{K0I} \approx (13 / 200^2) / (0.8 / 0.85^2) \approx 3 \times 10^{-4}$

(b) If both stars had the same atmospheric temperatures and opacities, what would be the approximate ratio of strengths of the Ca II K lines in their spectra? (The ionization potential of calcium is 6.1 eV.)

We are considering CaII K lines in both stars, so the ionization potential, χ_i, is the same. Since the temperature is the same for both stars, it is only the electron density (or electron pressure) that determines the line strength ratio. From the Saha equation (Equation 13-9; see also section 8-4b), the ionization ratio is inversely proportional to the electron density. Assume that $P_{electron} \propto P_{total}$. Use the form of the hydrostatic equilibrium (Equation 13-5):

$$P = (g / \kappa) \tau$$

assuming $\kappa_I = \kappa_V$, and $\tau_I = \tau_V$, we get the ratio of the the pressure in the dwarf (V) and supergiant (I) K stars:

$$P_I / P_V = [(g_I / \kappa_I) \tau_I] / [(g_V / \kappa_V) \tau_V]$$
$$= g_I / g_V = 3 \times 10^{-4} \quad \text{[from part (a)]}$$

So if line strengths are $\propto$ to 1/pressure (or $1/N_e$), then

$$(\text{line strength})_I / (\text{line strength})_V \approx 1 / 3 \times 10^{-4} \approx 3 \times 10^3$$

(c) What assumptions did you make to answer part (b)?
The assumptions used in part (b) are stated there.

(d) What is the ratio of the mean densities of these two stars?
The mean densities can be computed from the masses and radii (volume) of the stars:

$$<\rho_I> / <\rho_V> = \{M_I / [(4/3) \pi R_I^3]\} / \{M_V / [(4/3) \pi R_V^3]\}$$
$$= (13 / 200^3) / (0.8 / 0.85^3) = 1.2 \times 10^{-6}$$

(e) If the atmospheric electron densities of these stars were directly proportional to their mean densities, would your answer to (b) remain unchanged? Why?
If N_e were directly proportional to the mean density, then the ratio of line strengths in

part (b) would be $\approx 10^6$ rather than $\approx 10^3$. In calculating line strengths we want to consider mean densities of the atmosphere, not the mean density of the entire star.

13-6. Why is the H-R diagram of stars in the solar neighborhood (within 500 pc) *not* an unambiguous two-dimensional plot?

The H-R diagram of the stars in the solar neighborhood does not have the strong selection effect in favor of more luminous stars that the H-R diagram of the brightest stars has. However, there are some ambiguities introduced by the Population II (low metal abundance) stars which are interloping into the solar neighborhood. Since the diagram thus includes both Population I and Population II stars -- which have slightly different characteristics for the same mass and age -- the consequence is a less well-defined main-sequence and evolutionary tracks on the H-R diagram. An unambiguous H-R diagram would thus include metal abundance as well as the spectral class (temperature, color index) and luminosity (absolute magnitude).

13-7. Why is an absolute magnitude - spectral type H-R diagram quite different (in principle) from an apparent magnitude - (B-V) H-R diagram?

The apparent magnitude - (B-V) H-R diagram is only meaningful if the stars plotted are at the same distance (such as in an open cluster, globular cluster, or association), whereas the absolute magnitude - spectral type H-R diagram can include all stars whose distances are known. In the cluster H-R diagram, all stars are the same age and have the same metal abundance, while these are not necessarily true for all stars which are otherwise selected for an H-R diagram.

The B-V color index is only approximately correlated with spectral type (or temperature), with abundance effects and stellar size (pressure) causing a spread in the color index - spectral type correlation. These effects produce quantitatively different H-R diagrams, depending upon the parameters selected for the axes.

13-8. In determining distances via the main-sequence fitting technique, why must we refrain from comparing the observed H-R diagram of a galactic cluster with the "calibrated" H-R diagram of M 3?

In determining distances via the main sequence fitting technique, we can not compare the observed H-R diagram of a galactic cluster with a calibrated H-R diagram of a globular cluster because the elemental abundances (metal abundances) are different. A (young) galactic cluster contains high metal abundance Population I stars, while an (old) globular cluster contains low metal abundance Population II stars. The compositional differences will shift the main sequence slightly. The widely different ages of the galactic and globular clusters will also make it difficult to compare the main sequences of the two clusters.

13-9. (a) It is sometimes said that the spectral type of a star depends only upon luminosity and surface temperature. Under what conditions is this statement approximately true?

In Equation 13-14, we saw that the line equivalent width depends upon not only the luminosity and surface temperature, but also the mass and chemical composition:

$$EW = EW(L, T, M, \mu).$$

The spectral type will also depend upon these stellar parameters. In order to state that the spectral type of star depends only on the luminosity and surface temperature, some assumptions need to be made: (i) the stars are main sequence stars; (ii) there is a unique mass-luminosity relationship; and (iii) the chemical composition is the same (or that chemical composition has little influence on the spectral type). Since these assumptions are not true, the statement is merely a generalization for a selected group of stellar types.

(b) Give three generic examples that violate the statement that stellar masses are uniquely determined by their colors.

Stellar masses may not be uniquely determined by their colors if:

(i) two stars of the same spectral class are of different luminosity classes. They will have the same color but different sizes and masses;

(ii) two stars are of different chemical composition. Stars of the same mass but different metal abundances will have different colors -- so stars of the same color need not have the same mass;

(iii) the star is just approaching or leaving the main sequence. Evolutionary effects during stages near the main sequence result in slightly different colors of a star; and

(iv) the stars have had their evolution influenced by companion stars. Due to mass exchange, the evolutionary tracks of stars are altered -- so a star's color may not be a realistic indication of its mass.

13-10. In Section 13-1A, we wrote the mean molecular weight μ for a fully ionized gas as

$$1/\mu = 2X + (3/4)Y + (1/2)Z$$

where, for example,

$X \equiv$ hydrogen mass fraction

= mass density of hydrogen / mass density of *all* constituents.

Derive this relationship for μ, indicating the assumptions and approximations used at each step.

The mean molecular weight, μ, is

μ = total atomic mass / total number of free particles.

To find μ in terms of mass fractions it is convenient to consider

$1/\mu$ = total number of free particles / total atomic mass.

If we want to consider mass rather than atomic mass we use

$1/(\mu m_H)$ = total number of free particles / total mass of sample.

Consider the contribution of each element individually, then sum the contributions:

$1/(\mu m_H) = \Sigma_z$ (total number of free particles contributed by element z) /

(total mass of sample)

where z $\equiv$ atomic number.

$1/(\mu m_H) = \Sigma_z \{$[(total number of free particles contributed by element z)
/ (mass of element z)] x [(mass of element z / volume)
/ (total mass of sample / volume)]$\}$

Now,

$X \equiv$ (mass density of H / mass density)

$Y \equiv$ (mass density of He / mass density)

$Z \equiv$ (mass density of heavy elements / mass density)

so in general,

$X_z \equiv$ mass fraction of element z

= [(mass of element z / volume) / (total mass of sample / volume)].

Thus, considering H, He, and heavy elements (z > 2) gives

$1 / (\mu m_H)$ = (total number of free particles from H / mass of H) x X
+ (total number of free particles from He / mass of He) x Y
+ (total number of free particles from others / mass of others) x Z

Assume complete ionization, so that:

Each H atom generates 2 particles (one proton and one electron) for one atomic mass unit, so:

total number of free particles from H / mass of H = $2 / m_H$

Each He atom generates 3 particles (a nucleus and 2 electrons), for 4 atomic mass units, so:

total number of free particles from He / mass of He = $3 / (4 m_H)$

For a heavy element, we assume that A = 2 Z (atomic mass is twice the atomic number). Deviations from this assumption are of negligible importance because the mass fraction of heavy elements is generally small (< 2%). Each atom produces Z + 1 free particles (the nucleus and z electrons) for about 2 Z atomic mass units, so:

total number of free particles from heavy elements / mass of heavy elements =
$(Z+1) / (A m_H) \approx (Z+1) / (2 Z m_H) \approx 1 / (2 m_H)$

Summing the individual terms:

$1 / (\mu m_H) = (2 / m_H) X + (3 / 4 m_H) Y + (1 / 2 m_H) Z$

$1 / \mu = 2 X + (3/4) Y + (1/2) Z$ for the case of complete ionization.

13-11. The number 5040 appears in Equations 13-10 and 13-11.

(a) Show where this number comes from (how it is derived).

The number 5040 in Equations 13-10 and 13-11 is not a number of mystical significance to a cult of astronomers, but rather arises from the numerical constants in the Boltzmann and Saha equations: $(1 / k) \log_{10} e = 5040$.

Taking the base 10 logarithm of the Boltzmann equation

$N_B / N_A \propto \exp [(E_A - E_B) / k T]$

gives

$\log_{10} (N_B / N_A) = [(E_A - E_B) / k T] (\log_{10} e) + $ constant
$= (0.4343 / k) (1 / T) (E_A - E_B) + $ constant

If E is in eV, then use $k = 8.61 \times 10^{-5}$ eV / K, so

$$\log_{10}(N_B/N_A) = (5040/T)(E_A + E_B) + \text{constant}.$$

The factor 5040 enters the Saha equation (Equation 13-11) in a similar manner.

(b) What are the units of this number in these equations?

Because k has units eV / K and $\log_{10} e$ is dimensionless, the quantity 5040 has units of K / eV (i.e. 5040 K / eV).

13-12. Use hydrostatic equilibrium to compare the central pressure of the Sun and

(a) a B0 V star

The equation of hydrostatic equilibrium is discussed in Section 4-3B, and Problems 4-11, 5-12 and 6-11 deal with the central pressures in solar system objects. (See also Section 16-1A, where the equation of hydrostatic equilibrium is discussed in relationship to stellar structure. However, the equation for the central pressure differs by a factor of two compared to the equation introduced in Section 4-3B.)

In Section 4-3B, the central pressure of a body is estimated by the hydrostatic equilibrium equation.

$$P_c = (2\pi/3) G <\rho>^2 R^2$$

If we assume (as is done in arriving at this equation) that the density is constant throughout the volume, then

$$<\rho> = (3/4\pi) M / R^3$$

substituting gives

$$P_c = [3/(8\pi)] G M^2 / R^4$$

or $P_c \propto M^2 / R^4$.

The central pressure for the Sun is

$$P_c = [3/(8\pi)] (6.67 \times 10^{-11} \text{ N m}^2/\text{kg}^2)(1.99 \times 10^{30} \text{ kg})^2 / (6.96 \times 10^8 \text{ m})^4$$

$$= 1.34 \times 10^{14} \text{ N/m}^2 = 1.34 \times 10^9 \text{ atm}$$

[Note that this differs from the central pressure for the Sun estimated in Section 16-1A where the equation for the central pressure differs by a factor of two.]

To compare other stars to the Sun, use $P_c \propto M^2 / R^4$.

From Appendix Table A4-3, we find (interpolating for the G2 stars):

Spectral Type	R	M
B0V	7.6 $R_\odot$	17 $M_\odot$
G2III	7.8	2.6
G2I	110.	11.2

For the B0V star:

$$P_c/P_{c\odot} = (M/M_\odot)^2 / (R/R_\odot)^4 = (17)^2 / (7.6)^4 = 0.087$$

(b) a G2 **III** star
For the G2III star:
$$P_c/P_{c\odot} = (2.6)^2/(7.8)^4 = 0.0018$$

(c) a G2 **I** star
For the G2I star:
$$P_c/P_{c\odot} = (11.2)^2/(110)^4 = 8.6 \times 10^{-7}$$

13-13. Using Figures 13-8 and 13-11, estimate the distance to an M Ib star of apparent magnitude +1.0.
From Figures 13-8C or 13-11B, an MIb star has $M_V = -4.5$. If $m_V = 1.0$, from Equation 11-6:
$$m - M = 5 \log(d_{pc}) - 5$$
$$1.0 - (-4.5) = 5 \log(d_{pc}) - 5$$
$$\log(d_{pc}) = 2.1$$
$$d = 126 \text{ pc}$$

13-14. Estimate the distances to the following clusters from their color magnitude figures (use Table A4-3 to convert B-V to spectral type):
(a) the Pleiades (Figure 13-9)
[Note that the problem incorrectly refers to Figure 13-10.]
From Appendix Table A4-3, for main sequence (luminosity class **V**) stars:

B - V	1.2	0.8	0.65	0.30	0.0
Spectral type	K5	K0	G0	F0	A0

We can now convert the spectral types in Figures 13-7 and 13-8 to B-V color index.

For the Pleiades (Figure 13-9) we read off the figure and compare to the calibrated H-R diagram (Figure 13-7):

Spectral Class	K5	K0	G0	F0	A0
m_V(observed)	13	11	10	8.5	6.5
M_V(calibrated)	9	7	5	4	1
$m_V - M_V$	4	4	5	4.5	5.5

mean $m_V - M_V \approx 4.6$

From Equation 11-6:
$$m_V - M_V = 5 \log(d_{pc}) - 5$$
$$4.6 = 5 \log(d_{pc}) - 5$$
$$\log(d_{pc}) = 1.92$$
$$d = 83 \text{ pc}$$

(b) M3 (Figure 13-10)
[Note that the problem incorrectly refers to Figure 13-12.]
For the globular cluster M3 (Figure 13-10) the estimate will be less accurate because of

differences between H - R diagrams for Population **I** and Population **II** stars which produce a small motion to the left of the main sequence. Proceeding nonetheless, we read off the figure and compare to the calibrated H- R diagrams:

Spectral Class	K0	G0	F0	
m_V(observed)	21	20	18	
M_V(calibrated)	7	5	4	
$m_V - M_V$	14	15	14	mean $m_V - M_V \approx 14.3$

$m_V - M_V = 5 \log (d_{pc}) - 5$
$14.3 = 5 \log (d_{pc}) - 5$
$\log (d_{pc}) = 3.86$
$d = 7{,}200 \text{ pc} = 7.2 \text{ kpc}$

13-15. Estimate the radii of both a main sequence M star (M**V**) and a red supergiant (M**I**) using information from the H-R diagrams in the text.

From the H-R diagrams in Figures 13-7 and 13-8
$T_{eff} \approx 3500 \text{ K}$ for an M star
$M_V \approx -6$ for an M0I star (rough average of M0Ia and M0Ib classes)
$M_V \approx 10$ for M0V star

[Note: students might get slightly different values using Appendix Table A4-3.]

The luminosity of a star is given by (Equation 13-13; also Section 13-1B)
$L = 4 \pi R^2 \sigma T_{eff}^4$
so
$R = (1 / T_{eff}^2)(L / (4 \pi \sigma))^{1/2}$.

We need to find the luminosities of the stars from their M_V using Equation 11-20b:
$\log (L_* / L_\odot) = 1.9 - 0.4 M_{bol*}$
where (Equation 11-21)
$M_{bol*} = M_V + BC$
with the bolometric correction, BC, for M0 stars being ≈ -1.2 (Appendix Table A4-3).

For the M0I star,
$\log (L_* / L_\odot) = 1.9 - 0.4 (-6.0 - 1.2) = 4.78$
$L_* = 6.0 \times 10^4 L_\odot$
$= (6.0 \times 10^4)(3.9 \times 10^{26} \text{ W}) = 2.3 \times 10^{31} \text{ W}$
so
$R = (1 / (3500 \text{ K})^2) [(2.3 \times 10^{31} \text{ W}) / (4 \pi \times 5.67 \times 10^{-8} \text{ W/m}^2 \text{ K}^4)]^{1/2}$
$= 4.6 \times 10^{11} \text{ m} = 4.6 \times 10^8 \text{ km}$

$$\approx 660 \, R_\odot$$
$$\approx 3 \, AU \quad !!$$

Supergiant M0I stars are thus hundreds of times the size of the Sun; in fact they are larger than the Earth's orbit.

For the M0V star,
$$\log(L_*/L_\odot) = 1.9 - 0.4(10.0 - 1.2) = -1.62$$
$$L_* = 2.4 \times 10^{-2} \, L_\odot$$
$$= (2.4 \times 10^{-2})(3.9 \times 10^{26} \, W) = 9.4 \times 10^{24} \, W$$

so
$$R = (1/(3500 \, K)^2)[(9.4 \times 10^{24} \, W)/(4\pi \times 5.67 \times 10^{-8} \, W/m^2 \, K^4)]^{1/2}$$
$$= 3.0 \times 10^8 \, m = 3.0 \times 10^5 \, km$$
$$\approx 0.43 \, R_\odot$$

Main sequence M0V stars are thus about half the size of the Sun.

13-16. Using the H-R diagram in Figure 13-7 and the relationship between temperature and spectral type in Figure 13-6, estimate how many times larger is Betelgeuse than

(a) Antares

Inspection of Figure 13-7B shows that Betelgeuse has a spectral type M2 and absolute magnitude -5.0, while Antares has a spectral type M1 and absolute magnitude -2.6 (Note that these values differs from the absolute magnitudes given in Appendix Table A4-2).

The difference in absolute magnitude corresponds to a luminosity ratio of (Equation 11-4)
$$L_{Antares}/L_{Betelgeuse} = 10^{0.4(M_{Betelgeuse} - M_{Antares})}$$
$$= 10^{0.4(-5.0 - (-2.6))} = 0.11$$

[Note we are ignoring minor effects due to bolometric corrections.]

From Figure 13-6, make the approximation that an M1 and M2 star have about the same temperature ≈ 3500 K.

The radius of a star is given by (Equation 13-13)
$$L \propto R^2 T^4 \quad \text{or} \quad R \propto L^{1/2}/T^2$$

so the ratio of the radii of Antares to Betelgeuse is
$$R_{Antares}/R_{Betelgeuse} = (L_{Antares}/L_{Betelgeuse})^{1/2}/(T_{Antares}/T_{Betelgeuse})^2$$
$$\approx (0.11)^{1/2}/(1)^2 = 0.33 \approx 1/3$$

(b) β Crucis

From Figure 13-7B, β Crucis is a B0 star of absolute magnitude -5.0. From Figure 13-6, a B0 star has temperature $\approx 25{,}000$ K $\approx 7.1 \, T_{Betelgeuse}$.

$$L_{\beta\ Crucis} / L_{Betelgeuse} = 10^{0.4(M_{Betelgeuse} - M_{\beta\ Crucis})}$$
$$= 10^{0.4(-5.0 - (-5.0))} = 1.0$$
$$R_{\beta\ Crucis} / R_{Betelgeuse} \approx (1)^{1/2} / (7.1)^2 = 2.0 \times 10^{-2} = 0.02$$

(c) α Centauri

From Figure 13-7A or 13-7B, α Centauri is a G2 star of absolute magnitude 4.5. From Figure 13-6, a G2 star has temperature ≈ 5800 K ≈ 1.7 $T_{Betelgeuse}$.

$$L_{\alpha\ Centauri} / L_{Betelgeuse} = 10^{0.4(M_{Betelgeuse} - M_{\alpha\ Centauri})}$$
$$= 10^{0.4(-5.0 - (5.0))} = 1.0 \times 10^{-4}$$
$$R_{\alpha\ Centauri} / R_{Betelgeuse} \approx (1.0 \times 10^{-4})^{1/2} / (1.7)^2 = 3.5 \times 10^{-3} \approx 0.0035$$

13-17. (a) Using the excitation and ionization potentials for calcium, magnesium, helium, and hydrogen (Table 8-3), explain the relative absorption line strengths of neutral and singly ionized atoms of these four elements for stars of different spectral types as shown in Figure 13-6.
Reviewing Table 8-3,

atom / ion	excitation potential	ionization potential
calcium (Ca)	1.9 eV	6.1 eV
magnesium (Mg)	2.7	7.6
hydrogen (H)	10.2	13.6
helium (He)	20.9	24.5

To understand how Figure 13-6 results from Table 8-3, note the order of excitation and ionization potentials for the four different elements. The order of the excitation and ionization potentials is directly related to the temperature at which the appropriate lines would be strongest. We'd expect the strongest of the lines to be from Ca**I**, Mg**I**, Ca**II**, Mg**II**, H**I**, H**I**, He**II** as the temperature increases (there are no H**II** lines).

At the lowest temperatures, neutral calcium (Ca**I**) will be excited, so we might expect absorption lines from these levels. As we go to hotter stars, we'd expect Mg**I** lines be be strong. The neutral calcium lines will diminish in strength rapidly with temperature as the ionization potential is sufficiently low; calcium atoms will be ionized (Ca**II**) before hydrogen atoms are excited. By the time a sufficient number of hydrogen atoms are excited to the n=2 level (for Balmer absorption lines to appear), the Ca**I** (and to some extent the Ca**II**) lines would have diminished in strength. We'd expect (erroneously) that the Mg**II** lines would be strong at temperatures just above those where Ca**II** lines become strong; the Mg**II** lines don't appear to follow our ordering of line strength. At even higher temperatures, the hydrogen would begin to be ionized before a large number of He**I** atoms are excited, so we'd expect the hydrogen Balmer lines to disappear before the He**I** lines reach maximum. The He**I** lines would diminish in strength at higher temperatures as the ionized He**II** line appear.

The argument presented is complicated somewhat in that multi-electron atoms are more complicated than hydrogen in their energy levels and resultant absorption line spectra.

(b) Why do both hot O stars and cool M stars have weak hydrogen absorption lines in their spectra? [Review Sections 8-2 and 8-4; Figure 13-6 shows the relative line strengths of different atoms in stars of different spectral class - temperature.] For Balmer-line absorptions to occur, the electron must be in the n = 2 energy level. Hot O stars have weak Balmer hydrogen absorption lines in their spectra because most hydrogen atoms are ionized (Saha equation; Equation 13-9), so there are few atoms with electrons in the n = 2 level. Cool M stars do have neutral atoms, but the temperature is too low to excite many electrons to the n = 2 level (most atoms are in the n = 1) energy level (Boltzmann equation; Equation 13-8).

13-18. With the unaided eye, away from light pollution, you can see stars with a visual apparent magnitude of 6 or brighter. For each of the main-sequence spectral types, calculate the maximum distance that you could observe a star of that type for this limiting magnitude.

The absolute magnitudes of the various spectral classes are found in Appendix Table A4-3. Consider main sequence stars (luminosity class **V**). [Alternately, the absolute magnitudes can be estimated from the H-R diagrams in the text, such as Figure 13-11B or Figure 13-8C.]

The distance that a star of given absolute magnitude and limiting apparent magnitude can be seen is given by Equation 11-7:

$$M = m + 5 - 5 \log(d_{pc})$$

For m = +6.0, solving for d

$$\log(d_{pc}) = (6.0 + 5 - M)/5 = (11 - M)/5$$

$$d_{pc} = 10^{-(M-11)/5}$$

Solving for an O5 star with absolute magnitude M = -6.0,

$$d_{pc} = 10^{-(-6.0-11)/5} = 10^{3.4} \approx 2500 \text{ pc}$$

Similarly for the other spectral types:

Spectral Class	M_V	Maximum Distance Visible to Naked Eye
O5	-6.0	2500 pc
B5	-1.1	260 pc
A5	+2.1	60 pc
F5	+3.4	33 pc
G5	+5.2	14 pc
K5	+8.0	4 pc
M5	+12.3	0.5 pc

Chapter 14: Our Galaxy: A Preview

14-1. Using a simple diagram, explain why our Galaxy appears as the Milky Way in the night sky.

The Milky Way Galaxy has a relatively flat disk shape as shown in the diagram (in actuality, the galactic plane is much narrower than shown; see also Figure 14-1B and Figure 14-8B).

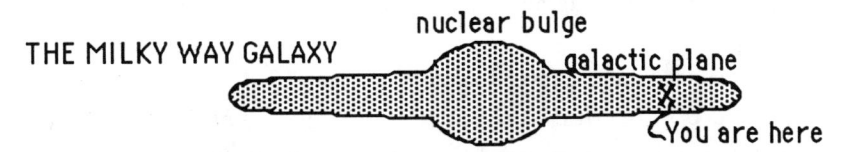

```
                           nuclear bulge
    THE MILKY WAY GALAXY   ⌒⌒⌒⌒⌒  galactic plane
         ≈≈≈≈≈≈≈≈≈≈≈≈≈≈≈≈≈≈≈≈≈≈≈≈≈≈≈≈≈≈
                                    ✕
                                    ↙ You are here
```

We are immersed in the disk about 2/3 of the way from the center to the edge. If we look along the plane of this disk (the galactic plane) we see a large number of stars per unit area. Hence we see the "Milky Way". If we look perpendicular to this plane, we see relatively few stars per unit area in the sky.

14-2. Using Figure 14-4, describe the various aspects of the Milky Way on evenings of different seasons for an observer in New Mexico. Use the altitude-azimuth horizon system of coordinates, and mention the tilt of the Milky Way to the horizon and the observability of Sagittarius.

[Note: The Figure 14-4 from the previous edition has been removed from this edition, thus making it difficult for the student to do this problem. The student is referred to Appendix A10-3 and Appendix Figure A-18, although it may be difficult for many students to effectively use this diagram to answer the question.]

Note that the galactic equator is inclined 63° to the celestial equator -- thus the "Milky Way" will pass through some circumpolar constellations (within 27° from the north celestial pole), will cross the celestial equator at a high inclination (63°), and extend to within 27° of the south celestial pole. The galactic poles are 27° from the celestial equator. The galactic plane cross the celestial equator at about 7^h and 19^h right ascension, and the galactic center is in Sagittarius at right ascension 17^h42^m, declination -29°. Using this information, we can deduce the visibility of the Milky Way for observers in the U.S.

Due to the high inclination of the galactic plane to the celestial equator, the plane is best visible during those months when that right ascension is best visible. The 7^h right ascension portion of the plane (near Orion - Monoceros) is best seen in the winter months (being near the meridian throughout the evening) while the 19^h right ascension portion of the plane (near Sagittarius - Scutum) is best seen during the summer months (near the meridian in the evening). If one is a morning observer, the best times to view these portions of the Milky Way are in fall and spring, respectively (the Milky Way being near the meridian at sunrise). For the northerly portion (near Cassiopeia) of the Milky Way, the best time to view is during the fall and winter months when Cassiopeia is highest in the sky during evening hours. The southerly portion of the Milky Way is not visible to a northern observer. Since the brightest part of the Milky Way is in the direction of the galactic center (Sagittarius), the best time to view this part of the Milky Way is during the summer months, when it will be low in the southern sky -- reaching an altitude of $\approx 35°$ in the southern U.S., and only $\approx 15°$ in the northern U.S. The best place to view this portion of the Milky Way is from the southern hemisphere.

14-3. The integral count ratio $N(m+1)/N(m) = 3.98$ with the assumption that stars are uniformly distributed in space. Make a schematic diagram plotting log N(m) versus m and show what effect each of the following would produce:

(a) a uniform distributed interstellar obscuration of 1 mag/kpc.

General galactic obscuration will cause us to see fewer stars than we would without obscuration. The absorption would increase with distance from the observer, so the incremental number of stars seen would be progressively less as we look at fainter (more distant) stars.

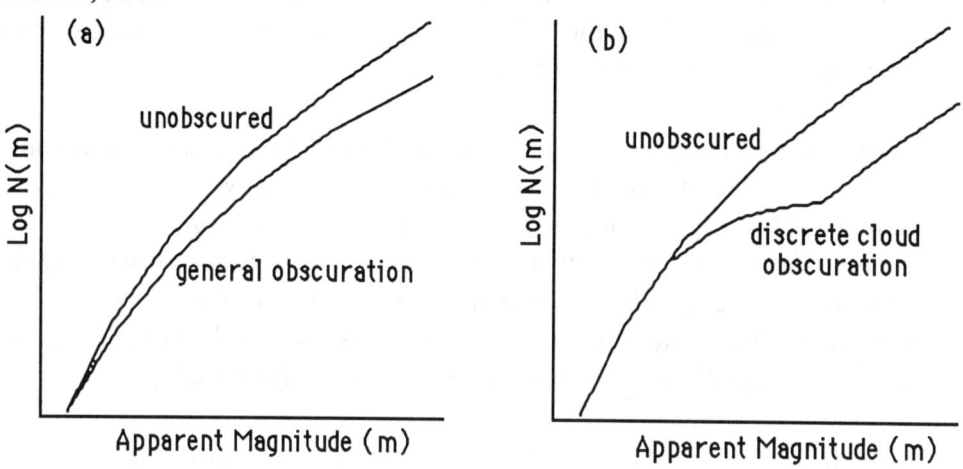

(b) a strongly absorbing interstellar dust cloud localized at the apparent magnitude m_{cloud}.

For discrete clouds of dust, those stars in front of the cloud are unobscured. At the apparent magnitude of the cloud, m_{cloud}, there is an abrupt change in the number of stars seen at fainter magnitude since light from stars that would have been seen without the obscuration is heavily absorbed. For stars beyond the cloud, the curves are again parallel, since these stars' light undergoes the same absorption in the cloud.

14-4. What stellar population would you expect to find (and why)

(a) in the nucleus of our Galaxy

Nuclei of galaxies contain Population **II** stars, so we would expect the same for the nucleus of the Milky Way. We have, however, observed molecular clouds near the galactic center suggesting the possibility of some Population **I** stars.

(b) in the spiral arms of the Andromeda galaxy

Stars in the spiral arms of M31 would be similar to those of our own galaxy, containing primarily Population **I** stars. In the disk we would see Population **I** and Disk Population stars.

(c) in the Pleiades star cluster

The Pleiades star cluster is a young open star cluster located in the spiral arms. Hence we would expect to see Population **I** stars.

(d) in intergalactic space (beyond the halo)

In intergalactic space, beyond the halo, we would expect to see only old extreme Population II stars, or possibly Population III stars (which have been predicted, but not yet conclusively observed to exist). There is insufficient gas and dust clouds in this region for star formation, so any stars seen would be very old -- having escaped from the galaxy.

(e) in the galactic bulge

In the galactic bulge we'd expect to see Population II stars. The stars are old, with little gas and dust for new star formation.

14-5. (a) A globular cluster is in an elliptical orbit (e = 0.9) about the center of our Galaxy, reaching *apogalacticon* (farthest distance from the center) at the distance of 40 kpc. What is the *perigalacticon* (nearest) distance, and how long will this cluster require to complete one orbit?

For elliptical orbits (Equation 1-3), the distance from the focus to a point on the ellipse is given by:

$$r = a(1 - e^2) / (1 + e \cos(\theta))$$

where $\theta = 0°$ at perigalacticon. For e = 0.9, r = 40 kpc, we have at apogalacticon ($\theta = 180°$):

$$40 \text{ kpc} = a(1 - 0.9^2) / (1 + 0.9 \cos(180°)) = 1.90 \, a$$
$$a = 21.0 \text{ kpc}$$

So at perigalacticon,

$$r_{perigalacticon} = 21.0 \text{ kpc} (1 - 0.9^2) / (1 + 0.9 \cos(0°)) = 2.1 \text{ kpc}$$

From Kepler's Third Law (Harmonic Law) (Equation 1-24 or Equation 12-1):

$$P^2 = a^3 / (M_{galaxy} + M_{cluster})$$

where P in years, a in A.U., M in $M_\odot$.

Taking $M_{galaxy} \approx 10^{12} \, M_\odot \gg M_{cluster}$ (Table 14-1),

and a = (40 + 2.1) / 2 = 21 kpc = 4.3×10^9 A.U.

$$P^2 = (4.3 \times 10^9)^3 / (10^{12}) = 8.0 \times 10^{16} \text{ yr}^2$$
$$P = 2.8 \times 10^8 \text{ yr} \approx 300 \text{ million yr}$$

Alternately, a total galactic mass of $7 \times 10^{11} \, M_\odot$, as listed in Appendix 7, could have been used in this calculation, giving a period of $\approx 3.4 \times 10^8$ yr.

[Note, however, that we really want to use the galactic mass internal to the globular cluster's orbit.]

(b) What is the approximate speed of escape from our Galaxy in the solar neighborhood if the Sun's circular orbital speed about the galactic center is 220 km/s?

The escape speed is (Equation 2-8):

$$v_{esc}^2 = 2GM/R$$

Applied to the Sun's position in the galaxy ($R_\odot$ is the Sun's distance from the galactic center):

$$v_{esc\odot}^2 = 2GM_{Galaxy}/R_\odot$$

From Equation 14-6 we have the relationship between the Sun's speed about the galaxy and the galaxy's gravitational attraction:

$$v_\odot^2 = GM_{Galaxy}/R_\odot$$

Combining these two equations,

$$v_{esc\odot}^2 = 2v_\odot^2$$

or

$$v_{esc\odot} = (\sqrt{2})v_\odot$$
$$= (1.414)(220 \text{ km/s})$$
$$= 310 \text{ km/s}$$

14-6. Consider the center of our Galaxy to be a spherical star cluster of radius R and uniform mass density ρ.
(a) What is the total mass M of this cluster?

The total mass of the cluster is: $M_{tot} = V\rho = [(4/3)\pi R^3]\rho$.

(b) What is the mass contained within the sphere of radius $r < R$?
If the density is uniform, the mass within a radius $r < R$ is given by

$$M_r = (4/3)\pi r^3 \rho = M_{tot}(r/R)^3$$

(c) In terms of M, what is the angular speed ω of a star in circular orbit at a distance r ($<R$) from the cluster's center? Note that $\omega = v/r$, where v is the circular orbital speed of the star in kilometers per second.

For a star in a circular orbit at a distance r ($< R$) from the center of the cluster, only the mass interior to r (M_r) contributes to the gravitational force; the mass external to radius r does not contribute. For a circular orbit, setting the centripetal force equal to the gravitational force (Equation 14-6):

$$F_{centripetal} = F_{gravitational}$$
$$m_{star} v^2/r = GM_r m_{star}/r^2$$
$$v^2 = GM_r/r$$

Using the result from step b,

$$v^2 = GMr^2/R^3$$
$$v = (GM_r/r)^{1/2} = r(GM/R^3)^{1/2}$$

The angular speed is given by

$$\omega = v/r = (GM_r/r^3)^{1/2} = (GM/R^3)^{1/2}$$

Note that the angular speed is independent of distance from the center of the cluster!

(d) Using an analogy, explain why we refer to the result of part (c) as *rigid-body rotation*.

Since the angular speed of a star within a uniform mass distribution is independent of radius from the center, all stars within the cluster complete (circular) orbits in the same time -- as if the cluster was rotating as a solid body. We can thus refer to this as "rigid-body rotation".

14-7. Assume that the galactic disk may be approximated by a plane-parallel slab 500 pc thick and having the galactic plane at its midline. Take the Sun to be located in the galactic plane.

(a) How many magnitudes of absorption are there at b = 90°?

For the Sun on the galactic plane, at b = 90° we are looking through half the thickness of the 500 pc slab, or through 250 pc. Using an average visual absorption of 1 magnitude per kpc (Section 14-2A), we get 0.25 magnitude of absorption looking directly out of the plane of the galaxy.

(b) How many magnitudes of absorption are there at the general galactic latitude b?

At a general galactic latitude, b, the distance, d, through which we are looking in the slab is given by: sin(b) = 250 pc / d .

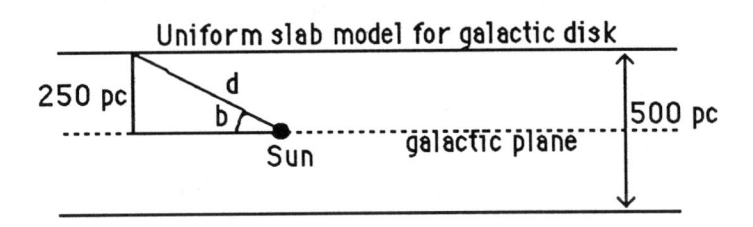

Uniform slab model for galactic disk

<figure: diagram showing 250 pc above galactic plane, 500 pc total slab thickness, Sun on galactic plane, distance d at angle b>

So, d = 250 pc / sin(b) = 250 pc x cosecant(b) ,
and the absorption is: A = 0.25 magnitude / sin(b) .

(c) Explain why the region b ≤ 10° is called the "zone of avoidance" (essentially total obscuration in terms of apparent magnitudes).

At |b| ≤ 10°, A ≥ 0.25 mag. / sin(b) ≥ 1.4
so the general obscuration amounts to more than 1.4 magnitudes. Due to the clumpy nature of the galactic plane, many discrete interstellar clouds will produce an even higher absorption. Thus, many distant stars as well as external galaxies beyond our Milky Way are not visible through the interstellar absorbing medium along the plane of the galaxy -- albeit, the "zone of avoidance".

14-8. In the direction perpendicular to the galactic plane, approximately how thick (in kiloparsecs) is the galactic bulge?

Using Figure 14-8, where the distance of the Sun from the galactic center is 8.5 kpc, the galactic bulge has a thickness perpendicular to the plane of the galaxy of ≈ 5 kpc.

14-9. This chapter suggests that stars formed when disk materials (gas and dust) catch up with a density wave. Assume that newborn stars continue about the galactic center with a circular speed appropriate to their distance.

(a) How far will an O star move from the spiral arm of its birth in 1 million years?

At approximately the Sun's distance from the galactic center, stars orbit the galactic center with a speed of ≈ 220 km / s (Section 14-4; see also Figure 19-10). In 1 million years, a newly formed O star will have traveled

$$d = (220 \text{ km/s}) \times (1.0 \times 10^6 \text{ yr}) \times (3.16 \times 10^7 \text{ s/yr})$$
$$= 7 \times 10^{15} \text{ km} = 225 \text{ pc} \approx 1/4 \text{ kpc}$$

The O star will not have moved far from the spiral arm in which it was created.

(b) Are you surprised to find our Sun in a spiral arm (or near one, at least)?
In the 4.6 billion years since the Sun was formed, it has moved

$$d = (220 \text{ km/s}) \times (4.6 \times 10^9 \text{ yr}) \times (3.16 \times 10^7 \text{ s/yr})$$
$$= 3.2 \times 10^{19} \text{ km} \approx 1.0 \times 10^6 \text{ pc} = 1.0 \times 10^3 \text{ kpc}$$

This corresponds to ≈ 20 orbits about the galactic center, at the Sun's current orbital radius. The fact that the Sun is near a spiral arm is mere coincidence -- it could just as well be located well away from an arm, or within one (the probability depending on the width of the arms and inter-arm distances at the Sun's location in the galaxy).

14-10. If the interstellar absorption average 1 mag/kpc, calculate the optical depth for this absorption in the galactic plane.

Optical depth, τ, is defined by (Equation 10-4):

$$F = F_o e^{-\tau} \quad \text{or} \quad F/F_o = e^{-\tau}$$

The brightness ratio F/F_o is expressed in magnitudes by (Equation 11-3):

$$F/F_o = 10^{0.4(m_o - m)}$$

where m_o is the magnitude before extinction, m is the magnitude after extinction.
Equating the two expressions for F/F_o:

$$e^{-\tau} = 10^{0.4(m_o - m)}$$

where for the left hand side: $\log_{10}(e^{-\tau}) = -\tau \log_{10}(e) = -\tau (0.434)$

and for the right hand side: $\log_{10}(10^{0.4(m_o - m)}) = 0.4 (m_o - m)$

so $\quad -(0.434) \tau = 0.4 (m_o - m)$

$$\tau = 0.922 (m - m_o)$$

Since interstellar absorption is 1 magnitude per kpc ($m - m_o = 1$),

$$\tau = 0.922 / \text{kpc}$$

or, 1 magnitude of absorption per kiloparsec corresponds to an optical depth of 0.922.

14-11. For each of the bins in Figure 14-6, convert absolute visual magnitude to luminosity (in solar units). Multiply the luminosity by the number of stars for each bin. Make a table of the product number of stars x luminosity versus absolute visual magnitude. If these stars are representative of the Galaxy, at what absolute visual magnitude (counting upward) is 95% of the Galaxy's luminosity accounted for? Roughly what range of spectral types contributes the bulk of the Galaxy's luminosity?
Convert the absolute visual magnitude to luminosity (ignoring the bolometric

correction) using Equation 11-20b:
$$\log(L/L_\odot) = 1.9 - 0.4 M_{bol}.$$
Use of this formula and Figure 14-4 (the 50 nearest stars) gives the following table.

Mag (in bin)	1.25	2.75	4.25	5.75	6.25	7.25	8.25	8.75	10.25	10.75
L (in $L_\odot$)	25.1	6.31	1.58	0.40	0.25	0.100	0.040	0.025	0.006	0.004
#stars	1	1	1	3	1	1	2	1	4	2
L x #stars	25.1	6.31	1.58	1.20	0.25	0.100	0.080	0.025	0.024	0.008

table (continued)

11.25	11.75	12.25	12.75	13.25	13.75	14.25	14.75	15.25	15.75	16.75
.0025	.0016	.0010	.0006	.0004	.0003	.0002	.0001	.0001	.00004	.00002
5	3	3	3	6	1	2	3	2	1	2
.0125	.0048	.0030	.0018	.0024	.0004	.0003	.0004	.0001	.00004	.00004

The total luminosity in the above table, from 48 stars, is 34.7 $L_\odot$. If these stars are representative of our galaxy, then 95% (or 33.0 $L_\odot$ in this example) of the Galaxy's luminosity is accounted by stars of brighter than absolute magnitude 5.0 ± 0.5. This is roughly the absolute magnitude of the Sun. So stars in the spectral type from O through F (and the earlier G stars) account for the bulk of the Galaxy's luminosity.

14-12. Repeat the previous exercise except convert luminosity to mass by the rough formula $M \approx L^{1/3}$, where M and L are in solar units. Is most of the mass contained in high-, low-, or intermediate-mass stars? Roughly what range of spectral types contribute most of the mass of the Galaxy?

Using the luminosities deduced in Problem 14-11, and the more accurate mass-luminosity relationship for main sequence stars given by Equations 12-5b & c:

$$M/M_\odot \approx (L/L_\odot)^{1/4} \qquad \text{for } M > 0.43 M_\odot$$
$$M/M_\odot \approx (4.3 L/L_\odot)^{0.43} \qquad \text{for } M < 0.43 M_\odot$$

calculate the mass contribution of these stars (ignoring, again, the bolometric correction). Assume that all these stars are main sequence stars.

Mag (in bin)	1.25	2.75	4.25	5.75	6.25	7.25	8.25	8.75	10.25	10.75
M (in $M_\odot$)	2.24	1.59	1.12	0.80	0.71	0.56	0.45	0.38	0.21	0.17
#stars	1	1	1	3	1	1	2	1	4	2
M x #stars	2.24	1.59	1.12	2.40	0.71	0.56	0.90	0.38	0.84	0.34

table (continued)

11.25	11.75	12.25	12.75	13.25	13.75	14.25	14.75	15.25	15.75	16.75
0.14	0.12	0.10	0.08	0.07	0.06	0.05	0.04	0.04	0.02	0.02
5	3	3	3	6	1	2	3	2	1	2
0.70	0.36	0.30	0.24	0.42	0.06	0.10	0.12	0.08	0.02	0.04

The total mass in the above table, from 48 stars, is 13.5 $M_\odot$. If these stars are representative of our galaxy, then ≈ one-third (or 4.95 $M_\odot$ in this example) of the Galaxy's mass is accounted by stars of brighter than absolute magnitude 5.0 ± 0.5 (that is, a mass greater than ≈1.5 $M_\odot$). Thus, the stars that contribute the most luminosity (Problem 14-11) do **not** contribute a proportionate amount of the mass. So stars of all spectral types contribute to the Galaxy's mass. The mass of a galaxy is much more evenly distributed amongst the different spectral classes than is true of the luminosities (Problem 14-11). High, low, and intermediate mass stars all contribute to the total mass, with intermediate mass stars comprising a significant contribution.

14-13. The star Deneb, in the constellation Cygnus, is one of the most luminous stars visible with the unaided eye. Stars like Deneb are therefore the most distant stars we can see in the Galaxy.
(a) Given its apparent magnitude of 1.3 and distance of 430 pc, calculate the absolute visual magnitude of Deneb (ignore interstellar absorption).
The absolute magnitude is given by Equation 11-7
$$M = m + 5 - 5 \log(d_{pc})$$
$$= 1.3 + 5 - 5 \log(430)$$
$$= -6.9$$
(This agrees with the absolute magnitude shown in Figure 13-7B and Appendix Table A4-2.)

(b) Ignoring interstellar absorption, how distant could a star like Deneb be and still be visible to the unaided eye (m = 6)?
The distance at which Deneb could be seen is also given by Equation 11-7,
$$M_{Deneb} = m_{limiting\ mag} + 5 - 5 \log(d_{pc})$$
$$-6.9 = 6 + 5 - 5 \log(d_{pc})$$
$$\log(d_{pc}) = 3.58$$
$$d = 3.8 \times 10^3 \text{ pc} = 3.8 \text{ kpc}$$

(c) Deneb lies close to the galactic plane. Assuming a constant interstellar absorption of 1.0 magnitude per kiloparsec, estimate the absorption towards Deneb in magnitudes.
At a distance of 0.43 kpc, the light from Deneb would suffer 0.43 magnitude of absorption.

(d) Assuming interstellar absorption, how distant could a star like Deneb be and still be visible with the unaided eye?
The light from a star viewed through the galaxy is absorbed by $(d_{pc})/10^3$ magnitudes (or d_{kpc} magnitudes), so the distance to which we could see Deneb in the plane of the Galaxy is given by Equation 11-7 modified by this absorption:
$$M_{Deneb} + (d_{pc})/10^3 = m_{limiting\ mag} + 5 - 5 \log(d_{pc})$$
$$-6.9 + (d_{pc})/10^3 = 6 + 5 - 5 \log(d_{pc})$$

$$5 \log (10^3 d_{kpc}) + (10^3 d_{kpc}) / 10^3 = 17.9$$
$$5 \log (10^3) + 5 \log (d_{kpc}) + d_{kpc} = 17.9$$
$$5 \log (d_{kpc}) + d_{kpc} = 2.9$$

Using a numerical guessing trial and error procedure gives
$$d = 1.7 \text{ kpc}$$

Deneb could be seen at a distance of 1.7 kpc if it suffered 1 magnitude of absorption per kiloparsec. Comparing this result to step (a), where an unabsorbed Deneb could be seen to a distance of 3.8 kpc, reveals the effect which interstellar gas and dust have on our view of the Galaxy and Universe.

14-14. Assume that the interstellar absorption in the galactic plane has a roughly constant value of 1 magnitude per kiloparsec.

(a) What is the absolute magnitude of the faintest star that could be detected at the galactic center by a telescope with a limiting magnitude of 23.5?

The galactic center is 8.5 kpc distant (Table 14-1), so light from a star would suffer 8.5 magnitudes of extinction for an interstellar absorption of 1 magnitude per kpc. A star at the galactic center would have to have an apparent magnitude of 23.5 - 8.5 = 15.0 without absorption in order to just be seen with a telescope of limiting magnitude 23.5. From Equation 11-7, such a star would have an absolute magnitude

$$M = m + 5 - 5 \log (d_{pc})$$
$$= 15.0 + 5 - 5 \log(8.5 \times 10^3)$$
$$= 0.35$$

(b) Compare this result to the value if there were no interstellar extinction.

If there were no interstellar absorption,
$$M = m + 5 - 5 \log (d_{pc}) = 23.5 + 5 - 5 \log(8.5 \times 10^3) = 8.85$$

(c) In each case, what stars on the H-R diagram have the required absolute magnitude?

For the case of interstellar absorption, from inspection of Figure 13-8C, all stars of luminosity class I, II, the M stars of class III, and only those of spectral type O and B of classes IV and V would be visible.

For the case of no interstellar absorption, from inspection of Figure 13-8C, all stars of luminosity class I, II, III, and IV and those of class V with spectral type O to K would be visible. Only the main sequence M stars would be undetectable.

(d) These stars at the galactic center cannot actually be observed at visual wavelengths. Assuming that the stars do indeed exist, why do you think they are not visible?

These stars in the galactic center cannot actually be observed since the interstellar absorption is not uniform with 1 magnitude per kpc extinction. In the plane of the galaxy, particularly in the inner parts of the galaxy, the extinction is much greater due to dense clumps of interstellar gas and dust.

14-15. (a) Assuming a constant interstellar extinction of 1 magnitude per kiloparsec, what is the maximum distance to which we could see a bright globular cluster in our Galaxy using a telescope with a limiting visual magnitude of 23.5?

The brightest globular clusters have absolute magnitude M = -10 (Section 14-2B) [Note: some students may select an absolute magnitude between -10 and -4 for a typical globular cluster.]. The effective limiting magnitude with extinction taken into account is

$$m_{eff} = m_{limiting} - (d_{pc})/10^3 .$$

We can modify equation 11-7 to

$$M = m_{eff} + 5 - 5 \log(d_{pc}) = m_{limiting} - (d_{pc}/10^3) + 5 - 5 \log(d_{pc})$$

which becomes, for the limiting magnitude m = 23.5,

$$-10 = 23.5 - (d_{pc}/10^3) + 5 - 5 \log(d_{pc})$$

$$5 \log(d_{pc}) + (d_{pc}/10^3) = 38.5$$

converting distances to kpc

$$5 \log(10^3 d_{kpc}) + (10^3 d_{kpc}/10^3) = 38.5$$

$$5 \log(10^3) + 5 \log(d_{kpc}) + d_{kpc} = 38.5$$

$$5 \log(d_{kpc}) + d_{kpc} = 23.5$$

Using a numerical guessing trial and error procedure gives
   d = 17.3 kpc.
We should be able to see luminous globular clusters on the other side of our Galaxy.

(b) Globular clusters are observed around the Andromeda galaxy (M31), a nearby spiral galaxy 960 kpc from our Galaxy. Reconcile this observation with your answer in (a).

We can see globular clusters in the galaxy M31 because we are looking out of the galactic plane, so the path length through the galaxy is small. There is no extinction between the galaxies.

To see how far we can see a globular cluster without extinction,

$$M = m_{limiting} + 5 - 5 \log(d_{pc})$$

$$d_{pc} = 10^{(m_{limiting} + 5 - M)/5}$$

$$= 10^{(23.5 + 5 - (-10))/5} = 5 \times 10^7 \text{ pc} = 5{,}000 \text{ kpc}$$

Luminous globular clusters could be seen well beyond (more than five times more distant) the Andromeda galaxy.

Chapter 15: The Interstellar Medium and Star Birth

15-1. An A0 V star has an apparent visual magnitude of 12.5 and an apparent blue magnitude of 13.2.

(a) What is the color excess for this star?

An A0 V star has an intrinsic color index of B - V = 0 (Appendix Table A4-3). Using Equation 15-3, the color excess is:

$$CE = CI_{observed} - CI_{intrinsic} = (13.3 - 12.5) - 0.0 = 0.8$$

(b) What is the visual absorption in front of this star?

The visual absorption in front of this star is given by Equation 15-4,

$$A_V \approx 3 \,(CE) = 3\,(0.8) = 2.4$$

(c) Calculate the distance of the star (in parsecs).

The distance can be calculated using Equation 15-2, and $M_V = +0.6$ (Appendix Table A4-3):

$$m_V - M_V = 5 \log(d_{pc}) - 5 + A_V$$
$$12.5 - 0.6 = 5 \log(d_{pc}) - 5 + 2.4$$
$$\log(d_{pc}) = 2.9$$
$$d \approx 790 \text{ pc}$$

(d) What error would have been introduced if you had neglected interstellar absorption?

If we had ignored the interstellar absorption, we would have made a considerable error (about a factor of three) in the distance calculation.

$$m_V - M_V = 5 \log(d_{pc}) - 5$$
$$12.5 - 0.6 = 5 \log(d_{pc}) - 5$$
$$\log(d_{pc}) = 3.38$$
$$d \approx 2400 \text{ pc} = 2.4 \text{ kpc}$$

$$d_{without\ absorption} / d_{with\ absorption} = 2400 \text{ pc} / 790 \text{ pc} \approx 3.0$$

or $\quad d_{without\ absorption} / d_{with\ absorption} = 10^{A_V/5} = 10^{2.4/5} = 3.0$

15-2. (a) How does interstellar reddening alter the Planck spectral energy curve of a star? Sketch approximate curves for an A star (10,000 K) in the visible part of the spectrum, both with and without reddening.

The interstellar reddening absorbs more in the blue than in the red; so the Planck curve will be suppressed more at blue wavelengths than at the red. The total energy will be less and the wavelength of peak emission will shift slightly to the red. For a 10,000K blackbody, the unreddened and reddened Planck curves would qualitatively appear as shown. The unreddened curve is from Figure 8-14.

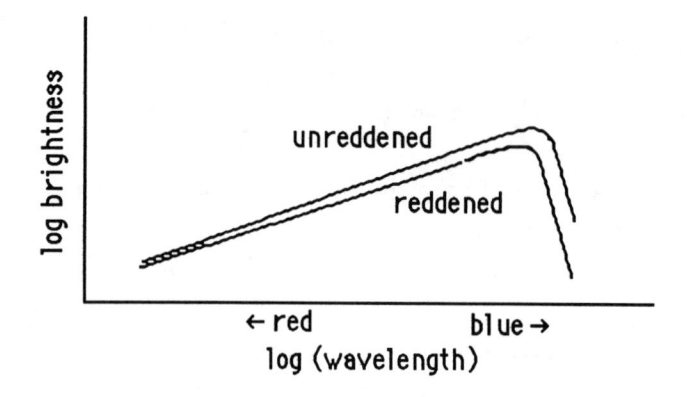

```
        log brightness
                        ┌─── unreddened
                        │   ┌── reddened
                        │   │
        ← red        blue →
        log (wavelength)
```

(b) What effect would interstellar reddening have on the color-magnitude (H-R) diagram of a star cluster? Draw a diagram to support your answer.

The interstellar reddening has the effect of making a star appear redder and fainter. A star will thus be placed slightly cooler and less luminous on the H-R diagram than it actually is. The main sequence will be shifted to the right and down. The effect will be greater for blue stars than for red stars.

15-3. What are the observational clues to the nature of the interstellar dust grains, and how have these clues been interpreted in terms of models?

Some clues as to the nature of the interstellar grains are: the general galactic extinction, wavelength dependence of the extinction (reddening), interstellar polarization, and infrared spectral features (particularly the silicate and ice features at 9.7 μm and 3.07 μm). The dust grains seem to be made up of silicate (or possible iron or graphite) cores with ice mantles. The grains seem to also come in a range of sizes and compositions. The interstellar polarization suggests that the grains (at least some of them) are elongated and magnetic.

15-4. Several galactic nebulae have been photographed in color. Two different types occur, reddish and bluish.

(a) Explain the physics of these two types of nebulae.

The bluish nebulae are reflection nebulae. We see starlight that is reflected by the dust in the nebulae. The red light is not reflected (scattered) as much because the grain sizes are such that the red light can more readily pass through the dust cloud. If we were to view a star through the dust cloud it would appear reddened.

The reddish nebulae are emission nebulae -- HII regions. The red color is due to H-α emission. UV photons from embedded stars ionize the hydrogen atoms, producing the HII region. The electrons recombine to an excited atomic state, then cascade down to the ground state emitting lower energy Balmer series photons. The low energy photons easily escape the HII region -- so we see these photons. The dominant Balmer line is the H-α line.

(b) Briefly describe the spectra you would expect to observe for these two types of nebulae.

Because the bluish reflection nebulae are due to scattered starlight, we would expect to see a highly dereddened (blue) stellar spectrum -- continuum emission with stellar absorption features. The reddish emission nebulae are due to recombination and de-excitation of atoms (predominantly hydrogen, though other elements do contribute to the emission). We would see an emission line spectrum in emission nebulae -- thus their name.

15-5. Why are none of the hydrogen Balmer lines seen as interstellar absorption lines, even though hydrogen is the most abundant element in the Universe? (*Hint:* Recall the energy level diagram of hydrogen.) [Review Section 8-2.] To see the Balmer series absorption lines we need the hydrogen atoms to be excited to the second (n = 2) level -- the first excited level. The interstellar gas is of sufficiently low density and temperature that collisional excitation to n = 2 is negligible -- most atoms are in the ground state. The relatively low interstellar ultraviolet flux makes radiative excitation unimportant except in HII regions surrounding the luminous O and B stars. We therefore do not see Hydrogen Balmer absorption lines. (For a further discussion of excitation, see the discussions of the Boltzmann and Saha equations in Section 8-4.)

15-6. (a) Name the two factors that primarily determine the size of an HII region.
The ultraviolet flux (particularly at wavelengths shorter than that corresponding to the ionization energy of hydrogen -- wavelengths less than 91.2 nm) is an important factor in determining the size of the HII region. The greater the UV flux, the greater the number of ionizations which occur, so the larger the HII region. This UV flux is ultimately determined by the surface temperature (spectral type) of the exciting star.
The hydrogen density is also important. The higher the density, the smaller the HII region will be since absorption of the UV photons will occur closer to the star. Density clumps of hydrogen gas will also produce a non-spherical Strömgren "sphere".

(b) Explain the physical basis for your answer to part (a) in terms of the Strömgren sphere.
The UV flux ($\lambda \leq 91.2$ nm) is required to ionize the hydrogen atoms. The greater the number of UV photons the greater the number of ionizations which can occur (the hotter the star, the more UV photons are produced). However, some recombinations will occur, with equilibrium occurring when the ionization and recombination rates are equal. The continued production of UV photons from the star maintains the equilibrium -- resulting in a stable Strömgren sphere size. For smaller gas densities, the UV photons can travel a longer distance from the exciting star before being absorbed by an H atom, so the Strömgren sphere will be larger.

(c) The brightness of an HII region depends only on the gas density of the nebula. Explain this phenomenon in terms of what you know about the hydrogen atom. (*Hint:* Remember the Saha equation.)
The brightness of an HII region depends on the number of recombinations (since most recombined electrons cascade to the ground state -- emitting photons -- before they are subsequently re-ionized). From the Saha equation (Equation 8-30 and Figure 8-13), we see that as the hydrogen density increases, the number density of ionized hydrogen

and electrons increases -- so the recombination rate increases.

15-7. Forbidden lines are observed both in the solar corona and in gaseous nebulae, but they are not the same forbidden lines.
(a) How do the lines differ in the two cases?
As a result of the solar corona's high temperature ($\approx$ 2 million K), the forbidden lines are from highly ionized atoms such as Fe **X** (9 electrons ionized). The gaseous nebulae are at much lower temperatures ($\approx$ 10,000 K), so the forbidden lines are from atoms in lower ionization states, such as O **III**.

(b) Why are nebular forbidden lines not emitted by the solar corona? What about coronal forbidden lines for nebulae?
The highly ionized atoms found in the solar corona require high temperatures (few million K) to produce the required collisional excitations. The lower temperature (few thousand K) nebulae do not have enough collisional ionizations to produce the coronal forbidden lines.
The solar corona has such a high collision rate that the atoms in the metastable states required to produce the nebular forbidden lines suffer a collision before the required transition can occur.

15-8. Somewhere in our Galaxy resides a cloud of neutral hydrogen gas with a radius of 10 pc. The gas density is 10^7 atoms/m^3.
(a) How many 21-cm photons does the cloud emit every second?
According to the information in the chapter (Section 15-2e), about 75% of the neutral hydrogen atoms in an interstellar cloud are in the excited hyperfine state (spins aligned). For an atom, on the average there is a spontaneous downward (spin-flip) transition for each atom every few million years. However, collisional excitations or de-excitation occurs every 400 years. Because of the rapid (relative to a few million years!) collisional excitation and de-excitation rates, an equilibrium is established in the hyperfine state populations with 75% in the spin parallel state.

The number of 21-cm photons emitted per second is the number of H atoms in the cloud which are in the excited hyperfine state divided by the average time between transitions. Since the size of the cloud is given as 10 pc = 3.1 x 10^{17} m, and the time between transitions is few million years $\approx 10^{14}$ s,
$$\text{\# of 21-cm photons} = (0.75) \times n_H \times \text{Volume}_{cloud} / \text{(time between transitions)}$$
$$= (0.75) (10^7 \text{ m}^{-3}) [(4/3) \pi (3.1 \times 10^{17} \text{ m})^3] / 10^{14} \text{ s}$$
$$= 9.4 \times 10^{45} / \text{s} \approx 10^{46} / \text{s}$$

(b) If the cloud is 100 pc from the Sun, what is the energy flux of this radiation (in W / m^2) at the Sun?
The energy of a 21-cm photon is (Equation 8-15):
$$E = h\nu = hc/\lambda$$

$$E = (6.63 \times 10^{-34} \text{ J s}) (3.0 \times 10^8 \text{ m/s}) / (0.21 \text{ m})$$
$$= 9.5 \times 10^{-25} \text{ J}$$

so the total energy produced per second by the cloud is

$$L = (9.5 \times 10^{-25} \text{ J / transition}) \times (9.4 \times 10^{45} \text{ transitions / s})$$
$$= 8.9 \times 10^{21} \text{ J/s} \approx 9 \times 10^{21} \text{ J/s}$$

At a distance of 100 pc (= 3.1×10^{18} m), the energy flux is (Equation 8-17):

$$F = L / (4 \pi d^2)$$
$$= 8.9 \times 10^{21} \text{ W} / [4 \pi (3.1 \times 10^{18} \text{ m})^2]$$
$$= 7.4 \times 10^{-17} \text{ W/m}^2 \approx 10^{-16} \text{ W/m}^2$$

15-9. What are the energies (in electron volts) of the photons that characterize the
(a) Lyman continuum limit (91.2 nm)
The energy of a photon is given by (Equation 8-15):

$$E = h\nu = hc/\lambda$$
$$= (6.625 \times 10^{-34} \text{ J s}) (3.0 \times 10^8 \text{ m/s}) (1 \text{ eV} / 1.60 \times 10^{-19} \text{ J}) / \lambda$$
$$= 1.24 \times 10^{-6} \text{ eV m} / \lambda$$

so, the Lyman continuum,

$$\lambda = 91.2 \text{ nm} = 9.12 \times 10^{-8} \text{ m} \implies E = 13.6 \text{ eV}$$

(b) nebular line of [O III] at 500.7 nm

$$\lambda = 500.7 \text{ nm} = 5.007 \times 10^{-7} \text{ m} \implies E = 2.48 \text{ eV}$$

c) neutral hydrogen line (21 cm)

$$\lambda = 21 \text{ cm} = 2.1 \times 10^{-1} \text{ m} \implies E = 5.92 \times 10^{-6} \text{ eV}$$

d) ammonia (NH_3) emission (1.3 cm)

$$\lambda = 1.3 \text{ cm} = 1.3 \times 10^{-2} \text{ m} \implies E = 9.54 \times 10^{-5} \text{ eV}$$

e) hydrogen Balmer line, H-α (656.3 nm)

$$\lambda = 656.3 \text{ nm} = 6.563 \times 10^{-7} \text{ m} \implies E = 1.89 \text{ eV}$$

15-10. Using Bohr's formula for the wavelengths of hydrogen lines, calculate the wavelength and frequency of the radio recombination line with upper level 93 and lower level 92.
The wavelengths of transitional lines in hydrogen atoms is given by (Equation 8-25):

$$1/\lambda_{ab} = R(1/n_b^2 - 1/n_a^2) \quad \text{where } R = 1.096776 \times 10^7 \text{ m}^{-1}$$

For $n_a = 93$, and $n_b = 92$,

$$1/\lambda_{ab} = 1.10 \times 10^7 \text{ m}^{-1} [1/(92)^2 - 1/(93)^2]$$
$$= 1.10 \times 10^7 \text{ m}^{-1} (2.53 \times 10^{-6}) = 2.77 \times 10^1 \text{ m}^{-1}$$

$$\lambda_{ab} = 3.61 \times 10^{-2} \text{ m} = 3.61 \text{ cm}$$

$$\nu = c/\lambda = (3.0 \times 10^8 \text{ m/s}) / (3.61 \times 10^{-2} \text{ m})$$
$$= 8.32 \times 10^9 \text{ Hz} = 8.32 \text{ GHz}$$

This transition is in the radio portion of the electromagnetic spectrum.

15-11. A pure hydrogen gas cloud of number density $n = 10^7$ atoms/m^3 surrounds an O star that generates 10^{49} photons/s at wavelengths shorter than 91.2 nm. The rate at which such recombinations occur is $\alpha = (2 \times 10^{-19}) n^2 / \text{m}^3 \cdot \text{s}$.

(a) Balance the number of ionizations with the number of recombinations to determine the Strömgren radius of the resultant H II region.

The recombination rate, α, (where $\alpha(2)$ is the recombination coefficient defined in the text Section 15-2B; and n [number density of atoms] $\approx n_e$ [e.g. number density of electrons] $\approx n_H$ [e.g. number density of hydrogen nuclei]) is

$$\alpha = \alpha(2) n_e n_H = (2 \times 10^{-19}) n^2 / \text{m}^3 \text{ s}$$
$$= (2 \times 10^{-19})(10^7)^2 / \text{m}^3 \text{ s}$$
$$= 2 \times 10^{-5} / \text{m}^3 \text{ s}$$

This is the number of recombinations per unit volume per second.

The recombinations occur throughout the Strömgren sphere of radius r_s, so the total number of recombinations is this rate times the volume of the sphere, V_s. For equilibrium, the number of recombinations per second equals the number of ionizations per second -- which equals the number of ionizing photons per second, N, emitted by the ionizing star. Equating recombinations with ionizations:

$$V_s \alpha = N_{UV}$$
$$[(4/3) \pi (r_s)^3] [\alpha(2) n_e n_H] = N_{UV}$$
$$[(4/3) \pi (r_s)^3] (2 \times 10^{-5} / \text{m}^3 \text{ s}) = 10^{49} / \text{s}$$
$$r_s^3 = 1.19 \times 10^{53} \text{ m}^3$$
$$r_s = 4.9 \times 10^{17} \text{ m} = 16 \text{ pc}$$

(b) The Sun produces about 5×10^{23} photons/s with $\lambda \leq 91.2$ nm. Calculate its Strömgren radius in astronomical units for an interplanetary medium density of 10^9 atoms/m^3.

For the Sun, with $N = 5 \times 10^{23}$ photons/s, and the hypothetical interplanetary density of $n = 10^9$ atoms/m^3, balancing recombinations and ionizations gives:

$$[(4/3) \pi r_s^3] [(2 \times 10^{-19})(10^9)^2 / \text{m}^3 \text{ s}] = 5 \times 10^{23} / \text{s}$$
$$r_s^3 = 5.97 \times 10^{23} \text{ m}^3$$

$$r_s = 8.4 \times 10^7 \text{ m} = 5.6 \times 10^{-4} \text{ AU}$$

This is less than the Sun's radius, so to do the problem correctly we need to consider the volume of the shell around the Sun rather than a sphere. However, the Strömgren sphere for the Sun is so small as to be insignificant!

A G star inside a cloud of density $10^9 / \text{m}^3$ (more like an intracloud density than a real interplanetary density) would thus be unable to produce an **HII** region. The density in the solar system is actually about a factor of 1/1000 the density posed in the problem, so the size of any ionized region around the Sun is insignificant.

15-12. Approximate the Galaxy as a uniform disk of constant thickness. Show that the optical depth of extinction by interstellar dust should approximately obey the law $\tau_\lambda \propto \csc b$ (except for small b), where b is galactic latitude. At what galactic latitudes would an astronomer want to observe other galaxies without having to worry much about extinction?
(See also Problem 14-7)

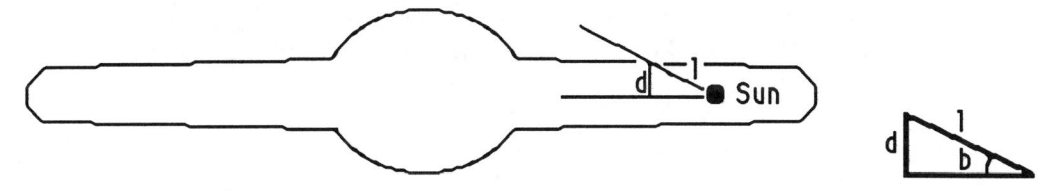

We are located sufficiently far away from the galactic center that the galactic plane can be approximated as a disk of uniform thickness (for galactic latitudes above a few degrees, where we are not viewing through the galactic bulge). For those latitudes for which we look only through the uniform thickness slab of the galactic plane, the path length that we look through the galaxy, l, is related to the half-thickness of the galaxy, d, and the galactic latitude, b (assuming the Sun is on the galactic plane):
$\sin(b) = d / l$ or $l = d / \sin(b) = d \csc(b)$.

Now, the optical depth, τ_λ, is proportional to the path length through which we view, assuming a uniform disk, so $\tau_\lambda \propto l$, so $\tau_\lambda \propto d \csc(b)$.

The latitude above which an astronomer would want to observe galaxies depends on the proportionality constant and the galactic half-thickness. However, the astronomer would want the extinction to be at most a few magnitudes. For the galactic parameters given in Chapter 14 (see Problem 14-7), specifically half-thickness of the galactic plane of 250 pc and general extinction of 1.0 magnitudes per kpc, the interstellar absorption is given by:
$A_\lambda = 1.086 \tau_\lambda = 0.25 \csc(b)$ magnitudes.
If we want to keep $A_\lambda < 1$ magnitude, then $\csc(b) < 4.0$; so we want to observe above $b = 15°$ (or 7° if we can tolerate 2 magnitudes extinction).

15-13. An open cluster of stars is found to contain main-sequence O5 stars with observed color indices (B - V) of 0.4. These O stars are observed to have apparent magnitudes of 10.0.

(a) Use Appendix Table A4-3 to calculate the distance to the cluster. Include the effects of extinction by dust.

The intrinsic color index of main sequence O5 stars is, from Appendix Table A4-3,
$$CI = B - V = -0.32.$$
These stars have $M_V = -6.0$. To correct for dust absorption, use Equation 15-3:
$$CE = CI_{observed} - CI_{intrinsic}$$
$$= 0.4 - (-0.32) = 0.72$$
The interstellar absorption is given by Equation 15-4:
$$A_V \approx 3 \, (CE) = 3 \, (0.72) = 2.16$$
The distance can be determined using Equation 15-2:
$$m_V - M_V = 5 \log(d_{pc}) - 5 + A_V$$
$$10.0 - (-6.0) = 5 \log(d_{pc}) - 5 + 2.2$$
$$\log(d_{pc}) = 3.76$$
$$d = 5.75 \times 10^3 \text{ pc} = 5.8 \text{ kpc}$$

(b) Determine the apparent magnitude of a G0 main-sequence star in the cluster.

A G0V star has $M_V = +4.4$ (Appendix Table A4-3). Applying Equation 15-2 as above,
$$m_V = M_V + 5 \log(d_{pc}) - 5 + A_V$$
$$= 4.4 + 5 \log(5.75 \times 10^3) - 5 + 2.2$$
$$= 20.4$$

15-14. Consider an extended region filled with neutral hydrogen gas at a density of nm_H and temperature T, where m_H is the mass of the hydrogen atom. Now consider a small spherical volume of radius R inside this region where the density is slightly higher than in the surrounding region.

(a) Show that the radius of this volume must satisfy the condition
$R \geq (3 \, kT / 2 \pi G n \, m_H^2)^{1/2}$ in order for gravitational collapse of the denser regions to occur.

(*Hint:* Obtain an expression for gravitational pressure by integrating Equation 16-1 over r.)

We will integrate the hydrostatic equilibrium equation (Equation 16-1 or Equation 4-1) to derive the gravitational pressure.
$$dP/dr = - G \, M(r) \, \rho(r) / r^2$$
where the density is
$$\rho(r) = n \, m_H \equiv \rho = \text{constant}$$
and the mass is given by
$$M(r) = (4/3) \, \pi \, \rho \, r^3$$
Substituting,
$$dP = - (4/3) \, \pi \, G \, (n \, m_H)^2 \, r^3 \, dr / r^2$$
$$\int dP = - (4/3) \, \pi \, G \, (n \, m_H)^2 \int r \, dr$$
Integrating from the center ($r = 0$, $P = 0$) to the surface ($r = R$, $P = P_R$),
$$P_R = - (4/3) \, \pi \, G \, (n \, m_H)^2 \, (R^2 / 2)$$

The negative sign indicates that the gravitational pressure is inward.

For collapse to occur the gravitational pressure must be greater than the gas pressure given by the ideal gas law (Equation 13-1), or

$|P_{grav}| \geq |P_{gas}| = nkT$

$(2/3) \pi G (n m_H)^2 R^2 \geq nkT$

$R \geq (3kT / 2\pi G n m_H^2)^{1/2}$

(b) Obtain a lower limit for the mass of the collapsing volume.
The lower limit to the mass of the collapsing volume is:

$M \geq (4/3) \pi R^3 \rho$

$M \geq (4/3) \pi (3kT / 2\pi G n m_H^2)^{3/2} n m_H$

$M \geq [(4/3) \pi (3kT / 2\pi G)^{3/2}] n^{-1/2} m_H^{-2}$

(c) For the general interstellar medium, we can take the approximate values $n = 10^6 / m^3$ and $T = 100$ K. Obtain a numerical value for the mass limit in solar masses.
Using these values in the last equation,

$M \geq (4/3) \pi [3 (1.38 \times 10^{-23} \text{ J/K}) (100K)$
$\qquad\qquad / 2\pi (6.67 \times 10^{-11} \text{ N m}^2/\text{kg}^2)]^{3/2}$
$\qquad\qquad \times (10^6/\text{m}^3)^{-1/2} \times (1.67 \times 10^{-27} \text{ kg})^{-2}$

$M \geq (4.19) (9.88 \times 10^{-12} \text{ m}^{-1} \text{ kg}^2)^{3/2} (10^{-3} \text{ m}^{3/2}) (3.59 \times 10^{53} \text{ kg}^{-2})$

$M \geq 4.67 \times 10^{34} \text{ kg} = 2.34 \times 10^4 M_\odot$

(d) Is a collapsing region of the general interstellar medium likely to produce a single star like the Sun?
The minimum mass is larger than the mass of the Sun. Thus, if this cloud is to form the Sun, either much of the mass must be lost in star formation or the cloud must divide into sub-clouds which forms many stars. If we require that a single cloud contract to form a solar-like star, then the mass necessary for self-gravitational collapse must be less -- either the gas temperature is lower or the density must be greater. Inspection of the answer in part (b) shows that the temperature can not be sufficiently low to lower the minimum mass requirement to the order of 1 $M_\odot$; therefore the density must be higher.

15-15. A star count is made on a photograph that contains a dark nebula. Ten times fewer stars are counted in front of the cloud than are found on a region away from the cloud with the same solid angle. For ease of calculation, assume that the dark cloud is completely opaque, that the stars are uniformly distributed in space, that all stars have absolute magnitudes of 5.0, and that the limiting magnitude of the photograph is 15.0. Calculate an approximate distance to the cloud.
Assuming that all the stars have M = 5.0 and the limiting magnitude of the photograph

is m = 15.0, we can compute the maximum distance observed in the unobscured photograph using (Equation 11-6 or Equation 15-1):

$$m - M = 5 \log(d_{pc}) - 5$$
$$\log(d_{pc}) = (15 - 5 + 5) / 5 = 3$$
$$d = 1,000 \text{ pc}$$

We see unobscured to a distance of 1000 pc. Since there are less stars visible in the line of sight to the cloud, the cloud must be closer than this distance.

To calculate the cloud's distance, take the ratio of the volumes observed in the obscured and unobscured regions. For a given solid angle, Ω,

$$\text{volume} \propto \text{Area} \times r \propto (\Omega r^2) r \propto r^3, \text{ so}$$

$$V_{\text{not obscured}} / V_{\text{obscured}} \approx (d_{\text{not obscured}} / d_{\text{cloud}})^3$$

If the stars are uniformly distributed, then since 10 times more stars are seen in the unobscured region,

$$V_{\text{not obscured}} / V_{\text{obscured}} = 10 \approx (1000 \text{ pc} / d_{\text{cloud}})^3$$
$$d_{\text{cloud}} \approx 1000 \text{ pc} / (10^{1/3}) \approx 460 \text{ pc}$$

15-16. There is indirect evidence in L 1551 - IRS 5 for a roughly Solar System-sized disk surrounding the protostar. Consider the feasibility of directly observing a solar system size disk ($\approx$ 50 AU) at a distance of 150 pc by estimating the angular size of such a disk. Comment on your result.

Using the small angle approximation, $\theta \approx \tan(\theta) = d / D$, where θ is the observed angle in radians of an object of linear size d at a distance D, or for θ in arc seconds,

$$\theta \approx \tan(\theta) = 206,265 \, (d / D) \text{ arcsec}.$$

$$\theta = 2.06 \times 10^5 \, (50 \text{ AU}) / (150 \text{ pc} \times 2.06 \times 10^5 \text{ AU/pc}) \text{ arcsec}$$
$$= 0.33 \text{ "}$$

The angular size of L 1551 - IRS 5 would be 1/3 arcsec. This is right at the limit of resolution (due to atmospheric seeing; see Section 9-1 for a discussion of seeing) from the best of sites on the Earth; it would also be just resolvable by the Hubble telescope.

15-17. Using Figure 15-22 and an assumed average outflow velocity of 50 km/s, estimate the age of the molecular outflow associated with IRc2 in the Orion Nebula.

From Figure 15-21 (not Figure 15-22 as stated in the problem), we see that the bipolar outflow extends roughly 9000 AU from the center region at IRc2 to the edge of the lobes. If the nebula has been expanding at a speed 50 km/s, the age of the expanding region is

$$t = d / v$$
$$= (9 \times 10^3 \text{ AU}) (1.496 \times 10^8 \text{ km/AU}) / (50 \text{ km/s}) = 2.7 \times 10^{10} \text{ s}$$
$$= 8.5 \times 10^2 \text{ years} = 850 \text{ years}$$

15-18. Let's examine again what conditions would result in the formation of the Sun.

(a) Consider the following: A core of a giant molecular cloud has a radius of about 1 pc, a temperature between 30 and 100 K, and a number density of some $10^{10}/m^3$. Is it likely to collapse to form a solar-type star?

See Section 15-3A. Some insight into the possibility of collapse to form a solar-type star can be found by estimating the mass of the cloud. The mass of a spherical cloud is given by

$$M = (4/3) \pi R^3 \rho$$
$$= [(4/3) \pi (1 \text{ pc} \times 3.086 \times 10^{16} \text{ m/pc})^3] \rho = (1.23 \times 10^{50} \text{ m}^3) \rho$$

If the cloud is molecular hydrogen, the mass density is

$$\rho = 2 m_H N$$
$$= 2 (1.67 \times 10^{-27} \text{ kg}) (10^{10}/m^3) = 3.34 \times 10^{-17} \text{ kg}/m^3$$

where N is the number density.
The mass of the cloud before collapse is therefore

$$M = (1.23 \times 10^{50} \text{ m}^3)(3.34 \times 10^{-17} \text{ kg}/m^3)$$
$$\approx 4 \times 10^{33} \text{ kg} \approx 2000 \, M_\odot$$

For this cloud to collapse to a solar-type star, it must lose 99.95 of its initial mass.

Equation 15-9 can also provide some insight into the right size scale, L, of the collapsing region (for L in m, T in K, and ρ in kg/m^3),

$$L \approx 10^7 (T/\rho)^{1/2}$$
$$= 10^7 [30/(3.34 \times 10^{-17})]^{1/2} \quad \text{to} \quad = 10^7 [100/(3.34 \times 10^{-17})]^{1/2}$$
$$= 9.5 \times 10^{15} \text{ m} \approx 0.3 \text{ pc} \qquad\qquad = 1.7 \times 10^{16} \text{ m} \approx 0.55 \text{ pc}$$

This is about one-third the actual size of the cloud.
Thus this cloud is a bit large and hot to be likely to collapse to a solar-type star.

(b) Small knots within cores of giant molecular clouds have radii of about 0.1 pc, temperatures between 30 and 200 K, and number densities of some $10^{12}/m^3$. Is one more likely to collapse to form a solar-type star than the case in part (a)?

Follow the arguments from part (a), but for a smaller knot in a cloud.

$$M = [(4/3) \pi (0.1 \text{ pc} \times 3.086 \times 10^{16} \text{ m/pc})^3] \rho = (1.23 \times 10^{47} \text{ m}^3) \rho$$
$$\rho = 2 (1.67 \times 10^{-27} \text{ kg})(10^{12}/m^3) = 3.34 \times 10^{-15} \text{ kg}/m^3$$

The mass of the knot in the cloud before collapse is

$$M = (1.23 \times 10^{47} \text{ m}^3)(3.34 \times 10^{-15} \text{ kg}/m^3)$$
$$\approx 4 \times 10^{32} \text{ kg} \approx 200 \, M_\odot$$

For this knot to collapse to a solar-type star, it must lose 99.5 of its initial mass.

The size scale of the collapsing region is

$$L \approx 10^7 (T/\rho)^{1/2}$$

$$L = 10^7 [30 / (3.34 \times 10^{-15})]^{1/2} \quad \text{to} \quad = 10^7 [200 / (3.34 \times 10^{-15})]^{1/2}$$
$$= 9.5 \times 10^{14} \text{ m} \approx 0.03 \text{ pc} \qquad = 2.4 \times 10^{15} \text{ m} = 0.08 \text{ pc}$$

This is about two-thirds the actual size of the knot.

Thus this knot in a cloud is more likely to collapse to form a solar-type star than is the cloud in part (a), since it must lose a smaller amount of its initial mass and its size is more comparable to the size scale of a collapsing cloud.

(c) For the more likely case, what is the free-fall time of the collapse?

The time scale for the collapse of the knot in part (b) is given by Equation 15-10 (for ρ_o in kg/m^3),

$$t_{ff} = (6.44 \times 10^4) / \rho_o^{1/2} \text{ s}$$
$$= (6.44 \times 10^4) / (3.34 \times 10^{-15})^{1/2} \text{ s}$$
$$= 1.1 \times 10^{12} \text{ s} = 3.5 \times 10^4 \text{ yr}$$

(d) For the more likely case, what is the average luminosity radiated by the protostar during its collapse phase?

From the virial theorem, the energy available to radiate during gravitational collapse for the knot in part (b) is half the available gravitational potential energy, U, (Section 15-3A)

$$E_{radiation} = (1/2) U \approx -(1/2) G M^2 / L$$

Using 0.1 pc = 3.086×10^{15} m for the size scale of the collapsing region,

$$E_{radiation} = -(1/2)(6.67 \times 10^{-11} \text{ N m}^2/\text{kg}^2)(4 \times 10^{32} \text{ kg})^2 / (3.086 \times 10^{15} \text{ m})$$
$$= -1.7 \times 10^{39} \text{ J}$$

Using the time scale of collapse from part (c), the average luminosity during the collapse is

$$\langle L \rangle = E_{radiation} / t_{ff}$$
$$= (-1.7 \times 10^{39} \text{ J J}) / (1.1 \times 10^{12} \text{ s}) = 1.6 \times 10^{27} \text{ W} \approx 4 L_\odot$$

This is reasonable. Of course, the collapse is not uniform in rate, so the luminosity will vary during the collapse.

15-19. Estimate the initial size of the cloud required to form a 50, 10, and 0.5 $M_\odot$ star. Compare your result for that of a 1 $M_\odot$ star and the size of the Solar System.

From Section 15-3A, for a collapsing cloud of mass M and size L,
$$k T / m_H \approx G M / L$$
so
$$L \approx G M m_H / k T$$
$$\approx (6.67 \times 10^{-11} \text{ N m}^2/\text{kg}^2)(1.67 \times 10^{-27} \text{ kg}) / (1.38 \times 10^{-23} \text{ J/K}) \times (M/T)$$

$$\approx (8.07 \times 10^{-15} \text{ m K / kg}) M / T$$

For the mass in solar masses

$$L \approx (1.61 \times 10^{16} \text{ m K}) (M / M_\odot) / T$$

For an interstellar cloud of initial temperature $T \approx 10$K, the size of a cloud required to form a 50 $M_\odot$ star is

$$L \approx (1.61 \times 10^{16} \text{ m K}) (50) / (10 \text{ K})$$
$$\approx 8.0 \times 10^{16} \text{ m}$$
$$\approx 8.0 \times 10^{16} \text{ m} / (1.496 \times 10^{11} \text{ m / AU}) \approx 5.4 \times 10^5 \text{ AU}$$
$$\approx 5.4 \times 10^5 \text{ AU} / (2.06 \times 10^5 \text{ AU / pc}) \approx 2.6 \text{ pc}$$

To form a 10 $M_\odot$ star

$$L \approx (1.61 \times 10^{16} \text{ m K}) (10) / (10 \text{ K})$$
$$\approx 1.6 \times 10^{16} \text{ m}$$
$$\approx 1.6 \times 10^{16} \text{ m} / (1.496 \times 10^{11} \text{ m / AU}) \approx 1.1 \times 10^5 \text{ AU}$$
$$\approx 1.1 \times 10^5 \text{ AU} / (2.06 \times 10^5 \text{ AU / pc}) \approx 0.5 \text{ pc}$$

To form a 1 $M_\odot$ star

$$L \approx (1.61 \times 10^{16} \text{ m K}) (1) / (10 \text{ K})$$
$$\approx 1.6 \times 10^{15} \text{ m}$$
$$\approx 1.6 \times 10^{15} \text{ m} / (1.496 \times 10^{11} \text{ m / AU}) \approx 1.1 \times 10^4 \text{ AU}$$
$$\approx 1.1 \times 10^4 \text{ AU} / (2.06 \times 10^5 \text{ AU / pc}) \approx 0.05 \text{ pc}$$

To form a 0.5 $M_\odot$ star

$$L \approx (1.61 \times 10^{16} \text{ m K}) (0.5) / (10 \text{ K})$$
$$\approx 8.0 \times 10^{14} \text{ m}$$
$$\approx 8.0 \times 10^{14} \text{ m} / (1.496 \times 10^{11} \text{ m / AU}) \approx 5.4 \times 10^3 \text{ AU}$$
$$\approx 5.4 \times 10^3 \text{ AU} / (2.06 \times 10^5 \text{ AU / pc}) \approx 0.03 \text{ pc}$$

The size of the solar system is ≈ 50 AU, so to form 50, 10, 1, and 0.5 $M_\odot$ stars the sizes of the original cloud are approximately 10,000 times, 2,000 times, 200 times, and 100 times, respectively, the size of the solar system. Even for forming low mass stars, the initial cloud is quite large.

15-20. A typical T Tauri star has a stellar wind moving at 100 km/s and a mass loss rate of about 10^{-7} to 10^{-8} $M_\odot$/year.

(a) What is the momentum lost per second by the wind?

The momentum (Section 1-3A), p = m v, lost per second is given by
$$\text{momentum loss} / s = (\text{mass loss} / s)(v)$$
For a mass loss rate of 10^{-8} $M_\odot$/year

momentum loss / s
$$= (10^{-8}\ M_\odot/\text{year})(1.99 \times 10^{30}\ \text{kg}/M_\odot)(1\ \text{year} / 3.16 \times 10^7\ \text{s})(10^5\ \text{m/s})$$
$$= (6.3 \times 10^{14}\ \text{kg}/\text{s})(10^5\ \text{m/s})$$
$$= 6.3 \times 10^{19}\ \text{kg·m/s} / \text{s}$$

A mass loss rate of 10^{-8} to 10^{-7} $M_\odot$/year corresponds to a momentum loss of 6.3×10^{19} to 6.3×10^{20} kg·m/s / s.

(b) The winds sweep out surrounding dust until, by conservation of momentum, it slows to about 10 km/s. What is the rate at which dust can be driven away?

Assuming the dust surrounding the T Tauri star is initially at rest, it will be accelerated from 0 km/s to 10 km/s by the stellar wind -- essentially an inelastic collision. Applying the conservation of momentum, $d\mathbf{p}/dt = 0$ (Equation 1-11), in every second for a mass loss rate of 10^{-8} $M_\odot$/year:

$$m_{wi} v_{wi} + m_{di} v_{di} = (m_w + m_d) v_f$$
$$(6.3 \times 10^{19}\ \text{kg·m/s}) + 0\ \text{kg·m/s} = (6.3 \times 10^{14}\ \text{kg} + m_d)(10^4\ \text{m/s})$$
$$m_d = 6.3 \times 10^{15}\ \text{kg} - 6.3 \times 10^{14}\ \text{kg} \approx 5.7 \times 10^{15}\ \text{kg}$$

for a mass loss rate of 10^{-7} $M_\odot$/year:

$$m_{wi} v_{wi} + m_{di} v_{di} = (m_w + m_d) v_f$$
$$(6.3 \times 10^{20}\ \text{kg·m/s}) + 0\ \text{kg·m/s} = (6.3 \times 10^{15}\ \text{kg} + m_d)(10^4\ \text{m/s})$$
$$m_d = 6.3 \times 10^{16}\ \text{kg} - 6.3 \times 10^{15}\ \text{kg} \approx 5.7 \times 10^{16}\ \text{kg}$$

The stellar wind of 10^{-8} to 10^{-7} $M_\odot$/year from the T Tauri star can sweep up 5.7×10^{15} to 5.7×10^{16} kg of dust per second, or 1.8×10^{23} to 1.8×10^{24} kg per year. That's about 3% to 30% the mass of the Earth per year; or about 10^{-7} to 10^{-6} $M_\odot$ per year.

Chapter 16: The Evolution of Stars

16-1. Verify that about 6 x 10^{11} kg of hydrogen is converted to helium in our Sun every second.

The primary nuclear reaction in the Sun is the proton-proton chain. In each chain, 4 protons (H nucleus) are converted into 1 Helium nucleus, with the release of energy. This energy is equal to the equivalent energy of the mass loss in this chain. The mass of 4 protons is 4.0312 amu = 6.692 x 10^{-27} kg; the mass of 1 helium nucleus is 4.0026 amu = 6.644 x 10^{-27} kg. The mass loss per chain is thus

$$\Delta m = 4.0312 \text{ amu} - 4.0026 \text{ amu} = 0.0286 \text{ amu} = 4.75 \times 10^{-29} \text{ kg}$$

This mass loss per chain corresponds to an energy (Equation 16-16):

$$E = \Delta m c^2 = 4.75 \times 10^{-29} \text{ kg} \times (3.00 \times 10^8 \text{ m/s})^2$$
$$= 4.27 \times 10^{-12} \text{ J}.$$

The solar luminosity (Appendix Table A7-1) is 3.90 x 10^{26} J/s, so there must be

$$(3.90 \times 10^{26} \text{ J/s}) / (4.27 \times 10^{-12} \text{ J/chain}) = 9.13 \times 10^{37} \text{ chains/s}$$

occurring in the core of the Sun.

Since each chain consumes 4 hydrogen atoms, 3.65 x 10^{38} hydrogen atoms are consumed per second, corresponding to 6.1 x 10^{11} kg of hydrogen being consumed per second.

16-2. (a) If a star is characterized by M = 2 x 10^{32} kg and L = 4 x 10^{32} W, how long can it shine at that luminosity if it is 100% hydrogen and converts all the H to He?

A stellar mass of 2 x 10^{32} kg (= 100 M$_\odot$) corresponds to:

$$(2 \times 10^{32} \text{ kg}) / (1.67 \times 10^{-27} \text{ kg/H nucleus}) = 1.20 \times 10^{59} \text{ H nuclei},$$

if the star is 100% H. Since 4 H nuclei are used in each proton-proton chain, there can be 3.0 x 10^{58} chains if all this mass is converted.

In each chain, 4.27 x 10^{-12} J of energy is released (Section 16-1D), so the star can produce a total of 1.28 x 10^{47} J of energy if the entire amount of H is consumed in the P-P chain.

If the luminosity of the star is 4 x 10^{32} W, then the lifetime of the star is:

$$t = E_{total} / L = (1.28 \times 10^{47} \text{ J}) / (4 \times 10^{32} \text{ J/s})$$
$$= 3.20 \times 10^{14} \text{ s} = 1.0 \times 10^7 \text{ yr} = 10 \text{ million yr}.$$

This is a short life-span -- the star is a massive star (100 M$_\odot$), with a high luminosity (10^6 L$_\odot$).

If, more realistically, only 10% of the H is converted to He, then this star will live for 1 million yr (after which time other fusion reactions may produce some of the stellar luminosity).

(b) Do a similar calculation for a star of mass 10^{30} kg and luminosity 4×10^{25} W.

The lifetime is proportional to the available energy (which is proportional to its mass, in the examples here) divided by the luminosity. Comparing this star to the star in part (a):

$t_{starb} / t_{stara} = (E_{starb} / E_{stara}) / (L_{starb} / L_{stara})$

$t_{starb} / 1 \times 10^7$ yr $= (1.0 \times 10^{30}$ kg $/ 2 \times 10^{32}) / (4 \times 10^{25} / 4 \times 10^{32})$

$t_{starb} = [(5.0 \times 10^{-3}) / (1.0 \times 10^{-7})] \times (1.0 \times 10^7$ yr$)$

$= 5.0 \times 10^{11}$ yr $= 500$ billion yr

This is a long life-span -- the star is a low mass star (0.5 $M_\odot$), with a low luminosity (0.1 $L_\odot$).

If, more realistically, only 10% of the H is converted to He, then this star will live (converting H into He) for 50 billion yr.

16-3. Briefly describe the evolution of the following stars from a cloud of gas and dust to their demise:
(a) M = 10 $M_\odot$, (b) M = 0.1 $M_\odot$. Clearly indicate which stages of the evolution are highly uncertain.

Rather than repeat a long description, the reader is referred to Section 16-3B, which describes evolutionary stages in stars of several different masses. Figures 16-3 to 16-7 and Table 16-1 are particularly useful for summarizing the major stages in evolution of stars of several different masses.

The major points to keep in mind when comparing evolution of stars of different masses are: the time required for collapse (shorter for more massive stars); the path of the protostar and pre-main sequence star on the H-R diagram (horizontal to left for massive star -- little luminosity change but major temperature changes; mostly vertical downward for low mass star -- little temperature change, but major luminosity change); the energy generation process during the main sequence stage in a star's life (proton-proton in low mass star with cooler core; CNO cycle in hotter core of high mass stars [Figure 16-2]); the lifetime on the main sequence (few million yr for massive star; 20+ billion yr for low mass star); the post main sequence evolution on the H-R diagram (see Figure 16-4) in terms of luminosity, temperature, and radius changes of surface; the evolution of the core (composition, fusion reactions, whether degeneracy -- or helium flashes --occurs in the core) during the post main sequence phase; the ejection of shell material (supernova remnant for massive star; planetary nebula for low mass star); and the ultimate fate of the core (neutron star, black hole, or white dwarf for massive star; white dwarf for low mass star).

16-4. Using Figure 16-8 and the data in Table 16-1, sketch the H-R diagrams for star clusters of ages 10^7, 10^8 and 10^9 years (these are *constant-time* lines!). Clearly label the axes and comment upon the significance of your results (turnoff points).

Figure 16-8 shows H-R diagrams for several clusters of various ages. The evolutionary paths (Figures 16-4 and 16-7) for various stellar masses can be used to follow the evolution of cluster stars. The general evolution of a cluster is shown in the diagram below, showing the zero age main sequence and the H-R diagrams of a theoretical cluster at various ages. Note that the main sequence turn-off moves down (to lower mass, lower temperature, lower luminosity stars) as the cluster ages. The gap in the 10^8 yr old cluster is due to the rapid evolutionary stages -- few stars will be in these stages at any given time.

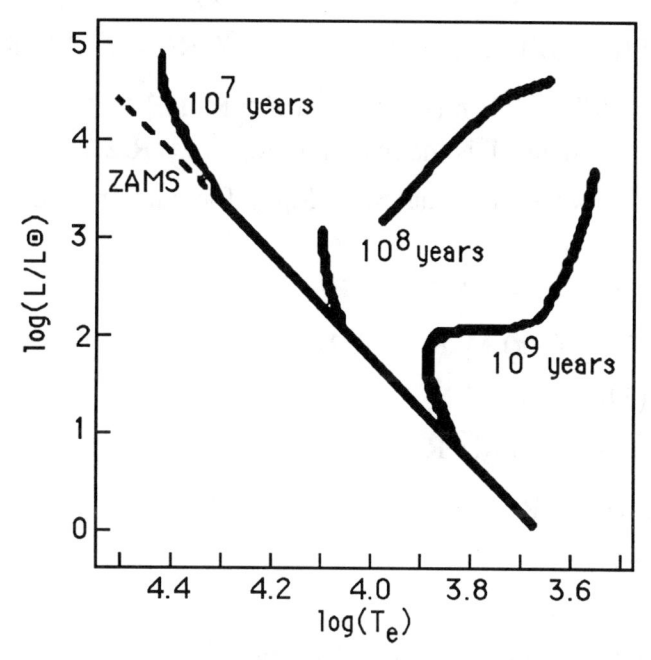

16-5. Although detailed models of stellar structure require the use of complex computer codes, simple scalings can be obtained by making rough approximations. For a variable x, we can substitute $\Delta x/\Delta r$ for dx/dr in order to obtain a crude result. (This method is a rough version of a numerical technique called finite differences.)

(a) Use the equation of hydrostatic equilibrium (16-1) to show that the central pressure scales as $P_c \propto M^2/R^4$. Substitute $\Delta P/\Delta r$ for dP/dr and take this difference between r=0 and r=R/2, that is, $\Delta P/\Delta r \approx [P(r=R/2) - P_c]/(R/2 - 0)$. You may assume that $P(r=R/2)$ is negligible compared with P_c. Also, substitute the mean density of the star $\langle\rho\rangle$ for $\rho(r)$.

[*Note:* For a good presentation of a model for stellar (solar) structure which can be calculated by students, the instructor is referred to Donald D. Clayton's article "Solar structure without computers", which appeared in Am.J.Phys., 54 (4), April 1986, p.354.]

Note, also, that this problem can be done less accurately using an outer limit of r = R (rather than r = R/2). The formalism is the same, only constant factors differ (see Sections 16-1A and 16-2B). The model being done here is a two-layer model for stellar structure, with the outer layer (from r = R/2 to r = R) showing little change in T,

P, or $\langle\rho\rangle$.

The Hydrostatic equation is (Equation 16-1):
$$dP/dr = -G M(r) \rho(r) / r^2.$$

Let $dP/dr \rightarrow \Delta P / \Delta r$
$$\Delta r = (R/2) - 0 = R/2$$
$$\Delta P = P(r = R/2) - P(r = 0) \approx -P_c$$
$$\text{since } P(r = R/2) \ll P(r = 0) \equiv P_c$$
$$\Delta P / \Delta r = [P(r = R/2) - P_c] / [R/2 - 0] \approx -P_c / (R/2) = -2 P_c R^{-1}$$
$$\rho(r) \approx \langle\rho\rangle = M' / [(4/3) \pi (R/2)^3] = (6/\pi) M' R^{-3}$$
where M' is the mass internal to $r = R/2$
(we assume that $M' \propto M_{total}$ for stars of different masses)

So,
$$\Delta P / \Delta r = -G M' \langle\rho\rangle / r^2$$
$$= -G M' (6/\pi) M' R^{-3} / (R/2)^2$$
$$= -(24/\pi) G M'^2 R^{-5}$$
$$-2 P_c R^{-1} = -(24/\pi) G M'^2 R^{-5}$$
$$P_c = (12/\pi) G M'^2 R^{-4}$$
$$P_c \propto M'^2 R^{-4}$$

(b) Now use the radiative transport equation and the same method as in part (a) to approximate dT/dr and obtain the theoretical mass-luminosity relationship, $L \propto M^3$. Assume that $\kappa(r)$ is constant and that $T(r) \propto T_c$.

The radiative transport equation (Equation 16-12a) is:
$$dT/dr = \{-3 \kappa(r) \rho(r) / [64 \pi \sigma r^2 T^3(r)]\} L(r).$$

Following procedure as in part (a), let
$$\Delta r \approx R/2 - 0 = R/2$$
$$\Delta T = T(r = R/2) - T(r = 0) \approx 0 - T_c = -T_c \text{ [where } T(r = R/2) \ll T(r = 0) \equiv T_c]$$
$$\kappa(r) = \kappa = \text{constant} \quad \text{(opacity)}$$
$$\rho(r) \approx \langle\rho\rangle$$
$$L(r) \approx L$$

so the equation of radiative transport becomes
$$\Delta T = \{-3 \kappa \langle\rho\rangle / [64 \pi \sigma (R/2)^2 T_c^3]\} L \Delta r$$
$$-T_c = \{-3 \kappa (M' / [(4/3) \pi (R/2)^3]) / [64 \pi \sigma (R/2)^2 T_c^3]\} L (R/2)$$
$$T_c = (3)(3/4)(8)(4)(1/2)(1/64\pi^2) \kappa \sigma^{-1} M' R^{-3} R^{-2} T_c^{-3} L R$$

$$T_c^4 = (9/16\pi)\, \kappa\, \sigma^{-1}\, M'\, R^{-4}\, L$$

$$L = [(16\pi/9)\, \sigma\, \kappa^{-1}]\, T_c^4\, R^4\, M'^{-1}$$

$$L \propto T_c^4\, R^4\, M'^{-1}$$

Use the equation of state (the perfect gas law, Equation 16-7)

$$P(r) = k\, \rho(r)\, T(r) / [\mu(r)\, m_H]$$

in a similar fashion to above, we get

$$P_c = k <\rho> T_c / \mu\, m_H$$

$$T_c = (\mu\, m_H / k)\, P_c / <\rho>$$

$$= (\mu\, m_H / k)\, [(12/\pi)\, G\, M'^2\, R^{-4}] / [(6/\pi)\, M'\, R^{-3}]$$

$$= (\mu\, m_H / k)\, (2G)\, M'\, R^{-1}$$

$$T_c \propto M'\, R^{-1}$$

Substituting this in the expression we found above for L:

$$L = [(16\pi/9)\, \sigma\, \kappa^{-1}]\, [(\mu\, m_H / k)\, (2G)\, M'\, R^{-1}]^4\, R^4\, M'^{-1}$$

$$= [(256\pi/9)\, \sigma\, \kappa^{-1}(G\, \mu\, m_H / k)^4\, \kappa^{-1}]\, M'^3$$

$$L \propto M'^3\, \kappa^{-1}$$

The opacity, κ, is a function of T and ρ, which, if included, would give $L \propto M^{3.3}$. If we ignore the opacity,

$$L \propto M'^3$$

(c) Use the same method to obtain a rough relationship between the effective temperature and mass of a main-sequence star. (No other variables should appear in the proportionality.)

Relating the mass to the effective surface temperature of the star, we know that the luminosity of the star must be proportional to the surface area times T^4 (the Stefan-Boltzmann Law, Equation 8-40 or 13-7):

$$L \propto R^2\, (T_{surface})^4$$

From part (b) we have $L \propto M^3$, so

$$T_s^4 \propto M^3\, R^{-2}$$

$$T_s \propto M^{3/4}\, R^{-1/2}$$

Using the empirical M-R relationship

$$R \propto M^{0.6}$$

$$T_s \propto M^{0.75}\, (M^{-0.3}) \propto M^{0.45}$$ -- very close to the T_c-M relationship above

(d) Combine your answers from parts (b) and (c) to obtain a relationship between T and L. Use the H-R diagrams (Figures 13-7 and 13-9) and Appendix Table A4-3 to compare the temperature of a star of $L = 10 L_\odot$ with that of the Sun. How does this compare with your theoretical T-L relationship (be quantitative)?

Combining: $L \propto M^3$ and $T_s \propto M^{0.45}$

$M \propto L^{1/3}$

$T_s \propto (L^{1/3})^{0.45} \propto L^{0.15}$

Comparing this with the Sun, the prediction is

$T_* / T_\odot = (10 L_\odot / L_\odot)^{0.15}$

$T_* = 1.4 T_\odot = 1.4 \times 5800K = 8,000K$

A star which is 10 times the luminosity of the Sun has an absolute magnitude which is 2.5 magnitude brighter than the Sun ($M_* = 2.3$). From Appendix Table A4-3, a main sequence star with this absolute magnitude is of spectral type A7, and has an effective surface temperature of 7,800K. Thus, our estimate is very close!

16-6. Assume that the amount of hydrogen mass available for nuclear reactions in the core of a star is $M_c \approx$ 0.5 M. Further assume for simplicity that the only energy-generating nuclear reaction is $4^1H \rightarrow {}^4He$ + energy (ignore the fact that some of the energy is in the form of positrons and neutrinos). Obtain an expression for the hydrogen-burning lifetime of a star in years as a function of mass in solar units. (Assume L ≈ constant during the hydrogen-burning phase and use the mass-luminosity relationship $L \approx M^3$ in solar units.)

[Note: The instructor may wish to have the students use a value of the core mass of 0.1 M, rather than the 0.5 M which is posed in the problem. As will be seen in the solution below, using $M_c = 0.5$ M results in calculated main sequence lifetimes which are a factor of 5 times too long.]

This is similar to Problem 16-2, except here only 50% of the hydrogen in the star is fused to helium via the proton-proton chain.

Consider first the Sun. In each chain, 4 protons (H nucleus) are converted into 1 helium nucleus, with the release of energy. This energy is equal to the equivalent energy of the mass loss in this chain. The mass of 4 protons is 4.0312 amu = 6.692×10^{-27} kg; the mass of 1 helium nucleus is 4.0026 amu = 6.644×10^{-27} kg. The mass loss per chain is thus

$\Delta m = 4.0312$ amu $- 4.0026$ amu $= 0.0286$ amu $= 4.75 \times 10^{-29}$ kg

This mass loss per chain corresponds to an energy (Equation 16-16):

$E = \Delta m c^2 = 4.75 \times 10^{-29}$ kg $\times (3.00 \times 10^8$ m/s$)^2$

$= 4.27 \times 10^{-12}$ J

The solar luminosity (Appendix Table A7-1) is 3.90×10^{26} J/s, so there must be

$(3.90 \times 10^{26}$ J/s$) / (4.27 \times 10^{-12}$ J/chain$) = 9.13 \times 10^{37}$ chains/s

occurring in the core of the Sun.

Since each chain consumes 4 hydrogen atoms, 3.65×10^{38} hydrogen atoms are consumed per second, corresponding to 6.1×10^{11} kg of hydrogen being consumed per second.

If half the mass of the Sun ($0.5 \times 2.0 \times 10^{30}$ kg = 1.0×10^{30} kg) is available for fusion, then the Sun can live in its hydrogen burning stage for

$\quad$ t = $(1.0 \times 10^{30}$ kg$) / (6.1 \times 10^{11}$ kg$/$s$)$
$\quad\quad$ = 1.64×10^{18} s = 5.2×10^{10} yr = 52 billion yr.

This is high compared to the accepted value of ≈ 10 billion yr for the main sequence life of the Sun. It is more reasonable to expect that 10% (not 50%) of the Sun's hydrogen will be utilized during the main sequence lifetime of a solar mass star. This would result in a main sequence lifetime of 10 billion yr (rather than the extremely high value of 52 billion yr calculated above).

Using the mass-luminosity relationship from Section 16-3B, $L \approx M^{3.3}$ (in solar units), the main-sequence lifetime is:

$\quad$ lifetime = mass available / luminosity = $M / L = M / M^{3.3} = M^{-2.3}$

so

$\quad$ lifetime$_*$ = (52 billion yr) $/ (M_* / M_\odot)^{2.3}$ = (52 billion yr) $(M_\odot / M_*)^{2.3}$,

or if we use the more reasonable lifetime for the Sun of 10 billion yr,

$\quad$ lifetime$_*$ = (10 billion yr) $/ (M_* / M_\odot)^{2.3}$ = (10 billion yr) $(M_\odot / M_*)^{2.3}$.

16-7. (a) Estimate the central pressure for 0.5, 10, and 50 $M_\odot$ stars. Compare these pressures to the central pressure of the Sun.

By analogy to the Sun (Section 16-1A):

$\quad P_c = G M \rho / R$

and if $\rho = 3 M / 4 \pi R^3$

$\quad P_c = 3 G M^2 / 4 \pi R^4$
$\quad\quad$ = $[3 (6.67 \times 10^{-11}$ N m^2 / kg$^2) / 4 \pi]$ M^2 / R^4
$\quad\quad$ = $(1.59 \times 10^{-11}$ N m^2 / kg$^2) M^2 / R^4$

To estimate the radius of the various stars use the main sequence empirical mass-radius relationship, $R \propto M^{0.6}$ (from Figure 12-11A), or $R = R_\odot (M / M_\odot)^{0.6}$, so

$\quad M^2 / R^4 = M^2 / [(R_\odot / M_\odot^{0.6}) M^{0.6}]^4$
$\quad\quad$ = $(6.96 \times 10^8$ m$)^{-4}$ $(1.99 \times 10^{30}$ kg$)^{2.4}$ $M^{-0.4}$
$\quad\quad$ = 2.22×10^{37} m^{-4} kg$^{2.4}$ $M^{-0.4}$

so $\quad P_c$ = $(3.53 \times 10^{26}$ N m^{-2} kg$^{0.4}) M^{-0.4}$ $\quad$ where M is in kg.

Or, simply,
$$P_c / P_{c\odot} = (M/M_\odot)^{-0.4}$$
$$P_c \approx (2.7 \times 10^9 \text{ atm}) (M/M_\odot)^{-0.4}$$

where we have used the Sun's central pressure from Section 16-1A. We will use the ratio to calculate central pressures, though the equation
$$P_c = (3.53 \times 10^{26} \text{ N m}^{-2} \text{ kg}^{0.4}) M^{-0.4}$$
could also be used, with the same results.

For a 0.5 $M_\odot$ star,
$$P_c = (2.7 \times 10^9 \text{ atm}) (0.5)^{-0.4} = 1.32 \ P_{c\odot} = 3.6 \times 10^9 \text{ atm}$$

For a 10 $M_\odot$ star,
$$P_c = (2.7 \times 10^9 \text{ atm}) (10)^{-0.4} = 0.40 \ P_{c\odot} = 1.1 \times 10^9 \text{ atm}$$

For a 50 $M_\odot$ star,
$$P_c = (2.7 \times 10^9 \text{ atm}) (50)^{-0.4} = 0.21 \ P_{c\odot} = 5.6 \times 10^8 \text{ atm}$$

The central pressure decreases slightly with stellar mass.

(b) Estimate the central temperature for 0.5, 10, and 50 $M_\odot$ stars. Compare these temperatures to the central temperature of the Sun.

Again, by analogy to the Sun (Section 16-1B; Equation 16-8):
$$T_c \approx P_c \mu m_H / \langle \rho \rangle k$$

For $\rho = 3M / 4\pi R^3$ and assuming $\mu = 1/2$,
$$T_c \approx (1/2)(1.67 \times 10^{-27} \text{ kg}) P_c / [(3M / 4\pi R^3)(1.38 \times 10^{-23} \text{ J/K})]$$
$$= (2.53 \times 10^{-4} \text{ kg K/J}) P_c R^3 / M$$

Using $P_c = (3.53 \times 10^{26} \text{ N m}^{-2} \text{ kg}^{0.4}) M^{-0.4}$ from part (a), and using the empirical mass-radius relationship, $R \propto M^{0.6}$ or $R = R_\odot (M/M_\odot)^{0.6}$ (from Figure 12-11A)

$$T_c = (2.53 \times 10^{-4} \text{ kg K/J})(3.53 \times 10^{26} \text{ N m}^{-2} \text{ kg}^{0.4}) M^{-0.4}$$
$$\times [(6.96 \times 10^8 \text{ m})\{M / (1.99 \times 10^{30} \text{ kg})\}^{0.6}]^3 / M$$
$$= (8.93 \times 10^{22} \text{ m}^{-3} \text{ K kg}^{1.4})(3.37 \times 10^{26} \text{ m}^3)(2.99 \times 10^{-55} \text{ kg}^{-1.8})$$
$$\times M^{-0.4+1.8-1}$$
$$= (9.01 \times 10^{-6} \text{ K kg}^{-0.4}) M^{0.4}$$

Or, simply,
$$T_c / T_{c\odot} = (M/M_\odot)^{0.4}$$

$$T_c = T_{c\odot}(M/M_\odot)^{0.4} = 12 \times 10^6 \text{ K } (M/M_\odot)^{0.4}$$

where we have used the central temperature of the Sun from Equation 16-9. We will use the ratio to calculate central pressures, through the equation

$$T_c = (9.01 \times 10^{-6} \text{ K kg}^{-0.4}) M^{0.4}$$

could also be used, with the same results.

For a 0.5 $M_\odot$ star,

$$T_c = 12 \times 10^6 \text{ K } (0.5)^{0.4} = 0.76 \; T_{c\odot} = 9.1 \times 10^6 \text{ K}$$

For a 10 $M_\odot$ star,

$$T_c = 12 \times 10^6 \text{ K } (10)^{0.4} = 2.5 \; T_{c\odot} = 3.0 \times 10^7 \text{ K}$$

For a 50 $M_\odot$ star,

$$T_c = 12 \times 10^6 \text{ K } (50)^{0.4} = 4.8 \; T_{c\odot} = 5.7 \times 10^7 \text{ K}$$

The central temperature increases slightly with stellar mass, but significantly enough to influence what form of hydrogen (p-p chain or CNO cycle) burning occurs in the star.

16-8. Calculate the main-sequence lifetimes for stars of 0.5, 1, 5, 10, and 25 $M_\odot$.

This problem makes use of the result of Problem 16-6. From that solution, assuming 10% of the star's hydrogen is in the core and converted to helium, we found an approximate relationship

$$\text{lifetime}_* = (10 \text{ billion yr}) (M_\odot / M_*)^{2.3}$$

M_*	lifetime$_*$
0.5 $M_\odot$	50 billion years
1.0 $M_\odot$	10 billion years
5. $M_\odot$	250 million years
10. $M_\odot$	50 million years
25. $M_\odot$	6 million years

These estimates could be improved with a more accurate mass-luminosity relationship than assumed here and in Problem 16-6. (That is, instead of assuming $L \approx M^{3.3}$ from Section 16-2B which was used in Problem 16-6, we could use the relationships in Section 12-2B where the exponent depends on the mass range.)

16-9. Assuming that a star radiates as a blackbody during all phases of its evolution, use the Stefan-Boltzmann Law to determine the radius (in units of $R_\odot$) of a 1$M_\odot$ star at all main stages in Figure 16-4. (*Hint:* Both the temperature and luminosity axes are logarithmic.)

The luminosity of a blackbody is given by Equation 13-13:

$$L = 4\pi R^2 \sigma T_{eff}^4$$

Comparing the radius to that of the main sequence Sun,

$$R/R_\odot = (L/L_\odot)^{1/2} / (T/T_\odot)^2.$$

At the different stages shown in Figure 16-4, approximate values are:

stage	log(L/L$_\odot$)	L/L$_\odot$	log(Teff)	Teff	T/T$_\odot$	R/R$_\odot$
main sequence	0	1	3.75	5600	1.0	1
first dredge-up	0.5	3	3.65	4500	0.80	2.7
RGB	1	10	3.6	4000	0.71	6.3
Helium flash	3	1000	3.5	3200	0.57	97
He -> C+O	1.8	60	3.6	4000	0.71	15
AGB	2.2	160	3.55	3500	0.62	33
Thermal Pulses	3.2	1600	3.5	3200	0.57	120
PN ejection	3.7	5000	3.55	3500	0.61	190
to white dwarf	4.0	10,000	>4.4	>25,000	>4.5	<4.9

16-10. Estimate the energy available and the lifetime for the helium-burning phase in a 1 M$_\odot$ star:

(a) Calculate the energy released per net reaction 3 ^{4}He → ^{12}C in the triple-alpha process. (*Note:* The weight of ^{4}He is 4.0026 and the weight of ^{12}C is 12.0000.)

The energy available is given by Equation 16-16:

$$E = \Delta m\, c^2$$
$$= (3 \times 4.0026\ \text{amu} - 12.0000\ \text{amu})(1.66 \times 10^{-27}\ \text{kg/amu})$$
$$\times (3.0 \times 10^8\ \text{m/s})^2$$
$$= 1.16 \times 10^{-12}\ \text{J per reaction.}$$

(b) What fraction of the available mass of 3 helium nuclei is liberated in the form of energy in the triple-alpha reaction? Compare this to the fraction of available mass liberated in the proton-proton reaction.

The fraction of mass liberated in energy in the triple-alpha reaction is the mass loss divided by the available mass in the 3 helium nuclei, in amu:

$$(3 \times 4.0026 - 12.0000) / (3 \times 4.0026) = 0.65 \times 10^{-4} = 0.00065$$

Contrasting that to the fraction liberated in the proton-proton reaction of 0.0071 (Section 16-1D), we see that the proton-proton chain converts 11 times more fractional mass into energy than the triple-alpha process.

(c) Assume that approximately 10% of the original mass of the star is in the form of ^{4}He in the stellar core during the helium-burning phase. Estimate the total energy available from the triple-alpha process.

The total energy available if 10% of a solar mass star takes part in the triple-alpha process

$$E_{total} = \text{(fraction of mass converted to energy)} \times (10\% \times \text{mass of Sun}) \times c^2$$
$$= (6.5 \times 10^{-4})(0.1 \times 1.99 \times 10^{30} \text{ kg})(3 \times 10^8 \text{ m/s})^2$$
$$= 1.2 \times 10^{43} \text{ J}$$

As expected from part (b), where we compared the efficiency of converting of mass to energy, the total energy available in the triple-alpha process is $\approx 1/11$ that of the proton-proton chain.

(d) During the helium-core-burning phase, some hydrogen burning is also occurring in a shell. Thus the star's luminosity is not due only to helium burning. Keeping this in mind, assume that the typical luminosity from helium burning is 10^2 L$_\odot$. Estimate the lifetime of the helium core burning phase.

The lifetime of the Sun in the helium burning phase is

$$\tau = E_{total}/L$$
$$= (1.2 \times 10^{43} \text{ J}) / (10^2 \times 3.9 \times 10^{26} \text{ W})$$
$$= 3.1 \times 10^{14} \text{ s}$$
$$= 9.7 \times 10^6 \text{ yr} \approx 10 \text{ million years}$$

This is 1/1000 the main-sequence lifetime of a 1 M$_\odot$ star, where 1/100 of this factor is due to the increased luminosity and $\approx 1/10$ is due to the lower fractional amount of mass converted to energy in the triple-alpha process compared to the proton-proton chain.

If part of the 100 L$_\odot$ is due to hydrogen burning, then the lifetime in this phase will be somewhat longer than the 10 million years calculated.

16-11. Consider a model star in which the density $\rho(r)$ goes as ρ_0 in the core ($r < r_0$), $\rho_0(r_0/r)^2$ in the region between the core and surface ($r_0 < r < R$), and 0 outside the surface ($r > R$).

(a) Find and expression for M(r).

The density $\rho = M/V$, so

$$M = \rho V.$$

For the core, where $r < r_0$, $\rho = \rho_0 = $ constant, so

$$M(r) = \rho_0 (4/3) \pi r^3 \qquad \text{for } r \le r_0$$

For the region outside the core, where $r_0 < r < R$, the total mass is the mass of the core plus the mass of the shell from r_0 to r, that is $\int_{r_0}^{r} \rho(r) 4\pi r^2 \, dr$.

$$M(r) = \rho_0 [(4/3) \pi r_0^3] + \int_{r_0}^{r} \rho_0 (r_0/r)^2 \, 4 \pi r^2 \, dr$$
$$= \rho_0 (4/3) \pi r_0^3 + \int_{r_0}^{r} \rho_0 r_0^2 \, 4 \pi \, dr$$
$$= \rho_0 (4/3) \pi r_0^3 + \rho_0 4 \pi r_0^2 \int_{r_0}^{r} dr$$

$$M(r) = \rho_0 (4/3) \pi r_0^3 + \rho_0 4 \pi r_0^2 (r - r_0) \quad \text{for } r_0 < r < R$$

and for the region outside the surface, the upper limit of the integral is R,

$$M(r) = \rho_0 (4/3) \pi r_0^3 + \rho_0 4 \pi r_0^2 (R - r_0) \quad \text{for } r > R$$

Summarizing,

$$M(r) = \rho_0 (4/3) \pi r^3 \quad \text{for } r \leq r_0$$
$$M(r) = \rho_0 (4/3) \pi r_0^3 + \rho_0 4 \pi r_0^2 (r - r_0) \quad \text{for } r_0 < r < R$$
$$M(r) = \rho_0 (4/3) \pi r_0^3 + \rho_0 4 \pi r_0^2 (R - r_0) \quad \text{for } r \geq R$$

(b) If the star's mass is 1 $M_\odot$ at $R = R_\odot$, and $r_0 = 0.1$ R, what is the value of ρ_0?
Assume in the question that: $r_0 = 0.1$ R should be $r_0 = 0.1$ $R_\odot$.
Using the last relationship above,

$$M(r) = 1 M_\odot = \rho_0 (4/3) \pi (0.1 R_\odot)^3 + \rho_0 4 \pi (0.1 R_\odot)^2 (R_\odot - 0.1 R_\odot)$$
$$= \rho_0 (4/3) \pi (0.001) R_\odot^3 + \rho_0 (4/3) \pi (3) (0.009) R_\odot^3$$
$$= \rho_0 (4/3) \pi (0.001 + 0.009 \times 3) R_\odot^3$$
$$= 0.028 \rho_0 (4/3) \pi R_\odot^3$$

$$\rho_0 = 35.7 \, M_\odot / [(4/3) \pi R_\odot^3]$$

Using solar values (Appendix Table A7-1)

$$\rho_0 = 35.7 \, (1.99 \times 10^{30} \text{ kg}) / [(4/3) \pi (6.96 \times 10^8 \text{ m})^3]$$
$$= 5.0 \times 10^4 \text{ kg}/\text{m}^3$$

This is a bit below the central density of 1.6×10^5 kg/m^3 given in Section 16-1A.

(c) Find an expression for P(r).
Using the equation of hydrostatic equilibrium (Equation 16-1)

$$dP = -G M(r) \rho(r) / r^2 \, dr$$

For the core region:

$$\int_0^r dP(r) = P(r) = -G \int_0^r \rho_0 (4/3) \pi r^3 \rho_0 / r^2 \, dr$$
$$= -G \rho_0^2 (4/3) \pi \int_0^r r \, dr$$
$$= -G \rho_0^2 (2/3) \pi r^2 \quad \text{for } r \leq r_0$$

For shells outside the core (the final integral is left to the reader):

$$P(r) = -G \rho_0^2 (2/3) \pi r_0^2 - G \int_{r_0}^r \rho_0 4 \pi r_0^2 (r - r_0) \rho_0 (r_0/r)^2 / r^2 \, dr$$
$$= -G \rho_0^2 (2/3) \pi r_0^2 - G \int_{r_0}^r \rho_0^2 4 \pi (r_0/r)^4 (r - r_0) \, dr$$
$$= -G \rho_0^2 (2/3) \pi r_0^2 - G \rho_0^2 4 \pi r_0^4 \int_{r_0}^r (1/r^3 - r_0/r^4) \, dr \quad \text{for } r_0 < r < R$$

16-12. The following table gives one model for the future evolution of the Sun:

Time (Gy)	Luminosity ($L_\odot$)	Radius ($R_\odot$)
5.5	1.08	1.04
6.6	1.19	1.08
7.7	1.32	1.14
8.8	1.50	1.22
9.8	1.76	1.36

For each future time, find the effective temperature of the Sun. Then plot an evolutionary track on an H-R diagram.

The effective temperature is given by Equation 11-19

$$L = 4\pi R^2 \sigma T_{eff}^4$$

$$T_{eff} = (L / 4\pi R^2 \sigma)^{0.25}$$

or in solar units

$$L/L_\odot = (R/R_\odot)^2 (T_{eff}/T_{\odot eff})^4$$

$$T_{eff}/T_{eff\odot} = (L/L_\odot)^{0.25} / (R/R_\odot)^{0.5}$$

Time (Gy)	Luminosity ($L_\odot$)	Radius ($R_\odot$)	Effective Temperature ($T_\odot$)
5.5	1.08	1.04	0.9996 = 5778 K
6.6	1.19	1.08	1.0050 = 5809 K
7.7	1.32	1.14	1.0039 = 5803 K
8.8	1.50	1.22	1.0019 = 5791 K
9.8	1.76	1.36	0.9877 = 5709 K

These values can be plotted an an H-R diagram -- note that the movement on the diagram is very small compared to later post-main sequence stages of evolution.

Chapter 17: Star Deaths

17-1. (a) A white dwarf has an apparent magnitude $m_v = 8.5$ and parallax $\pi = 0.2"$. Its bolometric correction is -2.1 mag, and $T_{eff} = 28,000$ K. Assume $A_V = 0$. Calculate the radius of the star. Compare your value with the radius of the Earth.

The apparent magnitude of the white dwarf and the parallax allow us to compute its absolute magnitude (Equation 11-8):

$$M_v = m_v + 5 + 5 \log(\pi")$$
$$= 8.5 + 5 + 5 \log(0.2)$$
$$= 10.0$$

Applying the bolometric correction gives the bolometric magnitude (Equation 11-21):

$$M_{bol} = M_v + BC$$
$$= 10.0 - 2.1 = 7.9$$

The bolometric magnitude allows us to compute the luminosity using Equation 11-20b:

$$\log(L/L_\odot) = 1.9 - 0.4 \, M_{bol}$$
$$= 1.9 - 0.4 \, (7.9) = -1.26$$
$$L = 0.055 \, L_\odot = 0.055 \, (3.90 \times 10^{26} \text{ J/s})$$
$$= 2.1 \times 10^{25} \text{ J/s}$$

Now, to determine the radius (Equation 11-19):

$$L = 4 \pi R^2 \sigma (T_{eff})^4$$
$$R^2 = [2.1 \times 10^{25} \text{ J/s}] / [4 \pi (5.67 \times 10^{-8} \text{ W m}^{-2} \text{ K}^{-4})(28,000 \text{ K})^4]$$
$$= 4.89 \times 10^{13} \text{ m}^2$$
$$R = 7.0 \times 10^6 \text{ m} = 7.0 \times 10^3 \text{ km} = 7,000 \text{ km}$$

This is about the radius of the Earth ($R_\oplus \approx 6,400$ km).

(b) A neutron star has $T_{eff} = 5 \times 10^5$ K and a radius of 10 km. What is its luminosity?

Using Equation 11-19:

$$L = 4 \pi R^2 \sigma (T_{eff})^4$$
$$= 4 \pi (1.0 \times 10^4 \text{ m})^2 (5.67 \times 10^{-8} \text{ W m}^{-2} \text{ K}^{-4})(5 \times 10^5 \text{ K})^4$$
$$= 4.45 \times 10^{24} \text{ J/s} = 0.011 \, L_\odot$$

(c) A protostellar cloud starts out with $T_{eff} = 15$ K and $R = 4 \times 10^4 \, R_\odot$. Determine the $L/L_\odot$ and the wavelength of the peak of the Planck curve.

Using Equation 11-19:

$$L/L_\odot = (R/R_\odot)^2 \, (T_{eff\text{-cloud}}/T_{eff\odot})^4$$
$$= (4 \times 10^4 \, R_\odot)^2 \, (15 \text{ K} / 5,800 \text{ K})^4$$
$$L = 7.2 \times 10^{-2} \, L_\odot = 0.072 \, L_\odot$$

The wavelength of the peak of the Planck curve can be found from Wien's Law (Equation 8-39):

$$\lambda_{max} T = 2.90 \times 10^{-3} \text{ m K}$$
$$\lambda_{max} = 2.90 \times 10^{-3} \text{ m K} / 15 \text{ K}$$
$$= 1.93 \times 10^{-4} \text{ m} = 193 \text{ µm}$$

17-2. Calculate the kinetic energy ($mv^2/2$) for each of the following:

(a) a nova outburst that accelerates a mass of 10^{-5} $M_\odot$ to a velocity of 10^3 km/s

$$\text{K.E.} = m v^2 / 2 = [10^{-5} \times (1.99 \times 10^{30} \text{ kg})] (10^6 \text{ m/s})^2 / 2 = 1.0 \times 10^{37} \text{ J}$$

(b) the formation of a planetary nebula in which a mass of 0.1 $M_\odot$ is accelerated to a velocity of 20 km/s

$$\text{K.E.} = [0.1 \times (1.99 \times 10^{30} \text{ kg})] (2.0 \times 10^4 \text{ m/s})^2 / 2 = 4.0 \times 10^{37} \text{ J}$$

(c) a supernova outburst that accelerates a mass of 1 $M_\odot$ to a velocity of 4×10^3 km/s

$$\text{K.E.} = [1.0 \times (1.99 \times 10^{30} \text{ kg})] (4.0 \times 10^6 \text{ m/s})^2 / 2 = 1.6 \times 10^{43} \text{ J}$$

How many years would it take the Sun to radiate away these energies?
The Sun's luminosity is $L_\odot = 3.9 \times 10^{26}$ J/s $= 1.23 \times 10^{34}$ J/yr.
Thus, for the Sun to radiate the energies given above would take:

Nova: $t = (9.95 \times 10^{36} \text{ J}) / (1.23 \times 10^{34} \text{ J/yr}) = 810$ yr

Planetary nebula: $t = (4.0 \times 10^{37} \text{ J}) / (1.23 \times 10^{34} \text{ J/yr}) = 3,300$ yr

Supernova: $t = (1.6 \times 10^{43} \text{ J}) / (1.23 \times 10^{34} \text{ J/yr}) = 1.3$ billion yr

17-3. How hot would a cloud of material become falling into a black hole? If it emitted as a blackbody, at what wavelength would it peak? (Ignore gravitational redshifts.)

The gravitational energy released as material falls into a black hole is the same as the kinetic energy that the material needs to escape from the same point. To escape from the Schwarzschild radius the material would have to travel at the speed of light, so to a rough approximation the energy released is $E \approx (1/2) m c^2$. From the virial theorem (Section 16-1D; Equation 21-1), half of this energy is radiated and half goes into thermal energy:

$$E_{thermal} \approx (1/2)(1/2) m c^2 = (1/4) m c^2$$

From the Boltzmann distribution, the thermal energy is related to the temperature:

$$E_{thermal} \approx (1/4) m c^2 \approx (3/2) k T$$

Since most of the particles falling into the black hole are ionized hydrogen (protons and electrons), the average mass per particle is $m \approx m_H / 2$, so:

$$T \approx (2/3)(1/4)(m_H/2)c^2/k$$
$$= (1/12)(1.67 \times 10^{-27} \text{ kg})(3.0 \times 10^8 \text{ m/s})^2/(1.38 \times 10^{-23} \text{ J/K})$$
$$= 9.1 \times 10^{11} \text{ K}$$

Using Wien's Law (Equation 8-39) to find the wavelength of peak emission:
$$\lambda_{max} = (2.9 \times 10^{-3} \text{ m K})/(9.1 \times 10^{11} \text{ K})$$
$$= 3.2 \times 10^{-15} \text{ m} = 3.2 \times 10^{-6} \text{ nm}$$

This wavelength corresponds to the γ-ray region of the electromagnetic spectrum (Table 8-1).

In practice, we would see this material in the accretion disk around the black hole, at a lower temperature (peaking in the x-ray portion of the spectrum).

17-4. Calculate the Schwarzschild radius for

(a) the Earth

The Schwarzschild radius is given by (Equations 17-7a and 17-7b):
$$R_s = 2GM/c^2 = 3 \text{ km} (M/M_\odot)$$

For the Earth, $M_\oplus = 6.0 \times 10^{24}$ kg $= 3.0 \times 10^{-6} M_\odot$, so

$R_{s\oplus} = 3$ km $(3.0 \times 10^{-6}) = 9 \times 10^{-6}$ km $= 9 \times 10^{-3}$ m $= 9$ mm -- pretty small

(b) the Sun

For the Sun, $R_{s\odot} = 3$ km

(c) a globular cluster

A globular cluster has a mass $\approx 10^5 M_\odot$, so

$R_{s\text{-globular cluster}} = 3 \times 10^5$ km $\approx 0.4 R_\odot$

(d) the Galaxy

The mass of the Milky Way Galaxy $\approx 10^{11} M_\odot$, so

$R_{s\text{-Galaxy}} = 3 \times 10^{11}$ km $\approx 2{,}000$ AU ≈ 0.01 pc

What trend do you notice?
The Schwarzschild radius increases linearly with the mass.

17-5. Consider stars of mass $1 M_\odot$. Compute the mean mass density for the following:

(a) our Sun ($R_\odot = 7 \times 10^5$ km)

The mean mass density of a sphere is (Equation 2-1):
$$\langle \rho \rangle = M/V = M/[(4/3)\pi R^3]$$

For a $1 M_\odot = 2.0 \times 10^{30}$ kg object, $\langle \rho \rangle = 4.8 \times 10^{29}$ kg $/R^3$

For the Sun
$$R_\odot = 7 \times 10^5 \text{ km} = 7 \times 10^8 \text{ m}$$
$$\langle \rho \rangle = 1.4 \times 10^3 \text{ kg/m}^3$$

(b) a white dwarf (R = 10^4 km)
$$R_{wd} = 10^4 \text{ km} = 10^7 \text{ m}$$
$$\langle \rho \rangle = 4.8 \times 10^8 \text{ kg/m}^3$$

(c) a neutron star (R = 10 km)
$$R_{ns} = 10 \text{ km} = 10^4 \text{ m}$$
$$\langle \rho \rangle = 4.8 \times 10^{17} \text{ kg/m}^3$$

Now consider a ^{12}C nucleus of radius r = 3 x 10^{-15} m and compute its mean density. Discuss the significance of all these results!
For a carbon nucleus,
$$m = 12 \, m_H = 12 \, (1.67 \times 10^{-27} \text{ kg}) = 2.0 \times 10^{-26} \text{ kg}$$
so
$$\langle \rho \rangle = (2.0 \times 10^{-26} \text{ kg}) / [(4/3) \pi (3 \times 10^{-15} \text{ m})^3]$$
$$= 1.8 \times 10^{17} \text{ kg/m}^3$$

Thus, the mean density of a neutron star is slightly greater than the mean density of a carbon nucleus! A neutron star can thus be considered as a "very large nucleus."

17-6. The oldest white dwarfs were formed about 10^{10} years ago with initial temperatures of about 10^9 K. Determine the current temperature of an old white dwarf of maximum mass 1.4 $M_\odot$, radius 7 x 10^6 m, and age 1 x 10^{10} years. Assume for simplicity that the density is constant throughout the star. What is the current wavelength of maximum intensity of this "white" dwarf? [*Hint:* Since the star cools by radiating as a blackbody, set L = - volume x number density of particles x k(dT/dt), where k is Boltzmann's constant. To solve this equation, separate the variables by putting all terms involving T on the left-hand side; then integrate.]
Using the hint given in the problem, the cooling rate is given by:
$$L = - V \times n \times k \, (dT/dt)$$
and we also know (Equation 11-19):
$$L = 4 \pi R^2 \sigma T^4$$
So
$$4 \pi R^2 \sigma T^4 = - [(4/3) \pi R^3] \, n \, k \, (dT/dt)$$
$$dT / T^4 = (3 \sigma / n k R) \, dt$$

$$\int (dT/T^4) = (3\sigma/nkR) \int dt$$

integrate from time $t = 0$ to t, with corresponding temperatures T_o and T_t

$$(1/3)\,[(1/T_t)^3 - (1/T_o)^3] = (3\sigma/nkR)(t - 0)$$

$$[(1/T_t)^3 - (1/T_o)^3] = 9\sigma t/nkR$$

We have,

$$T_o = 10^9 \text{ K}, \quad t = 10^{10} \text{ yr} = 3.16 \times 10^{17} \text{ s}, \quad R = 7 \times 10^6 \text{ m}$$

To calculate the number density, assume that the white dwarf is composed of carbon-12 nuclei. Each carbon nuclei generates 7 particles, the nucleus and 6 electrons, so the total number of particles can be estimated by dividing the total mass by the mass of a proton (a neutron) and multiplying by 7/12:

$$n = (7/12) \times (1.4 \, M_\odot)(2.0 \times 10^{30} \text{ kg}/M_\odot)$$
$$/\, [(4/3)\pi (7 \times 10^6 \text{ m})^3 (1.67 \times 10^{-27} \text{ kg})]$$
$$= 6.8 \times 10^{35} \text{ m}^{-3}$$

Evaluating,

$$[(1/T_t)^3 - (1/10^9 \text{ K})^3]$$
$$= 9\,(5.67 \times 10^{-8} \text{ J/s m}^2 \text{ K}^4)(3.16 \times 10^{17} \text{ s})$$
$$/\,[(1.38 \times 10^{-23} \text{ J/K})(6.8 \times 10^{35} \text{ m}^{-3})(7 \times 10^6 \text{ m})]$$

$$[(1/T_t)^3 - 10^{-27}] = 2.46 \times 10^{-9} \text{ K}^{-3}$$

$$T_t^3 = 4.07 \times 10^8 \text{ K}^3$$

$$T_t = 7.4 \times 10^2 \text{ K} = 740 \text{ K}$$

To calculate the wavelength of maximum emission, use Wien's Law (Equation 8-39):

$$\lambda_{max} = (2.9 \times 10^{-3} \text{ m K}) / 740 \text{ K} = 3.9 \times 10^{-6} \text{ m} = 3.9 \, \mu\text{m (infrared)}.$$

This "white dwarf" has cooled tremendously in the 10 billion years since its formation, and is now best detected in the infrared. It has about the same temperature as the day side of Mercury or the surface of Venus!

17-7. The speed of sound, given by $c_s = (5P/3\rho)^{1/2}$ for a nonrelativistic gas, where P is the pressure and ρ the density, is the speed at which a star will pulsate once oscillations are generated. Determine the speed of sound and the period of pulsations $\approx R/c_s$ as a function of mass (in solar units) for a nonrelativistic white dwarf (equation of state $P \approx 3.2 \times 10^6 \, \rho^{5/3}$ in SI units). Assume constant density and pressure. How does this time scale compare with the periods of the fastest pulsars? Is it possible for pulsars to be pulsating white dwarfs?

The pulsation speed is given by the speed of sound:

$$c_s = [(5/3)(P/\rho)]^{1/2}$$

For a relativistic white dwarf with equation of state

$$P \approx 3.2 \times 10^6 \rho^{5/3} \quad \text{(where P and } \rho \text{ are in SI units)}$$

the sound speed is

$$c_s = [(5/3)(3.2 \times 10^6 \rho^{5/3}/\rho)]^{1/2}$$

$$= (5.33 \times 10^6 \rho^{2/3})^{1/2} = 2.31 \times 10^3 \rho^{1/3}$$

Considering a typical white dwarf (Section 17-1B, Figure 17-4), with:

$$M = 0.7\, M_\odot,\ R \approx 0.01\, R_\odot = 7 \times 10^6\ m$$

So $\rho = M/V = M/[(4/3)\pi R^3]$

$$= 0.7 \times (2.0 \times 10^{30}\ kg)/[(4/3)\pi(7 \times 10^6\ m)^3]$$

$$= 1.0 \times 10^9\ kg/m^3$$

Note that for white dwarfs, $R \propto M^{-1/3}$ (Section 17-1A), so in solar mass units

$$\rho \propto M/(R^3) = M/M^{-1} = M^2$$

or

$$\rho = (1.0 \times 10^9\ kg/m^3)(M/0.7\,M_\odot)^2$$

$$= 2.0 \times 10^9\ kg/m^3\ (M/M_\odot)^2$$

[Note that the result depends upon the values selected above for the mass and radius of a "typical" white dwarf.]

Thus, the sound speed is

$$c_s = 2.31 \times 10^3\, \rho^{1/3}\ m/s$$

$$= (2.31 \times 10^3)[(2.0 \times 10^9)(M/M_\odot)^2]^{1/3}\ m/s$$

$$= 2.9 \times 10^6\, (M/M_\odot)^{2/3}\ m/s$$

$$= 9.7 \times 10^{-3}\, c\, (M/M_\odot)^{2/3}$$

Assuming $M = 0.7\, M_\odot$ and $R = 0.01\, R_\odot = 7 \times 10^6$ km, the period of pulsation is given by:

$$P \approx R/c_s \approx 10^6\ m/(2.3 \times 10^6\ m/s) \approx 0.4\ s.$$

The pulsation period of a white dwarf is of the order of seconds. The slowest pulsars (with periods of several seconds) are not inconsistent with pulsating white dwarfs. However, the fastest pulsars (with periods less than a second -- in some cases $\approx 1/100$ second) vary too rapidly to be pulsating white dwarfs.

17-8. The Crab Nebula pulsar radiates at a luminosity of about 1×10^{31} W and has a period of 0.033 s. If $M = 1.4\, M_\odot$ and $R = 1.1 \times 10^4$ m, determine the rate at which its period is increasing (dP/dt). How many years will it take for the period to double its present value? (*Hint:* You must integrate after isolating all the terms involving P on the left-hand side for the latter calculation.)

From Equation 17-6, the rate of change of a pulsar's period is given by:
$$dP/dt = (5/8\pi^2)(LP^3/MR^2).$$
For the Crab pulsar, $L \approx 10^{31}$ W, $P \approx 0.033$ s, $M \approx 1.4 M_\odot \approx 2.8 \times 10^{30}$ kg, and $R \approx 1.1 \times 10^4$ m, so
$$dP/dt = (6.33 \times 10^{-2})[(10^{31} \text{ W})(3.6 \times 10^{-6} \text{ s}^3)$$
$$/ (2.8 \times 10^{30} \text{ kg})(1.2 \times 10^8 \text{ m}^2)]$$
$$= 6.7 \times 10^{-14} \text{ s/s} = 2.1 \times 10^{-6} \text{ s/yr}$$

To determine the time it will take to double its period, rewrite the above equation:
$$dP/P^3 = (5/8\pi^2)(L/MR^2) dt$$
Assuming that M, R, and L remain the same (which is not true, since the luminosity will decrease with time),
$$\int (dP/P^3) = (5/8\pi^2) \int (L/MR^2) dt$$
Considering the limits, P = P (current, t = 0), P = 2P (future, t = t)
$$(-1/2)\{[1/(2P)^2] - [1/(P)^2]\} = (5/8\pi^2)(L/MR^2)[t - 0]$$
$$t = (-1/2)(1/4 - 1)[(5/8\pi^2)(L/MR^2) P^2]^{-1}$$
$$= (3/8)[(5/8\pi^2)(L/MR^2) P^2]^{-1}$$
Remembering, from above, that
$$dP/dt = (5/8\pi^2)(L/MR^2) P^3, \text{ we have}$$
$$(5/8\pi^2)(L/MR^2) P^2 = (dP/dt)/P$$
or $\quad P/(dP/dt) = [(5/8\pi^2)(L/MR^2) P^2]^{-1}$

So, the time for a pulsar's period to double is (assuming constant luminosity)
$$t = (3/8) P/(dP/dt)$$
This differs, by the factor (3/8) from the timescale approximated in the text.
Evaluating, for the Crab pulsar:
$$t = (3/8)(0.033 \text{ s})/(6.7 \times 10^{-14} \text{ s/s})$$
$$= 1.85 \times 10^{11} \text{ s} = 5.8 \times 10^3 \text{ yr} = 5,800 \text{ yr}$$

17-9. Equation 17-4 describes the redshift of electromagnetic waves emitted near a massive, compact object. Since time is in many respects the inverse of frequency, we can express the time-dilation effect by the formula $\Delta t'/\Delta t_{obs} = v_f/v_i$, where $\Delta t'$ is a time interval between two events (for example, consecutive ticks of a clock) in the source's frame and Δt_{obs} is the time interval between the same two events as measured by the observer. Notice that a clock placed in a strong gravitational field ticks more slowly than normal as observed by a distant observer, while a distant clock ticks more rapidly as observed by an observer in the gravitational field.

(a) How would a distant observer describe how the timing of events (time between events, how long events last, and so on) changes for an object falling into a black hole? Does the observer ever see the object cross the Schwarzschild radius? Comment.

A distant observer watching an object fall into a black hole sees the apparent time between events increase as the object nears the event horizon. The closer the object is to the event horizon, the stronger is the gravitational field and therefore the larger the apparent time between events. At the Schwarzschild radius, $v_f / v_i \gg 0$, so the time dilation effect $\Delta t' / \Delta t_{obs} \to \infty$. To the outside observer, the time interval between events (which are occurring near the black hole) increases, and events last longer. Hence a distant observer never actually observes the object close to (or crossing) the event horizon. The light emitted by the object falling into the black hole is progressively redshifted with time.

(b) Now describe the timing of distant events as observed by an observer falling into a black hole. What does the observer see as she or he crosses the Schwarzschild radius? Is there a paradox here? If so, can you resolve it?

An observer falling into the black hole sees the reverse. The time interval between external events is shorter. As the observer crosses the Schwarzschild radius, the time interval between distant events approaches zero.

17-10. A neutron decays into a proton, electron, and anti-neutrino via the weak interaction after about 15 min when it is outside the nucleus of an atom. Now imagine that a neutron is freed from its nucleus 3.00 km from the center of a black hole of mass 1 $M_\odot$. How long will it take the neutron to decay as measured by a distant observer? (Use the expression for time dilation given in the previous problem.)

The time dilation is given in Problem 17-9 and Equation 17-4:

$$\Delta t' / \Delta t_{obs} = v_f / v_i = (1 - 2GM/Rc^2)^{1/2} = [1 - (R_s/R)]^{1/2}$$

where $R_s = 2GM/c^2$, is the Schwarzschild radius of a black hole (Equation 17-7a).

For a 1 $M_\odot$ black hole,

$$R_s = 2\,(6.67 \times 10^{-11}\ N\,m^2/kg^2)\,(1.99 \times 10^{30}\ kg) / (3.00 \times 10^8\ m/s)^2$$

$$= 2.95 \times 10^3\ m = 2.95\ km$$

If the neutron is freed from the nucleus at the Schwarzschild radius, R_s, a distant observer would not see the neutron decay (time delay = ∞). However, if it was just outside (R = 3.00 km) the Schwarzschild radius (R = 2.95 km), then the decay will occur in:

$$15\ min / \Delta t_{obs} = [1 - (2.95\ km / 3.00\ km)]^{1/2} = 1.29 \times 10^{-1}$$

$$\Delta t_{obs} = 116\ min \approx 2\ hr$$

17-11. A star identical to the Sun is in a binary system with a black hole of mass M_H. Assume for simplicity that the density of the star is uniform and that the orbit is circular.

(a) Use Equation 3-9 to obtain the minimum separation from the black hole the star must have in order *not* to be torn apart by tidal forces.

The Roche limit is given by (Equation 3-9):

$$d = 2.44\,(\rho_M / \rho_m)^{1/3}\,R$$

The ratio of the densities is given by

$$\rho_M / \rho_m = (M_M / M_m) / (R_M / R_m)^3$$

For a black hole of mass M_{BH} in solar mass units, $R_{BH} = 3 \text{ km } M_{BH}$, while the solar-like star has $M_* = 1$, $R_* = 6.96 \times 10^5$ km, so

$$\rho_{BH} / \rho_* = (M_{BH} / 1) / [(3 \text{ km}) (M_{BH}) / 6.96 \times 10^5 \text{ km}]^3$$
$$= 1.25 \times 10^{16} \, M_{BH}^{-2}$$

Thus, the Roche limit for a solar-like star near a black hole of mass m_{BH} is

$$d = 2.44 \, (1.25 \times 10^{16} \, M_{BH}^{-2})^{1/3} \, (3 \text{ km } M_{BH})$$
$$= 1.7 \times 10^6 \text{ km } M_{BH}^{1/3}$$

If the solar-like star is greater than this distance from a black hole of mass M_{BH} (in solar masses), then the star will not be torn apart by tidal forces.

(b) At what black-hole mass is this minimum separation less than the Schwarzschild radius? Black holes *more* massive than this can swallow a star whole!

The Schwarzschild radius is: $R_s = 3 \text{ km } M_{BH}$, so the Roche limit in units of R_s is:

$$d / R_s = (1.7 \times 10^6 \text{ km } M_{BH}^{1/3}) / (3 \text{ km } M_{BH})$$
$$= 5.7 \times 10^5 \, M_{BH}^{-2/3}$$

For a star to be able to enter the Schwarzschild radius without being tidally disrupted,

$$d / R_s = 1 = 5.7 \times 10^5 \, M_{BH}^{-2/3}$$
$$M_{BH}^{-2/3} = 1.77 \times 10^{-6}$$
$$M_{BH} = 4.25 \times 10^8$$

If the black hole has a mass greater than 425 million solar masses, a solar-like star can be "swallowed" whole -- without being tidally disrupted. It is believed that supermassive black holes may exist in the center of some galaxies (see Chapter 24).

17-12. Assume a brown dwarf's luminosity derives from gravitational contraction. Its mass is 0.05 $M_\odot$, and its luminosity is $3 \times 10^{-5} \, L_\odot$. If we assume that its luminosity has been constant (even when the star had a much larger radius), how long can a star of this type radiate before the contraction is halted by electron degeneracy pressure (when $R \approx 9 \times 10^6 \, M^{-1/3}$ m, where M is in solar units)?

Electron degeneracy will halt the contraction when the star has reached a radius of

$$R \approx 9 \times 10^6 \, (M/M_\odot)^{-1/3} \text{ m} = 9 \times 10^6 \, (0.05)^{-1/3} \text{ m} = 2.44 \times 10^7 \text{ m}$$

The energy available per kg of stellar material from gravitational collapse is given by Equation 16-15:

$$E_{grav} = G M / 2 R$$
$$= (6.67 \times 10^{-11} \text{ N m}^2 / \text{kg}^2) \, (0.05 \times 2.0 \times 10^{30} \text{ kg}) / (2 \times 2.44 \times 10^7 \text{ m})$$

$$= 1.4 \times 10^{11} \text{ J/kg}$$

So the total energy available from gravitational contraction for a 0.05 $M_\odot$ (10^{28} kg) star is 1.4×10^{40} J.

The lifetime of the star to radiate from gravitational contraction is thus

$$t = E_{grav}/L = (1.4 \times 10^{40} \text{ J}) / [(3 \times 10^{-5}) \times (3.90 \times 10^{26} \text{ J/s})]$$

$$= 1.2 \times 10^{18} \text{ s} = 3.7 \times 10^{10} \text{ yr} = 37 \text{ billion years}$$

The brown dwarf can radiate (from energy derived from gravitational contraction) at its current luminosity for 37 billion years before its contraction is halted by electron degeneracy. It is thus possible that some of such a hypothetical brown dwarf's luminosity is due to gravitational energy.

17-13. A typical young neutron star has a radius of about 10^4 m and a temperature of roughly 10^6 K.

(a) What is the blackbody luminosity of such a neutron star?

The blackbody luminosity is given by (Equation 11-19):

$$L = 4\pi R^2 \sigma T^4.$$

For a neutron star with R = 10^4 m, and T = 10^6 K,

$$L = 4\pi (10^4 \text{ m})^2 (5.67 \times 10^{-8} \text{ W/m}^2 \text{ K}^4)(10^6 \text{ K})^4$$

$$= 7.1 \times 10^{25} \text{ W}$$

(b) What is the maximum distance a neutron star with these properties could have and still be detected by its optical blackbody radiation? Assume a limiting magnitude of 25 for detection and a bolometric correction of zero. Comment.

Assuming no bolometric correction (which is probably not correct), the absolute magnitude of this neutron star is (Equation 11-20b)

$$\log (L_*/L_\odot) = 1.9 - 0.4 M_{bol*}$$

$$M_{bol*} = [1.9 - \log(L_*/L_\odot)] / 0.4$$

$$= [1.9 - \log(7.1 \times 10^{25} \text{ W} / 3.9 \times 10^{26} \text{ W})] / 0.4$$

$$= [1.9 - (-0.74)] / 0.4$$

$$= 6.6$$

If this neutron star had an apparent magnitude of 25 (just barely detectable from Earth), from Equation 11-6,

$$m - M = 5 \log(d_{pc}) - 5$$

$$\log(d_{pc}) = [(25 - 6.6) + 5]/5 = 4.68$$

$$d = 4.8 \times 10^4 \text{ pc} = 48 \text{ kpc}$$

This is larger than the Galaxy. So, we should be able to see any such neutron stars in our Galaxy. From Wien's Law (Equation 13-6), we can show that this radiation would be seen at x-ray wavelengths,

λ_{max} = 2.9 x 10^{-3} m K / T ≈ 3 x 10^{-3} m K / 10^6 K ≈ 3 x 10^{-9} m = 3 nm,

so x-ray surveys of the sky might detect young neutron stars, though none have been discovered this way to date. The radio and optical luminosity which we observe from pulsars is from another radiation process (such as synchrotron radiation).

17-14. A cloud of hot gas is in a Keplerian circular orbit of radius 4.0 x 10^4 m about a black hole of mass $10 M_\odot$. Plot the ratio of observed-to-rest frequency or wavelength versus time over one orbital period of an emission line radiated by the cloud, as seen by an observer who views the orbital plane edge-on. Include both the gravitational and normal Doppler effects and use the relativistic formulas.

First find the speed of the gas in the cloud for a circular Keplerian orbit around the black hole (Equation 1-28):

$v_{orbit} \equiv v = (GM/r)^{1/2}$

$= [(6.67 \times 10^{-11}$ N m^2/kg^2) (10 x 2.0 x 10^{30} kg) / (4.0 x 10^4 m)$]^{1/2}$

$= 1.83 \times 10^8$ m / s = 0.6 c

If the orbital plane is seen edge on, the velocity component along the line of sight of the observer is given by v_r = v sin(θ) = 0.6 c sin(θ), where θ = 0° when the cloud is between the black hole and the observer.

Using the relativistic Doppler formula (Equation 8-14b), the frequency shift due to the orbital motion is:

$v_f/v_i = [(1 - v/c)/(1 + v/c)]^{1/2}$

The relativistic gravitational Doppler shift is given by Equation 17-4:

$v_f/v_i = [1 - 2GM/Rc^2]^{1/2}$

$= [1 - 2 (6.67 \times 10^{-11}$ N m^2/kg^2) (10 x 2.0 x 10^{30} kg)

$/ (4 \times 10^4$ m) (3.0 x 10^8 m/s$)^2]^{1/2}$

$= 0.51$

This gravitational redshift will modify the observed Doppler shift throughout the orbit.

The combined frequency shift is

$v_f/v_i = 0.51 [(1 - v/c)/(1 + v/c)]^{1/2}$

$= 0.51 [(1 - 0.6 \sin(\theta))/(1 + 0.6 \sin(\theta))]^{1/2}$

Evaluating for special cases:

for θ = 0°, 180° → sin(θ) = 0, so v_f/v_i = 0.51

for θ = 90° → sin(θ) = 1, so v_f/v_i = 0.51 $(0.4/1.6)^{1/2}$ = 0.25

for θ = 270° → sin(θ) = -1, so v_f/v_i = 0.51 $(1.6/0.4)^{1/2}$ = 1.02

The remainder of the values can be calculated, to produce the ratio of the observed to emitted frequency curve shown below. [Note that generally $\Delta v / v$ is plotted rather than

ν_f / ν_i, since the curve would then be sinusoidal in shape, rather than skewed as we see in the graph below.]

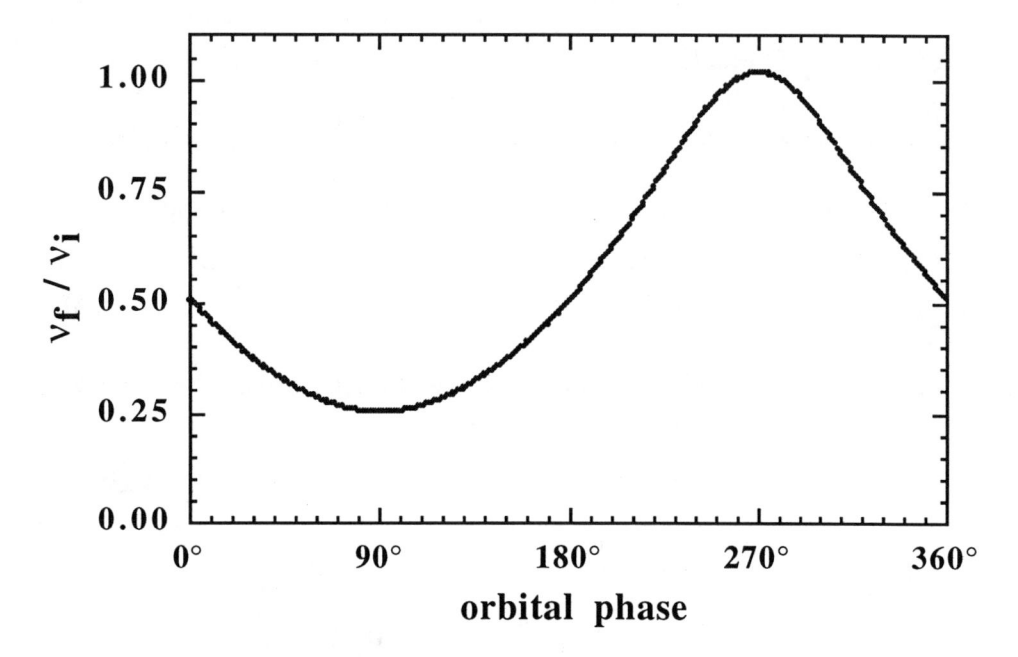

Similar calculations can be done for the observed wavelength shift (λ_f / λ_i), but this is left to the reader.

17-15. A magnetic dipole rotating with a period P radiates with a luminosity that is proportional to P^{-4}.

(a) Show that, long after its formation, a pulsar slows down according to the law $P \propto t^{1/2}$ (approximately).

For a magnetic dipole, the luminosity is related to the period by $L \propto P^{-4}$.

From Equation 17-6,

$$dP/dt = (5/8\pi^2)(LP^3/MR^2)$$

the luminosity of a pulsar is given by $L \propto P^{-3} dP/dt$, assuming the mass and radius of the neutron star remain constant. Combining these two equations,

$$L \propto P^{-4} \propto P^{-3} dP/dt$$

so

$$P \propto 1/(dP/dt)$$

or

$$dt \propto P\, dP$$

Integrating,

$$\int dt \propto \int P\, dP$$

$$t - t_0 \propto (1/2)[P^2 - P_0^2]$$

Long after formation, $t \gg t_0$, and $P^2 \gg P_0^2$, so $t \propto P^2$ or $P \propto t^{1/2}$.

203

(b) Using the result from part (a), show that the luminosity of the pulsar decreases approximately as $L \propto t^{-2}$.

Since $L \propto P^{-4}$ and $P \propto t^{1/2}$, we see that $L \propto t^{-2}$. The pulsar luminosity decreases quickly as a function of time.

17-16. For the Crab pulsar, calculate the difference between pulse arrival times at 430 and 196 MHz. Then compare this difference to that for the other pulsars listed in Table 17-1.

Note: The text gives the equation for time delay appropriate for cgs units. To use SI units, a factor of $1/4\pi\varepsilon_0$ must be included, where $\varepsilon_0 = 8.85 \times 10^{-12}$ $C^2/N\, m^2$. Also, the text has a factor of 2π in the numerator, which should be in the denominator.

From Section 17-2B, with the above mentioned corrections, the time delay between pulse arrivals is

$$t_2 - t_1 = (1/4\pi\varepsilon_0)[e^2/(2\pi m_e c)][(1/f_2^2) - (1/f_1^2)]\, DM$$

$$= [e^2/(8\pi^2 \varepsilon_0 m_e c)][(1/f_2^2) - (1/f_1^2)]\, DM$$

$$= [(1.6 \times 10^{-19}\, C)^2 / \{8\pi^2 (8.85 \times 10^{-12}\, C^2/N\, m^2)$$
$$\times (9.11 \times 10^{-31}\, kg)(3.00 \times 10^8\, m/s)\}] \times [(1/f_2^2) - (1/f_1^2)]\, DM$$

$$= (1.34 \times 10^{-7}\, m^2/s)[(1/f_2^2) - (1/f_1^2)]\, DM$$

The dispersion measure, DM, is usually specified in the weird units of pc/cm^3! The conversion factor between DM in these units and m/m^3 is

$$1\, pc/cm^3 = (3.086 \times 10^{16}\, m)/(10^{-2}\, m)^3 = 3.086 \times 10^{22}\, m^{-2}$$

so we can rewrite the equation for the time delay as

$$t_2 - t_1 = (1.34 \times 10^{-7}\, m^2/s)(3.086 \times 10^{22}\, cm^3/pc\, m^2)$$
$$\times [(1/f_2^2) - (1/f_1^2)]\, DM$$

$$= (4.11 \times 10^{15}\, cm^3/pc\, s)[(1/f_2^2) - (1/f_1^2)]\, DM$$

For the Crab pulsar (Table 17-1), $DM = 56.8\, pc/cm^3$, so

$$t_2 - t_1 = (4.11 \times 10^{15}\, cm^3/pc\, s)$$
$$\times [(1/(4.30 \times 10^8\, /s)^2) - (1/(1.96 \times 10^8\, /s)^2)] \times (56.8\, pc/cm^3)$$
$$= (4.11 \times 10^{15}\, cm^3/pc\, s)(-2.06 \times 10^{-17}\, s^2)(56.8\, pc/cm^3)$$
$$= -4.8\, s$$

Since the period of the Crab pulsar is 0.332 seconds, a pulse will be delayed by $4.8\, s / (0.332\, s/period) \approx 15$ periods between 430 MHz and 196 MHz.

17-17. Use the information in Section 17-2D and Kepler's third law to determine the sum of the masses of the components in the PSR +065564 binary system.

[*Note:* The fourth edition of the text no longer gives information about PSR +065564, so the instructor needs to give the following information to the student.]

The binary system PSR +065564 has a period of revolution 24 hour 41 minute = 8.886×10^4 s, and an orbital semi-major axis 7.5×10^5 km.

Using Kepler's third law (Equation 1-24)

$$P^2 = 4\pi^2 a^3 / G(M_1 + M_2)$$

$$\begin{aligned} M_1 + M_2 &= 4\pi^2 a^3 / G P^2 \\ &= 4\pi^2 (7.5 \times 10^8 \text{ m})^3 / (6.67 \times 10^{-11} \text{ N m}^2/\text{kg}^2)(8.886 \times 10^4 \text{ s})^2 \\ &= 3.2 \times 10^{28} \text{ kg} = 0.016 \text{ M}_\odot \end{aligned}$$

This does not agree with what one would expect (a sum of the masses on the order of one to two solar masses). Perhaps the parameters given in the third edition of the text were incorrect!

17-18. What would be the rotation period of the Sun if it collapsed to a radius of 10 km without losing angular momentum? Compare your result to the rotation periods of known pulsars.

Apply the conservation of angular momentum, L, to a collapsing spherical mass (see Section 17-2E for discussion of moment of interia):

$$L = I\omega = (2/5) M R^2 (2\pi/P)$$

so for conservation of angular momentum

$$L_\odot = L_{ns}$$

$$(2/5) M_\odot R_\odot^2 (2\pi/P_\odot) = (2/5) M_{ns} R_{ns}^2 (2\pi/P_{ns})$$

$$R_\odot^2 / P_\odot = R_{ns}^2 / P_{ns}$$

$$\begin{aligned} P_{ns} &= (R_{ns}/R_\odot)^2 P_\odot \\ &= [(10 \times 10^3 \text{ m})/(6.96 \times 10^8 \text{ m})]^2 (27 \text{ days} \times 8.64 \times 10^4 \text{ s/day}) \\ &= 5.2 \times 10^{-4} \text{ s} = 0.5 \text{ ms} \end{aligned}$$

This is comparable to the period of the fastest known pulsars (though very rapidly rotating pulsars are believed to have been spun-up after formation). It is reasonable that the rapid spin rate of young pulsars arises from the collapse of a star which obeys the conservation of angular momentum.

17-19. Compute the differential tidal force (see Chapter 3) on a person 3000 km from a 10 $M_\odot$ black hole. (Assume the person's mass is 90 kg and height is 2 m.)

From Equation 3-8, the differential tidal force is given by

$$\begin{aligned} dF &= -(2GMm/R^3) dR \\ &= -[2(6.67 \times 10^{-11} \text{ N m}^2/\text{kg}^2)(10 \times 1.99 \times 10^{30} \text{ kg})(9.0 \times 10^1 \text{ kg}) \\ &\qquad /(3 \times 10^6 \text{ m})^3] \times (2 \text{ m}) \end{aligned}$$

$$dF = -1.77 \times 10^4 \text{ N} \approx -1.8 \times 10^4 \text{ N}$$

A 90 kg person thus would experience a differential tidal force of $\approx 18{,}000$ N

17-20. By examining the gravitational redshift equation, you should see that the relativistic time dilation relates directly to it, if you think in terms of frequencies rather than wavelength. Consider two clocks, one keeping time T_1 at distance R_1 from a mass, the other T_2 at distance R_2. Then the time are related by

$$T_2/T_1 = [(1 - 2GM/R_2c^2)/(1 - 2GM/R_1c^2)]^{1/2}$$

(a) Compare clock, one at the surface of a neutron star, the other far away.
Using the approximate values for a neutron star,

$$M \approx 2 M_\odot \approx 4 \times 10^{30} \text{ kg}, R_2 \approx 10 \text{ km} = 10^4 \text{ m},$$

and take $R_1 = \infty$. Note for $R_1 = \infty$, the term $(1 - 2GM/R_1c^2)$ is equal to 1, so

$$T_2/T_1 = (1 - 2GM/R_2c^2)^{1/2}$$

$$T_2/T_1 = \{1 - [2\,(6.67 \times 10^{-11} \text{ Nm}^2/\text{kg}^2)\,(4 \times 10^{30} \text{ kg})$$
$$/\,(1 \times 10^4 \text{ m})\,(3 \times 10^8 \text{ m/s})^2)]\}^{1/2}$$

$$= 0.64$$

This is a considerable relativistic time dilation.

(b) Compare clocks, one at the surface of a white dwarf, the other far away.
Using the approximate values for a white dwarf star,

$$M \approx 0.7 M_\odot = 1.4 \times 10^{30} \text{ kg}, R_2 \approx 0.01 R_\odot = 7 \times 10^6 \text{ m},$$

and take $R_1 = \infty$, as above.

$$T_2/T_1 = \{1 - [2\,(6.67 \times 10^{-11} \text{ Nm}^2/\text{kg}^2)\,(1.4 \times 10^{30} \text{ kg})$$
$$/\,(7 \times 10^6 \text{ m})\,(3 \times 10^8 \text{ m/s})^2)]\}^{1/2}$$

$$= 0.99985$$

The relativistic time dilation is thus insignificant.

(c) Compare clocks, one just outside the Schwarzschild radius of a $3M_\odot$ black hole, the other far away.
Using the approximate values for a black hole,

$$M \approx 3 M_\odot = 6 \times 10^{30} \text{ kg}, R_2 = R_{\text{Schwarzschild}} = 3 \times 3 \text{ km} \approx 9 \times 10^3 \text{ m},$$

and take $R_1 = \infty$, as above.

$$T_2/T_1 = \{1 - [2\,(6.67 \times 10^{-11} \text{ Nm}^2/\text{kg}^2)\,(6 \times 10^{30} \text{ kg})$$
$$/\,(9 \times 10^3 \text{ m})\,(3 \times 10^8 \text{ m/s})^2)]\}^{1/2}$$

$$= 0.11$$

The relativistic time dilation is thus very significant for regions near black holes.

Chapter 18: Variable and Violent Stars

18-0. In a given constellation, the following designations have been assigned to variable stars: V502, SU, II, V956, XY, and AK. List these stars in the order of their discovery.
[*Note:* This is one of our favorite problems from the third edition, so we include it here even though it was omitted from the fourth edition.]
The student needs to understand the ordering of variable stars within a constellation discussed in Section 18-1. In order of their discovery, the stars were:
    SU, XY, AK, II, V502, V956.

18-1. Consider a Cepheid variable with a period of ten days, a mean radius of about 100 $R_\odot$, and a radial velocity averaging 15 km/s. What is the change in its radius, $\Delta R = R(2) - R(1)$?
The change in the radius will be the average velocity times (approximately) one half the period.

$$\Delta R = \langle v \rangle \times (P/2)$$
$$= (15 \text{ km/s}) [(10 \text{ days} \times 8.64 \times 10^4 \text{ s/day})/2]$$
$$= 6.5 \times 10^6 \text{ km} \approx 9.3 \, R_\odot$$

This is about 10% the mean radius of the star.

18-2. In your own words, describe and correlate the fluctuation profiles of the typical pulsating star illustrated in Figure 18-2.
The temperature of the variable star closely follows the brightness of the star. This correlation indicates that the magnitude change is at least in part due to surface temperature changes. As the brightness reaches a maximum, so does the temperature; however, the temperature minimum occurs slightly earlier than the brightness minimum -- so there also must be other effects causing the brightness variations. The changes in spectral type are in phase with the temperature fluctuations, as one would expect. The velocity/radius changes are slightly out of phase with the temperature and brightness changes. The radius is at a minimum shortly after the brightness is at a minimum, and the radius maximum lags the brightness peak. The radius profile is strongly asymmetric -- the changes at minimum radius are more rapid than at maximum radius. The brightness variations are caused in part by the radial variations as well as the temperature variations. The radial velocity variations are consistent with the expansion and contraction of the star.

18-3. A Cepheid variable in a hypothetical galaxy is observed to pulsate with a period of ten days, and its mean apparent visual magnitude is 18. It is not known whether this is a Population **I** or Population **II** Cepheid.
(a) What are the two possible distances to the galaxy (neglect interstellar absorption)?
Reading from the period luminosity relation (Figure 18-3), a Population **I** Cepheid with a 10 day period has $\log(L/L_\odot) = 3.5$, or $L = 3.2 \times 10^3 \, L_\odot$; a Population **II** Cepheid with the same period has $\log(L/L_\odot) = 2.66$, or $L = 4.6 \times 10^2 \, L_\odot$. (Student answers

will vary, depending on how well the logarithmic graph is read.) The absolute magnitudes are (Equation 11-5):

$$M_1 - M_2 = 2.5 \log (L_2/L_1)$$

so $M_\odot - M_I = 2.5 \log (3.2 \times 10^3) = 8.8$

$M_I = M_\odot - 8.8 = +4.8 - 8.8 = -4.0$

and $M_\odot - M_{II} = 2.5 \log (4.6 \times 10^2) = 6.7$

$M_{II} = M_\odot - 6.7 = +4.8 - 6.7 = -1.9$

For the Population **I** Cepheid, the distance is given by (Equation 11-6):

$\log (d_{pc}) = (1/5) (m - M + 5) = (1/5) (18 - (-4.0) + 5) = 5.4$

$d = 2.5 \times 10^5$ pc $= 250$ kpc

For the Population **II** Cepheid,

$\log (d_{pc}) = (1/5) (m - M + 5) = (1/5) (18 - (-1.9) + 5) = 4.98$

$d = 9.6 \times 10^4$ pc $= 95$ kpc

(b) What is the ratio of these distances?

The ratio of these distances in 250 kpc / 95 kpc = 2.6

(c) Would the ratio change if we considered other galaxies? Explain.

This ratio of the distances will not change if we considered Cepheid variables in other galaxies since the ratio of the distances is related to the differences in the absolute magnitudes of the stars (for an observed apparent magnitude). To see this, from part (a) we have:

$d_I / d_{II} = 10^{[0.2(m_I - M_I + 5)]} / 10^{[0.2(m_{II} - M_{II} + 5)]}$

$= 10^{[0.2(m_I - M_I) - (m_{II} - M_{II})]}$

For a given Cepheid star, the observed magnitude would be the same regardless of whether it was of Population **I** or Population **II**, so $m_I = m_{II}$:

$d_I / d_{II} = 10^{[0.2(M_{II} - M_I)]}$

Since, for a star of a given period, the difference between the absolute magnitude of a Population **I** and Population **II** Cepheid is independent of the distance, we have

$d_I / d_{II} =$ constant

Thus, if we make an error when interpreting observations of Cepheid variables in many galaxies by misidentifying Population **I** as Population **II** Cepheids, our estimate of the distance to each of these galaxies will be low by a multiplicative factor of 2.6 in distance -- our "scale of the Universe" will be incorrect by this factor. This is indeed what occurred during the early part of this century.

(d) Will this ratio differ for Cepheids of different periods?

To the extent that the P-L relation of Population **I** and **II** Cepheids are roughly parallel (the absolute magnitude difference between Population **I** and **II** is roughly the same for any period), the distance ratio calculated in part (b) will not be strongly dependent on the period of Cepheids which are used. However, a close inspection of Figure 18-3 reveals that the curves for Population I and II Cepheids are NOT parallel -- at the longer periods the curves diverge. Thus, the ratio of distances determined as in part (b) will increase for Cepheids of longer period.

18-4. If our telescope has a limiting magnitude of 22, what is the *maximum distance* to which we can see the stars indicated. How do these distances compare with the diameter of our Galaxy? (*Hint:* Consult Tables 18-1 to 18-3.)

Using Tables 18-1 and 18-3 for an estimate of the absolute magnitude at maximum of the different variable star classes, and Equation 11-6:

$$\log(d_{pc}) = (1/5)(m - M + 5)$$

we calculate the distances at which these stars can be observed to a limiting apparent magnitude of m = +22, assuming no interstellar absorption.

Type of Star	M_V maximum	maximum distance observable
a) RR Lyra stars	+0.5	200 kpc
b) Classical Cepheids	-6.0	4,000 kpc
c) W Virginis stars	-3.0	1,000 kpc
d) Novae (fast)	-9.2	17,000 kpc
Novae (slow)	-7.4	7,600 kpc
e) dwarf novae	+5.5	20 kpc
f) supernovae I	-20.0	2,500,000 kpc
supernovae II	-18.0	1,000,000 kpc

Note, for comparison, our Galaxy's diameter is ≈ 30 kpc. Except for the dwarf novae, we could see all these stars outside of our Galaxy. The Andromeda galaxy is ≈ 690 kpc distant (Table 23-1), so all but the dwarf novae and RR Lyrae stars could be seen in it. The supernovae are clearly the only objects which can be seen at extremely large distances.

18-5. The outburst of Nova Aquila (V603 Aql) occurred in June 1918, at which time it attained a brightness of -1.1 mag. Spectra showed Doppler-shifted absorption lines corresponding to a velocity of 1700 km/s. By 1926, the star was surrounded by a faint shell 16" of arc in diameter. Find the distance to Nova Aquilae in parsecs and the absolute magnitude at maximum.

The nova is expanding at a rate of 1,700 km / s. If it has a diameter of 16" (radius 8") after 8 yr, it is expanding (radially) at 1" / yr. Assuming the expansion is spherically symmetry, we can use an analogy to proper motion and use Equation 18-3

$$d\,[pc] = v\,[km/s] / 4.74\,\mu"$$
$$= 1,700 / (4.74 \times 1)$$
$$d = 360\text{ pc}$$

The absolute magnitude is found from Equation 11-6:

$$M = m + 5 - 5\log(d_{pc}) = -1.1 + 5 - 5\log(360) = -8.9.$$

This absolute magnitude tells us that Nova Aquilae was a fast novae (Table 18-3).

18-6. If a star becomes a supernova, by what amount does its luminosity change if it originally had an absolute magnitude of 5.0? of 2.0? (The absolute visual magnitude of a supernova at maximum is about -18.0.)

See Table 18-3 for typical absolute magnitudes of different classes of supernovae and other cataclysmic and eruptive variables.

The relationship between relative luminosities and magnitude differences is given by Equation 11-3:

$$L_{max}/L_{min} = 100^{-0.2(M_{max} - M_{min})} = 10^{-0.4(M_{max} - M_{min})}$$

gives

$$L_{max}/L_{min} = 10^{-0.4(-18-5)} = 10^{9.2} = 1.6 \times 10^9 \quad \text{for } M_{min} = +5.$$

and

$$L_{max}/L_{min} = 10^{-0.4(-18-2)} = 10^{8.0} = 1.0 \times 10^8 \quad \text{for } M_{min} = +2.$$

18-7. Consult Chapter 10 to find the energy output in watts of the most energetic solar flares. Referring to Chapter 13, compare this with the typical energy outputs of stars of the following spectral types: (a) F, (b) G, (c) K, (d) M. By what factor is the luminosity of each star increased during such a flare event? Which stars would you consider to be *observable* flare stars?

According to the information in Section 10-6C, the typical solar flares have a total energy output of 10^{25} J. [Note, some exceptional flares can be as large as 10^{30} J.] The largest flares take about 3 hours ($\approx 10^4$ s) to decay. The average flare luminosity is $L_f \approx 10^{25}$ J $/ 10^4$ s $\approx 10^{21}$ W $\approx 2.6 \times 10^{-6} L_\odot$.

Using the absolute magnitudes in Appendix Table A4-3, we can compute the bolometric magnitudes of main sequence stars of various spectral types (Equation 11-21):

$$M_{bol} = M_v + BC$$

and luminosities (Equation 11-20b):

$$\log (L_*/L_\odot) = 1.9 - 0.4 M_{bol*} \text{ ; where } L_\odot = 3.90 \times 10^{26} \text{ W}$$

where $L_f/L_* = (L_f/L_\odot) / (L_*/L_\odot) = 2.6 \times 10^{-6} L_\odot / (L_*/L_\odot)$.

Spectral Type	M_{bol}	$L_*/L_\odot$	L_f/L_*
Sun	+4.8 - 0.1 = +4.7	1.0	2.6×10^{-6}
F5	+3.4 -0.0 = +3.4	3.5	7.4×10^{-7}
G5	+5.2 -0.1 = +5.1	0.72	3.6×10^{-6}
K5	+8.0 -0.7 = +7.3	0.095	2.7×10^{-5}
M5	+12.3 - 2.1 = +10.2	0.0066	3.9×10^{-4}

Note that the fractional luminosity changes which we would observe for each stellar type is very small ($\approx 10^{-4}$ to 10^{-6}), beyond the detectability of Earth-based instruments (accuracy of measurement $\approx 0.1\% - 1\%$). Therefore, the flares that we detect in flare stars are considerably more energetic than is typical of solar flares. For comparable luminosity flares in stars of different spectral class, those flares in M and K stars would be the easiest to detect since the star's luminosity is the least.

18-8. Consider a rotating star with an extended atmosphere. If the stellar atmosphere is both rotating and expanding, draw the observed profile of a given spectral line (Figure 18-9) when the atmosphere rotates more slowly than the star.

For an extended atmosphere star that is both rotating and expanding (assume expansion velocity is less than rotational velocity), with the star rotating more rapidly than the atmosphere, the spectral line profile would include features seen in Figure 18-9. The profile would look similar to 18-9A, except that due to the expansion the emission feature from the envelope would be Doppler broadened.

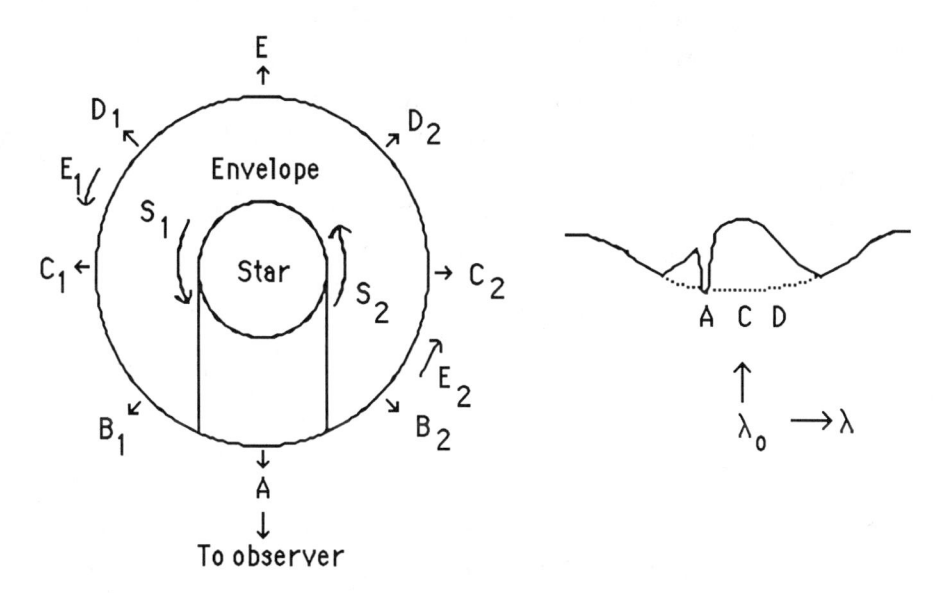

Also, since the envelope is expanding toward the observer at point A, the narrow absorption feature would be blue shifted. The profile would look as shown in the sketch above.

18-9. A white dwarf in a binary system accretes enough material to increase its mass beyond the Chandrasekhar limit (1.4 $M_\odot$) and collapse to the radius of a neutron star ($\approx 10^4$ m). Calculate the kinetic energy generated in such a collapse. Compare this with the value of approximately 10^{44} J relevant to a supernova explosion. Comment on the required efficiency of energy conversion.

According to the virial theorem, when the white dwarf collapses, half of the change in gravitational potential energy goes into kinetic energy and half goes into radiation. The increase in kinetic energy from collapse is thus (Equation 16-15)

$$KE = (1/2) \Delta PE = (G M m / 2 R_{final}) - (G M m / 2 R_{initial})$$
$$= (G M m / 2) (1 / R_{final} - 1 / R_{initial})$$
$$= [0.5 \times (6.67 \times 10^{-11} \text{ N m}^2 / \text{kg}^2) \times (1.4 \times 2.0 \times 10^{30} \text{ kg})^2]$$
$$\times [1 / (10^4 \text{ m}) - 1 / (7 \times 10^6 \text{ m})]$$
$$\approx (2.6 \times 10^{50} \text{ N m}^2) \times (10^{-4} \text{ m}^{-1})$$

211

$= 2.6 \times 10^{46}$ J

This kinetic energy is two orders of magnitude greater than that of a supernova explosion. Thus, an energy conversion efficiency of $\approx 1\%$ will suffice for white dwarfs in binary systems being responsible for some supernova explosions.

18-10. A certain contact-binary system contains a red giant and a neutron star. The neutron star has a mass of 1 $M_\odot$ and a radius of 10^4 m. The system radiates 10^{31} W in X-rays. Determine the rate of mass flow in solar masses per year from the red giant to the neutron star required to produce this luminosity. Assume that half of the change in gravitational potential energy of an accreted gas particle is converted to X-rays and that the separation of the two stars is much greater than the radius of the neutron star.
From the virial theorem, the energy available for radiation is given by half the gravitational potential energy (Equation 16-15)
$\quad$ E = (1/2) G M m / R
If the red giant is sufficiently far away, we can consider that the mass falls onto the neutron star from ∞. M is the mass of the neutron star and m is the mass of material falling onto the neutron star.
Taking the time derivative gives the luminosity

$\quad$ L $\equiv$ dE/dt = (1/2) G M (dm/dt) / R

dm/dt = $(10^{31}$ J / s) $(10^4$ m)
$\qquad$ / [(0.5) (6.67 $\times 10^{-11}$ N m^2 / kg^2) (2.0 $\times 10^{30}$ kg)]
$\quad$ = 1.5 $\times 10^{15}$ kg / s = 4.7 $\times 10^{22}$ kg / yr
$\quad$ = 2.4 $\times 10^{-8}$ $M_\odot$ / yr

18-11. The "luminosity" of a stellar wind is the rate at which kinetic energy is carried away by the wind
$\quad$ L_w = (1/2) (dm/dt) v^2

For the values given in the text, calculate the luminosities of
(a) T-Tauri stars

T Tauri stars have a mass loss rate of 10^{-7} to 10^{-8} $M_\odot$ per year (Section 18-3A). *The ejection velocity is not given in the chapter, but was given as $\approx$ 100 km/s in Problem 15-20.* For 10^{-8} $M_\odot$/yr,

$L_{\text{T Tauri}}$ = (1/2) $(10^{-8}$ $M_\odot$ / yr) (1.99 $\times 10^{30}$ kg/$M_\odot$) (3.16 $\times 10^7$ s/yr)$^{-1}$
$\qquad\qquad\qquad\qquad\qquad\qquad\qquad\qquad$ $\times$ (100 $\times 10^3$ m / s)^2
$\quad$ = (1/2) (6.3 $\times 10^{14}$ kg / s) (10^{10} m^2 / s^2)
$\quad$ = 3.1 $\times 10^{24}$ kg m^2 / s^3 = 3.1 $\times 10^{24}$ W

And for 10^{-7} $M_\odot$/yr, L = 3.1 $\times 10^{25}$ W.

(b) M giants

From Section 18-4C, we see that M giants have mass loss rates of 10^{-6} to 10^{-8} $M_\odot$ per year with expansion velocities of tens of km/s. (Let's interpret "tens" as thirty.) For 10^{-8} $M_\odot$/yr and a velocity of 30 km/s.

$$L_{M\ giants} = (1/2)\ (10^{-8}\ M_\odot\ /\ yr)\ (1.99 \times 10^{30}\ kg/M_\odot)\ (3.16 \times 10^7\ s/yr)^{-1}$$
$$\times\ (30 \times 10^3\ m\ /\ s)^2$$
$$= (1/2)\ (6.3 \times 10^{14}\ kg\ /\ s)\ (9 \times 10^8\ m^2\ /\ s^2)$$
$$= 2.8 \times 10^{23}\ W$$

And for 10^{-6} $M_\odot$/yr, $L = 2.8 \times 10^{25}$ W.

(c) M-supergiants

From Table 18-2, we see that M-supergiants have ejection velocities up to 26 km/s and mass loss rates of 5×10^{-6} to 5×10^{-9} $M_\odot$/year. For 5×10^{-9} $M_\odot$/year,

$$L_{M\ supergiants} = (1/2)\ (5 \times 10^{-9}\ M_\odot\ /\ yr)\ (1.99 \times 10^{30}\ kg/M_\odot)\ (3.16 \times 10^7\ s/yr)^{-1}$$
$$\times\ (26 \times 10^3\ m\ /\ s)^2$$
$$= (1/2)\ (3.15 \times 10^{14}\ kg\ /\ s)\ (6.76 \times 10^8\ m^2\ /\ s^2)$$
$$= 1.1 \times 10^{23}\ W$$

And for 5×10^{-6} $M_\odot$/yr, $L = 1.1 \times 10^{26}$ W.

How do they compare.
Summarizing,

$$L_{T\ Tauri} = 3.1 \times 10^{24}\ W\ to\ 3.1 \times 10^{25}\ W$$
$$L_{M\ giants} = 2.8 \times 10^{23}\ W\ to\ 2.8 \times 10^{25}\ W$$
$$L_{M\ supergiants} = 1.1 \times 10^{23}\ W\ to\ 1.1 \times 10^{26}\ W$$

These values are highly dependent on the velocities assumed. However, they are comparable to an order of magnitude. Note that the T Tauri stars are comparable to the M giant stars, partly due to their higher ejection speeds.

18-12. Another suggested mechanism for generating a type II supernova is explosive nuclear "burning" of the heavier elements, especially silicon. One model for a massive red supergiant predicts the presence of a shell of mass ≈ 2 $M_\odot$ and containing mostly ^{28}Si deep within the interior. Once this shell reaches ignition temperature, calculations show that the entire shell undergoes fusion within a fraction of a second. For simplicity, assume that the shell has a mass of 2 $M_\odot$, with ^4He making up half the nuclei and ^{28}Si the other half. The reaction is $^{28}Si + ^4He \rightarrow\ ^{32}S + \gamma$. Calculate the total energy released when the shell ignites and compare this with the value of about 10^{44} J required for a supernova explosion. (The masses of the nuclei are 27.9769 amu for ^{28}Si, 31.9721 amu for ^{32}S, and 4.0026 amu for ^4He.)

First compute the energy released per reaction for $^{28}Si + ^4He \rightarrow\ ^{32}S + \gamma$. The mass before the reaction is 27.9769 + 4.0026 = 31.9795 amu, while after the reaction the mass is 31.9721 amu. Thus 0.0074 amu of mass is lost (converted to energy) in each reaction (Equation 16-16):

$$\Delta E / \text{reaction} = 0.0074 \, m_H c^2$$
$$= 0.0074 \, (1.67 \times 10^{-27} \text{ kg}) (3.0 \times 10^8 \text{ m/s})^2$$
$$= 1.11 \times 10^{-12} \text{ J}$$

To estimate the number of reactions that take place assume that there are an equal number of He and Si nuclei, the number of reactions is found by dividing the total mass in the shell by the sum of the He and Si masses:

$$\text{\# of reactions} = 2 \, (2.0 \times 10^{30} \text{ kg}) / [(28 \text{ amu} + 4 \text{ amu}) (1.67 \times 10^{-27} \text{ kg/amu})]$$
$$= 7.5 \times 10^{55}$$

The total energy available is

$$E = (\Delta E / \text{reaction}) \times (\text{\# of reactions})$$
$$= (1.11 \times 10^{-12} \text{ J}) \times (7.5 \times 10^{55})$$
$$= 8.3 \times 10^{43} \text{ J}$$

A supernova explosion requires an energy of 10^{44} J, so the Si burning model is realistic.

18-13. Sketch the analogs to Figure 18-9 for the case of a cloud of hot gas falling in a spherical shell onto an even hotter star.

For hot gas falling onto a hotter star we would see a line profile analogous to that of Figure 18-9B (which is for an expanding cloud), but rather than the blueshift of the absorption line we would see a redshift -- since the infalling material is moving away from us. [See also Problem 18-8.]

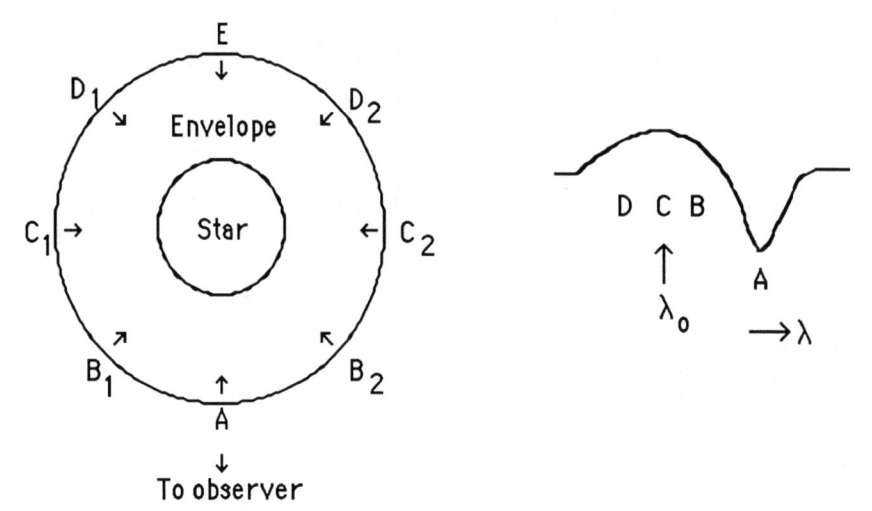

18-14. An O supergiant's mass can range from 15 to 40 $M_\odot$. Solve the mass function for Cyg X-1 for the mass of the X-ray source for this range of companion masses.

[*Note:* The equation in Section 18-6A for the mass function is incorrect. The equation *should have P instead of P^2 on the right hand side.]*

[*Note:* The easiest way to solve this equation is to use a spreadsheet. Input values of M_x from 1 to 40 $M_\odot$ into the 3rd order equation, and by trial and error (inspection) solve the equation. Of course, this problem could also be solved "by hand", or, more simply, the student could use $i = 90°$ as in Section 18-6A of the text and demonstrate that a value of $M_c \approx 33\ M_\odot$ is consistent with a M_x of 16 $M_\odot$.]

The mass function is given by Equation 18-6A (see also Section 12-3B) [Note the factor of P instead of P^2 as in the text.]:

$$f(M_x, M_c) = (M_x \sin i)^3 / (M_x + M_c)^2 = P(V_c \sin i)^3 / 2\pi G$$

where M_x is the mass of the x-ray source (Cyg X-1) and M_c is the mass of the companion (the star HD226868), with $15\ M_\odot < M_c < 40\ M_\odot$.

According to the text (Section 18-6A), the observed orbital period of Cyg X-1 is:

$P = 5.6$ days $= 4.8 \times 10^5$ s,

the observed projected orbital velocity of the companion is

$V_c \sin i = 76 \pm 1$ km/s $= 7.6 \times 10^4$ m/s,

and the mass function of Cyg X-1 is

$f(M_x, M_c) = 0.25 \pm 0.01\ M_\odot$.

[Let's check this

$$f(M_x, M_c) = P(V_c \sin i)^3 / 2\pi G$$
$$= (4.8 \times 10^5\ \text{s})(7.6 \times 10^4\ \text{m/s})^3 / 2\pi(6.67 \times 10^{-11}\ \text{N m}^2/\text{kg}^2)$$
$$= 5.0 \times 10^{29}\ \text{kg} = 0.25\ M_\odot \qquad \text{--- this checks}\ ]$$

The equation for the mass function (Equation 18-4) becomes

$$0.25\ M_\odot = (M_x \sin i)^3 / (M_x + M_c)^2$$

We do not know the inclination of the orbit, i, to our line of sight, so let's take a range of values. In reality, $i \neq 90°$ since we do not see eclipses and $i \neq 0°$ since that would imply $V_c \sin i = 0$. Let's take an approximate value of $i = 45°$,

For $M_c = 15\ M_\odot$, masses in solar masses

$$0.25 = (M_x \sin 45°)^3 / (M_x + 15\ M_\odot)^2 \approx (0.71\ M_x)^3 / (M_x + 15)^2$$
$$0.25\ M_x^2 + 7.5\ M_x + 56.25 = 0.35\ M_x^3$$
$$0.35\ M_x^3 - 0.25\ M_x^2 - 7.5\ M_x - 56.25 = 0$$
$$M_x \approx 7.0\ M_\odot \quad \text{(solved numerically)}$$

The x-ray object would be a black hole in this scenario since the maximum mass of a neutron star is $\approx 3\ M_\odot$.

For $M_c = 40 M_\odot$

$$0.25 = (M_x \sin 45°)^3 / (M_x + 40)^2$$

$$0.35 M_x^3 - 0.25 M_x^2 - 20 M_x - 400 = 0$$

$$M_x \approx 12.5 M_\odot \quad \text{(solved numerically)}$$

The x-ray object would also be a black hole in this scenario since the maximum mass of a neutron star is $\approx 3 M_\odot$.

Clearly the inferred mass for the compact x-ray emitting object depends on the orbital inclination and the mass of the companion star.

18-15. (a) Using the information provided in Section 18-5E and assuming a constant expansion rate, calculate the proper motion of the expanding nebula from Supernova 1987A.
Using Equation 18-3, a distance to Supernova 1987A of 52,000 pc (Section 18-5E) and expansion velocity of 17,000 km/s *(from the third edition of the text!)*:

$d = v_r / 4.74 \mu$ where v_r is in km/s, d in parsec, μ in arcsec per year

$$\mu = v_r / 4.74 \, d$$
$$= (1.7 \times 10^4) / [4.74 \times (5.2 \times 10^4)] \text{ ''/year}$$
$$= 6.9 \times 10^{-2} \text{ ''/year} = 0.069 \text{ ''/year}$$

Note, if one used a typical expansion velocity of Type II supernovae of $\approx 50,000$ km/s (Section 18-5B), the proper motion would be $\approx 2 \times 10^{-2}$ arcsec/year.

(b) How long must astronomers wait before the nebula has a diameter of 1"? Approximately what date will this be?
For a diameter of 1" (radius 0.5"), astronomers must wait

$$(0.5 \text{ ''}) / (0.069 \text{ ''/yr}) = 7.25 \text{ years} = 7 \, 1/4 \text{ years}$$

until mid 1994 (7 1/4 years after the explosion in the first quarter of 1987)

18-16. (a) Ignoring interstellar absorption, how bright (apparent magnitude) would a Type II supernova be if it exploded 1000 pc from the Earth? Compare this with the apparent magnitude of Venus (m = -3).
Using Equation 11-6, and an absolute magnitude of M = -18 for Type II supernovae (Table 18-3):

$$m = M + 5 \log(d_{pc}) - 5$$
$$= (-18) + 5 \log(1000) - 5$$
$$= -8$$

This is five magnitudes (or a factor of 100 in brightness) brighter than Venus. The supernova would easily be seen even during daylight, and would dominate the night sky (only the Moon would be brighter).

(b) Ignoring interstellar absorption, how distant could a Type II supernova be and still be visible with the unaided eye (m = 6)? Compare this with the diameter of the Galaxy.

Using Equation 11-6:
$$m = M + 5 \log(d_{pc}) - 5$$
$$d_{pc} = 10^{(m - M + 5)/5}$$
$$= 10^{(6 - (-18) + 5)/5} = 10^{5.8}$$
$$d = 6.3 \times 10^5 \text{ pc} = 630 \text{ kpc}$$

The supernova could be seen to a distance of 630 kpc. This is much larger than our Galaxy which is ≈ 100 kpc in diameter.

(c) Astronomers estimate that a supernova should occur on the average every 25 to 50 years in our Galaxy. Given that the last supernova in our Galaxy visible with the naked eye occurred nearly 400 years ago, comment on your result in (b).

Interstellar absorption cannot be ignored. Type II supernovae occur in the plane of the Galaxy, usually near spiral arms, since they are associated with low age massive stars located near the sites of their birth. The fact that we have not had a naked eye supernova in a period 8 to 16 times longer than the expected interval between explosions implies that interstellar absorption must be important.

Consider, as an example, a Type II supernova located 10 kpc away, assume 1 magnitude of extinction (absorption) per kpc (see Section 14-2 for a discussion of interstellar absorption). The supernova would thus appear 10 magnitudes fainter than it would if there were no interstellar absorption. The apparent magnitude would be (Equation 11-15)
$$m = M + 5 \log(d_{pc}) - 5 + A$$
$$= -18 + 5 \log(10{,}000) - 5 + 10 = 7$$

This supernova would be just fainter than the naked eye visibility limit. In actuality, it probably would appear even fainter since in the plane of the Galaxy localized clouds could result in even greater absorption.

(d) Ignoring interstellar absorption, how distant could a Type II supernova be and still be observed visually with a 16-inch telescope (m = 14)? Compare this with the distance to the Andromeda Galaxy (0.7 Mpc) and the nearest large clusters of galaxies, the Virgo cluster (15.7 Mpc).

Using Equation 11-6 (or Equation 11-15 with A = 0):
$$m = M + 5 \log(d_{pc}) - 5$$
$$d_{pc} = 10^{(m - M + 5)/5}$$
$$= 10^{(14 - (-18) + 5)/5} = 10^{7.4}$$
$$d = 2.5 \times 10^7 \text{ pc} = 25{,}000 \text{ kpc} = 25 \text{ Mpc}$$

A Type II supernova could thus be seen easily in the Andromeda Galaxy and without difficulty in galaxies in the Virgo cluster.

18-17. The spiral galaxy NGC 925 is one target of the HST Key Project on the Extragalactic Distance Scale. The following are V-band apparent magnitudes of selected Cepheids:

Cepheid Number	Period (Days)	V
5	48.5	23.68
8	37.3	24.67
15	30.1	24.85
36	20.2	25.35
77	10.8	25.90

Find the distance modulus and the distance to NGC 925 if $A_V = 0.42$.

We can correct the apparent magnitude for the interstellar extinction using $A_\lambda = 0.42$.

We can calculate the absolute magnitude from the period-luminosity relation for Cepheid variable stars (Equation 18-2; *note, the equation in the text is incorrect, the correct equation is given in the Key Equations at end of the chapter*),

$$M_V = -2.76 (\log P_{[days]} - 1.0) - 4.16$$

The distance modulus (Section 11-3), corrected for extinction, is given by Equation 15-5, and is added to the table,

$$(m_\lambda - A_\lambda) - M_\lambda = 5 \log(d) - 5$$

Cepheid#	Period (Days)	$V = m_V$	$m_V - A_V$	M_V	distance modulus
5	48.5	23.68	23.26	-6.05	29.31
8	37.3	24.67	24.25	-5.36	29.61
15	30.1	24.85	24.43	-5.48	29.91
36	20.2	25.35	24.93	-5.00	29.93
77	10.8	25.90	25.48	-4.25	29.73

The average (extinction corrected) distance modulus from these five stars is 29.70, so the distance to NGC 925 is

$$5 \log(d) - 5 = 29.70$$
$$d = 8.71 \times 10^6 \text{ pc} = 8.7 \text{ Mpc}$$

Chapter 19: Galactic Rotation: Stellar Motions

19-1. The Fe**I** emission lines (at 441.5 and 444.2 nm) in a comparison spectrum are located 15.00 and 15.43 mm, respectively, from an arbitrary reference point. If a stellar Ca**I** line (of rest wavelength 442.5 nm) is measured to be at 15.27 nm,

(a) what is the observed wavelength of the Ca**I** line?

To compute the observed wavelength of the Ca**I** line we assume that the relation between the wavelength and the observed position of the line in the spectrum is linear (in distance from the reference point) and interpolate the position of the Ca**I** line between the reference Fe**I** lines.

(444.2 nm $_{FeI}$ - 441.5 nm $_{FeI}$) / (15.43 mm $_{FeI}$ - 15.00 mm $_{FeI}$)

= (X nm $_{CaI}$ - 441.5 nm $_{FeI}$) / (15.27 mm $_{CaI}$ - 15.00 mm $_{FeI}$)

2.7 nm / 0.43 mm = (X - 441.5) nm / 0.27 mm

X = 441.5 nm + 1.7 nm

= 443.2 nm   --- the observed wavelength of the Ca**I** line

(b) what is the radial velocity of this star?

The radial velocity can be calculated from the non-relativistic Doppler formula (λ is the observed wavelength, λ_o is the rest frame wavelength) (Equation 19-1):

$$v_r = (\Delta\lambda / \lambda_o)\, c = [(\lambda - \lambda_o)/\lambda_o]\, c$$

$$= [(443.2\ \text{nm} - 442.5\ \text{nm}) / 442.5\ \text{nm}]\ (3.0 \times 10^5\ \text{km/s})$$

$$= (0.00158)(3.0 \times 10^5\ \text{km/s}) = 475\ \text{km/s}$$

Since the line is redshifted (to longer wavelengths), the star is receding with a velocity of 475 km / s.

19-2. By making the appropriate conversion in units, show that Equation 19-3 follows from Equation 19-2. Equation 19-2 relates the transverse velocity to the observed proper motion and distance of a star from an observer:

$$v_t = d \sin(\mu) \approx \mu\, d$$

where v_t is in pc/yr, μ is in radians/yr, and d is in pc.

Convert v_t from pc/yr to km/s,

1 pc = 3.086×10^{13} km, and 1 yr = 3.16×10^7 s.

Thus 1 pc/yr = 3.086×10^{13} km / 3.16×10^7 s = 9.77×10^5 km / s.

To convert μ from radians/yr to arcsec/yr, note 1 radian = 206,265".

Using these conversions:

$$v_t\ [\text{km/s}] = 9.77 \times 10^5\ v_t\ [\text{pc/yr}] = 9.77 \times 10^5\ \mu\ [\text{radians/yr}]\ d\ [\text{pc}]$$

$$= (9.77 \times 10^5)(1/206{,}265)\ \mu\ ["/\text{yr}]\ d\ [\text{pc}]$$

$$= 4.74\ \mu\ ["/\text{yr}]\ d\ [\text{pc}] = 4.74\ \mu\ ["/\text{yr}] / \pi\ ["]$$

where we have used the parallax of a star in seconds of arc is the reciprocal of the

distance in parsecs (Equation 11-2), $d\,[pc] = 1/\pi\,['']$.
Thus we have Equation 19-3:
$$v_t\,[km/s] = 4.74\,\mu['']\,d[pc] = 4.74\,(\mu['']/\pi[''])$$

19-3. A star located 90° from the solar antapex on the celestial sphere is at rest in the LSR 10 pc from the Sun. As seen from the Sun,

(a) by what angle (in arcsec) will this star appear to move on the celestial sphere in ten years?

The star is at rest w.r.t. the LSR, but the Sun has a speed of 19.5 km/s w.r.t. the LSR (Section 19-2). The star's proper motion (due to the Sun's relative motion) is given by (Equation 19-3):
$$\mu = 4.74\,d_{pc}/v_{\odot[km/s]} = 4.74\,(10)/19.5 = 2.43\,''/yr.$$
The star will move 24.3" in ten years.

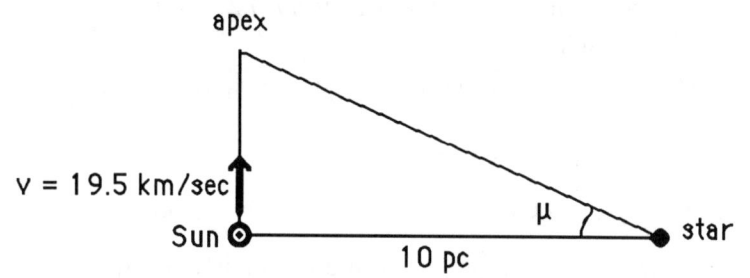

(b) in what direction will the star appear to move?

The star's apparent motion will be opposite to that of the Sun, or toward the solar antapex.

19-4. The star Delta Tauri is a member of the Taurus moving group. It is observed to have a proper motion of 0.115"/year and a radial velocity of 38.6 km/s and to lie 29.1° from the convergent point of the group.

(a) What is this star's parallax?

For the moving cluster method of determining stellar distance, we can use Equation 19-6
$$\pi'' = 4.74\,\mu''/(v_r\,\tan(\theta))$$
$$= (4.74) \times (0.115)/(38.6\,\tan(29.1°))$$
$$= 2.5 \times 10^{-2}\,'' = 0.025''$$

(b) What is its distance in parsecs?

From Equation 11-2: $d_{pc} = 1/\pi'' = 1/0.025 = 39.4$ pc

(c) Another star belonging to the same group lies only 20° from the convergent point. What are *its* proper motion and radial velocity?

For a star of the same group (i.e. the same distance) but with $\theta = 20°$, we find its radial velocity by assuming that all stars in the cluster have the same space velocity, V. Hence (Equation 19-5), $V = v_r/\cos(\theta)$. From the star in part (a), the space velocity of the cluster is:
$$V = (38.6\,km/s)/\cos(29.1°)$$

$$= 44.2 \text{ km/s}$$

Thus the radial velocity of a star 20° from the convergent point is:

$$v_r = V \cos(\theta)$$
$$= 44.2 \cos(20°) \text{ km/s}$$
$$= 41.5 \text{ km/s}$$

The proper motion can be found from Equation 19-6:

$$\mu'' = \pi'' v_r \tan(\theta) / 4.74$$
$$= (0.025)(41.5) \tan(20°) / 4.74$$
$$= 8.0 \times 10^{-2} \text{''} = .080 \text{''/yr}$$

19-5. Refer to the data given in Problem 19-4. Assume that a probable error of ±0.005"/year is associated with the proper motion of Delta Tauri. If we independently measure the trigonometric parallax of this star (with a probable error of ±0.005"), what are the uncertainties in parsecs in the distances determined from these two separate parallaxes?

If the proper motion measurement of Delta Tauri in Problem 19-4 is 0.115" ± 0.005", we have a 4.3% uncertainty in the measurement. Assume that the error in measuring the radial velocity or the angle to the convergent point are much less than 4.3%. Then, since the distance $\propto$ (proper motion)$^{-1}$ in the moving cluster method, the error in the distance determination will be 4.3%, or d = 39.4 ± 1.7 pc.

If the trigonometric parallax is measured to be π = 0.025" ± 0.005", we have a 20% uncertainty in the measurement. Since the distance $\propto$ (parallax)$^{-1}$ in the parallax method of determining distances, the error in the distance determination will be 20%, or d = 39.4 ± 7.9 pc.

Clearly the moving cluster method provides superior results in this case.

19-6. Assume that the mass of our Galaxy is 1.5×10^{11} solar masses and that it is *all* concentrated in a point at the galactic center.

(a) Plot the rotation curve (Θ versus R), with appropriate units and exemplary values along each axis, for this Keplerian case.

For the ideal case of circular Keplerian orbits, the orbital speed is given by (Section 19-4A):

$$V \equiv \Theta = (GM/R)^{1/2}$$

where R is the distance from the galactic center and M is the mass of the galaxy (concentrated in a point at the galactic center in this example)

$$M = 1.5 \times 10^{11} M_\odot = 2.99 \times 10^{41} \text{ kg}.$$

To express R in kpc and Θ in km/s, some unit conversions are necessary:

$$1 \text{ kpc} = 3.086 \times 10^{19} \text{ m}$$

gives $\Theta = \{(6.67 \times 10^{-11} \text{ N m}^2/\text{kg}^2)(2.99 \times 10^{41} \text{ kg}) / [(3.086 \times 10^{19} \text{ m})(R_{kpc})]\}^{1/2}$

$$= 8.04 \times 10^5 / (R_{kpc})^{1/2} \text{ m/s}$$

$$= 8.04 \times 10^2 / (R_{kpc})^{1/2} \text{ km/s} \qquad \text{where R is in kpc.}$$

Computing the circular orbital velocity for various distances from the galactic center:

R [kpc]	1	2	3	4	5	6	7	8	8.5	9	10	15	20
Θ [km/s]	804	569	464	402	360	328	304	284	276	268	254	208	180

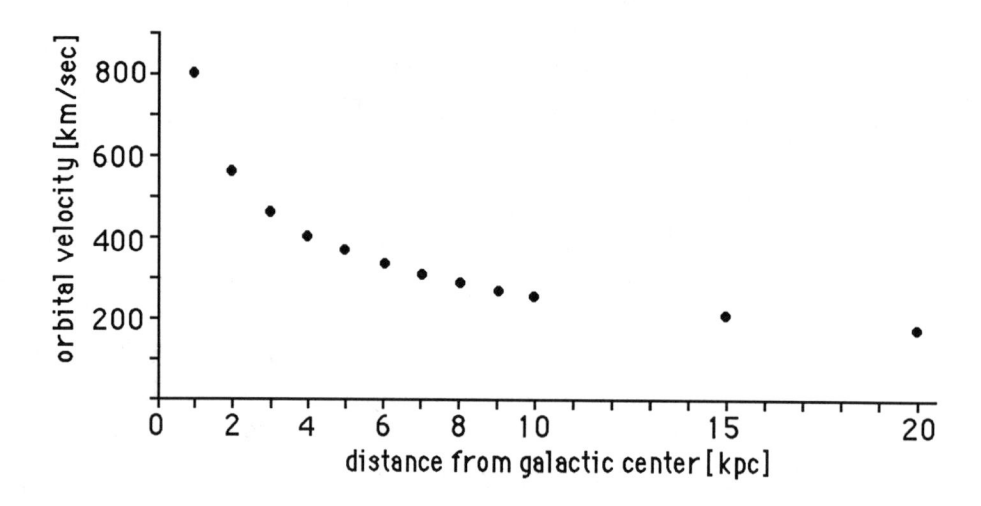

(b) Indicate the rotation period at R = 5, 8.5, and 20 kpc.
The rotation period is given by:

$$P = 2\pi R / V \equiv 2\pi R / \Theta.$$

P is conveniently expressed in yr, R in kpc, and Θ in km/s. Applying the appropriate conversion factors:

$$P_{yr} = 2\pi (3.086 \times 10^{16} R) / (3.16 \times 10^7 \Theta) = 6.14 \times 10^9 R_{kpc} / \Theta \text{ km/s}$$

At R = 5 kpc, P = 8.5×10^7 yr = 85 million yr

At R = 8.5 kpc, P = 1.9×10^8 yr = 189 million yr

At R = 20 kpc, P = 6.8×10^8 yr = 680 million yr.

The rotation period calculated for 8.5 kpc is smaller (the orbital velocity is larger) than the accepted value, since the Galaxy's mass distribution is not concentrated at the galactic center.

(c) What is the speed of escape from R = 8.5 kpc?
The escape speed is given by (Equation 2-8; see also Problem 14-5b)

$$v_{esc}^2 = (2GM/R) = 2\Theta^2$$

so $v_{esc} = (\sqrt{2})(GM/R)^{1/2} = (\sqrt{2})\Theta$

so at 8.5 kpc, $v_{esc} = (\sqrt{2}) \, 276 \text{ km/s} = 390 \text{ km/s}.$

19-7. A star in a galactic orbit of eccentricity 0.8 and semimajor axis 7 kpc moves through the solar neighborhood on its outward journey in the galactic plane. What is the velocity of this star with

respect to the LSR? Assume the Galaxy is a point mass (Keplerian motion).
The speed of a star in a Keplerian orbit is given by (Equation 1-28):

$$v^2 = G(M_{galaxy} + m_{star})(2/r - 1/a),$$

where the radial and tangential velocities are (Equations 1-26a and 1-26b):

$$v_r = (2\pi a / P)(e \sin(\theta))(1 - e^2)^{-1/2}$$

$$v_\Theta = (2\pi a / P)(1 + e \cos(\theta))(1 - e^2)^{-1/2}$$

the orbital period is given by Kepler's Harmonic Law (Equation 1-24):

$$P^2 = 4\pi^2 a^3 / G(M_{galaxy} + m_{star})$$

and the value θ can be calculated from Equation 1-3:

$$r = a(1 - e^2)/(1 + e\cos(\theta)).$$

Since $M_{galaxy} \gg m_{star}$, we can ignore m_{star} in the above equations.

Assume $M_{galaxy} = 1.0 \times 10^{11} M_\odot = 2.99 \times 10^{41}$ kg.

It is clear that the results will strongly depend on the mass of the galaxy that is selected, and the adopted distance of the Sun from the galactic center.

Expressing Kepler's Harmonic Law with P in yr, a in A.U., and M in $M_\odot$:

$$P^2 = (7000 \text{ pc} \times 206{,}265 \text{ A.U.} / \text{pc})^3 / (1.0 \times 10^{11} M_\odot)$$

$$= 3.01 \times 10^{16} \text{ yr}^2$$

$$P = 1.73 \times 10^8 \text{ yr}$$

Compute the value of θ, when the star's orbit crosses the Sun's galactic orbit:

$$8.5 \text{ kpc} = 7.0 \text{ kpc}(1 - 0.8^2)/(1 + 0.8\cos(\theta))$$

$$(1 + 0.8\cos(\theta)) = 2.96 \times 10^{-1}$$

$$\cos(\theta) = -0.88$$

$$\theta = 152°.$$

Computing the velocities:

$$v^2 = (6.67 \times 10^{-11} \text{ N m}^2/\text{kg}^2) \times (1.99 \times 10^{41} \text{ kg}) \times$$
$$\{[2/(2.62 \times 10^{20} \text{ m})] - [1/(2.16 \times 10^{20} \text{ m})]\}$$

$$= 3.99 \times 10^{10} \text{ m}^2/\text{s}^2$$

$$v = 2.00 \times 10^5 \text{ m/s} = 2.00 \times 10^2 \text{ km/s} = 200 \text{ km/s}$$

$$v_r = [2\pi(2.16 \times 10^{20} \text{ m})/(1.73 \times 10^8 \text{ yr} \times 3.16 \times 10^7 \text{ s/yr})]$$
$$\times (0.8 \times \sin(152°))(1 - 0.8^2)^{-1/2}$$

$$= (2.48 \times 10^5 \text{ m/s}) \times (3.76 \times 10^{-1}) \times (1.67)$$

$$= 1.56 \times 10^5 \text{ m/s} = 1.56 \times 10^2 \text{ km/s} = 156 \text{ km/s}$$
$$v_\Theta = [2\pi (2.16 \times 10^{20} \text{ m}) / (1.73 \times 10^8 \text{ yr} \times 3.16 \times 10^7 \text{ s/yr})]$$
$$\times (1 + 0.8 \cos(152°)) (1 - 0.8^2)^{-1/2}$$
$$= (2.48 \times 10^5 \text{ m/s}) \times (2.93 \times 10^{-1}) \times (1.67)$$
$$= 1.21 \times 10^5 \text{ m/s} = 1.21 \times 10^2 \text{ km/s} = 121 \text{ km/s}$$

As a check, does $v_r^2 + v_\Theta^2 = v^2$?
$$[(156)^2 + (121)^2]^{1/2} = 197 \approx 200 \quad \text{(it checks!)}$$

In the Sun's neighborhood, the LSR has $v_r = 0$, and $v_\Theta = 220$ km/s. The star thus has a velocity relative to the LSR of:

$v_r = 156$ km/s (toward us) and $v_\Theta = 99$ km/s (away from us).

19-8. Use Figure 19-10 to calculate the Galaxy's interior mass to the distance out to which the curve extends. (*Hint:* The Galaxy is *not* a point mass; the motion is *not* Keplerian!)

Using the rotation curve for the Milky Way Galaxy, we see that at a distance from the galactic center of 17 kpc, the rotational velocity is 235 km/s. The galactic mass interior to 17 kpc can be estimated by using Equation 14-6 (see also Section 19-4A):

$$M = V^2 R / G$$
$$= (2.35 \times 10^5 \text{ m/s})^2 \times (17 \times 3.086 \times 10^{19} \text{ m}) / (6.67 \times 10^{-11} \text{ N m}^2/\text{kg}^2)$$
$$= 4.29 \times 10^{41} \text{ kg} = 2.1 \times 10^{11} \text{ M}_\odot$$

19-9. The star BS 1828 has a proper motion of 0.24"/year along position angle 48° (east of north) and a parallax of 0.012". The H_β line ($\lambda_o = 486.1$ nm) appears at $\lambda = 485.9$ nm. What is the magnitude of the star's space velocity and what angle does the velocity make it to the line-of-sight (which points away from the Sun)?

Computing the tangential velocity of BS 1828 using Equation 19-3:
$$v_t = 4.74 (\mu''/\tau'') = 4.74 (0.24 / 0.012) = 94.8 \text{ km/s}$$

The radial velocity is computed from the Doppler shift (Equation 19-1):
$$v_r = (\Delta\lambda / \lambda_o) c = [(\lambda - \lambda_o) / \lambda_o] c$$
$$= [(485.9 \text{ nm} - 486.1 \text{ nm}) / 486.1 \text{ nm}] \, 3.0 \times 10^5 \text{ km/s}$$
$$= -1.23 \times 10^2 \text{ km/s} = -123 \text{ km/s} \quad \text{-- toward us}$$

To calculate the space velocity, use the Pythagorean theorem (Equation 19-4):
$$v_{space}^2 = v_r^2 + v_t^2 = (94.8 \text{ km/s})^2 + (123 \text{ km/s})^2$$
$$v_{space} = 155 \text{ km/s}$$

The angle of motion relative to the line of sight away from the Sun is given by:
$$\tan(\theta) = v_t / v_r = (94.8 \text{ km/s}) / (-123 \text{ km/s}) = -0.771$$
$$\theta = 142° \text{ (or 38° from a line pointing toward the Sun)}$$

19-10. Determine the proper motion (relative to the LSR) in arcseconds per year of a star in circular motion about the galactic center 4 kpc from the Sun and at a galactic longitude of 60°. Use the rotation curve given in Figure 19-10. (Expect your answers to be small!)

From the diagram below (like Figure 19-7), we can find the distance of the star

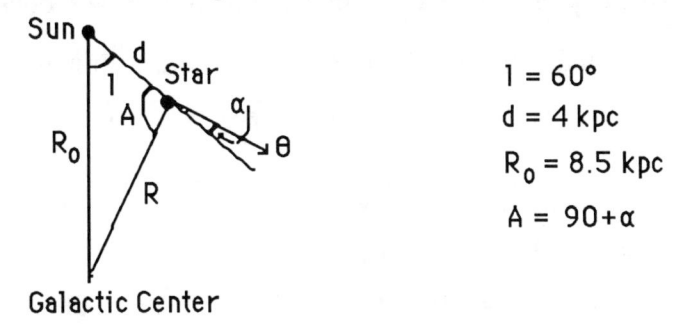

$l = 60°$
$d = 4$ kpc
$R_0 = 8.5$ kpc
$A = 90+\alpha$

from the galactic center, R, by using the law of cosines (distances in kpc):

$R^2 = R_0^2 + d^2 - 2 R_0 d \cos(l)$
$ = (8.5 \text{ kpc})^2 + (4 \text{ kpc})^2 - 2 (8.5 \text{ kpc}) (4 \text{ kpc}) \cos(60°)$
$ = 54.25 \text{ kpc}^2$
$R = 7.37$ kpc

From the galactic rotation curve, Figure 19-10, a star at this distance from the galactic center would have a circular orbital velocity of $\Theta = 230$ km / s.

The angle, α, between the star's circular velocity vector and our line of sight to the star can also be found from the law of cosines:

$R_0^2 = d^2 + R^2 - 2 d R \cos(90+\alpha)$
$8.5^2 = 4^2 + 7.37^2 - 2 (4) (7.37) \cos(90+\alpha)$
$\cos(90+\alpha) = -3.28 \times 10^{-2}$
$90+\alpha = 91.88°$
$\alpha = 1.88°$

The star's radial speed, v_r relative to the LSR is given by Equation 19-7:

$v_r = \Theta \cos(\alpha) - \Theta_0 \sin(l)$

where Θ_0 (Θ) is the LSR's (star's) circular orbital velocity about the galactic center,
so $v_r = (230 \text{ km / s}) \cos(1.88°) - (220 \text{ km /s}) \sin(60°)$
$ = 39$ km / s

and the star's tangential velocity relative to the LSR is given by Equation 19-10:

$v_t = \Theta \sin(\alpha) - \Theta_0 \cos(l)$
$ = (230 \text{ km / s}) \sin(1.88°) - (220 \text{ km /s}) \cos(60°)$
$ = -102$ km / s

Note that the values of v_t and v_r depend upon the orbital velocity of the star -- e.g. how well we have read the value off the rotation curve plot. But since the angle between the star's orbital velocity and our line of sight is small ($\approx 2°$), the first term (from the star's velocity) in the expression for v_t is small compared to the second term (from the LSR's velocity), so even an error of 20 or 30 km / s in our estimate of Θ will produce only a few percent error in v_t. However, v_r is more sensitive to the value we select for Θ.

The star's proper motion is given by Equation 19-3:
$$\mu["/yr] = v_t [km/s] / (4.74 \, d[pc])$$
$$= (102) / (4.74 \times 4000) = 5.4 \times 10^{-3} = 0.0054 \, "/yr$$
The star's proper motion is thus very small ($\approx 0.5"$ per century) -- due to the fact that the star is distant from the Sun.

19-11. The distance from the convergent point to the center of the Hyades cluster is 29.9°, and the cluster's radial velocity is 39.1 km/s. Use this information and that in the chapter to calculate a moving-cluster distance to the Hyades. If the error in the radial velocity determination is ±0.2 km/s, what is the error in the distance?

Using the moving cluster method to determine the distance to the Hyades (Equation 19-6) for d in pc, v_r in km / s, and μ in arcsec per yr:

$$d_{pc} = 1/\pi" = v_r \tan(\theta) / (4.74 \, \mu")$$
$$= (39.1)(\tan(29.9°)) / (4.74 \, \mu")$$
$$= 4.74 / \mu"$$

Neither the text nor the problem provide a value of the proper motion, μ. We can work backwards, and take the distance of 44.3 pc reported in the text, and compute the proper motion that must have been used in this calculation.
$$\mu" = 4.74 / 44.3 = 0.107"$$

If the radial velocity measurement is 39.1 ± 0.2 km / s, the error is 0.51%. The corresponding error in the distance (since distance $\propto$ radial velocity) is 0.51%, so the calculated distance is 44.3 ± 0.2 pc.

19-12. Derive Equation 19-6 for moving cluster parallaxes from Equations 19-3 and 19-5.
Equation 19-3:
$$v_t = 4.74 \, \mu" / \pi"$$
and Equation 19-5:
$$V = v_r / \cos(\theta) \text{ or } v_r = V \cos(\theta).$$
Since, from Equation 19-4,

$$V^2 = v_r^2 + v_t^2 = V^2 \cos^2(\theta) + v_t^2$$
$$v_t^2 = V^2(1 - \cos^2(\theta)) = V^2 \sin^2(\theta)$$
$$v_t = V \sin(\theta)$$

So, from Equation 19-3 and substituting,
$$\pi" = 4.74\, \mu"/v_t = 4.74\, \mu"/(V\sin(\theta)) = 4.74\, \mu"/[(v_r/\cos(\theta))\sin(\theta)]$$
$$= 4.74\, \mu"/[v_r \tan(\theta)] \quad \text{--- which is Equation 19-6}$$

19-13. Sections 14-5 and 19-4 state that the halo of our Galaxy contains nonluminous matter. If we can't see it, how do we know it's there?

Using the galactic rotation curve (Figure 19-10) and Kepler's third law we can estimate the mass of the Galaxy internal to that radius, including the halo. This dynamic mass is then compared to the luminous mass of the visible globular clusters and stars in the halo. The dynamic mass is greater than the luminous material, implying that there must be some non-luminous matter in the halo.

19-14. Barnard's star has a radial speed of -108 km/s, proper motion 10.34"/year, and parallax 0.546".

(a) What is the distance to Barnard's star in parsecs? In kilometers?

The distance is given by Equation 11-2:
$$d\,[\text{pc}] = 1/\pi["] = 1/0.546$$
$$d = 1.83\ \text{pc}$$
$$= 1.83\ \text{pc} \times (3.086 \times 10^{13}\ \text{km/pc}) = 5.65 \times 10^{13}\ \text{km}$$

(b) What is the tangential speed of Barnard's star?

The tangential speed is given by Equation 19-3:
$$v_t\,[\text{km/s}] = 4.74\, \mu["]/\pi["]$$
$$v_t = 4.74 \times 10.34 / 0.546\ \text{km/s}$$
$$= 89.8\ \text{km/s}$$

(c) What is the space velocity of Barnard's star and the angle that the space velocity makes with the line-of-sight?

The space velocity, V, is given by Equation 19-4:
$$V^2 = v_r^2 + v_t^2$$
$$= (-108\ \text{km/s})^2 + (89.8\ \text{km/s})^2 = 1.97 \times 10^4\ \text{km}^2/\text{s}^2$$
$$V = 140\ \text{km/s}$$

The angle to the line of sight is given by (Section 19-1D)
$$\tan(\theta) = v_t/v_r = 89.8\ \text{km/s}/(-108\ \text{km/s}) = -0.831$$
$$\theta = -39.7°$$

(d) At its closest approach, how distant in parsecs and light years will Barnard's star be? Compare this to the current distance to α Centauri.

The distance of closest approach is

$$X = (5.65 \times 10^{13} \text{ km}) \sin(39.7°) = 3.61 \times 10^{13} \text{ km}$$
$$= (3.61 \times 10^{13} \text{ km}) / (3.086 \times 10^{13} \text{ km/pc}) = 1.17 \text{ pc}$$
$$= 1.17 \text{ pc} \times (3.26 \text{ ly/pc}) = 3.81 \text{ ly}$$

Alpha Centauri is presently 1.31 pc distant, so Barnard's star will be closer in 10,000 years than is α Centauri today.

(e) In how many years will Barnard's star be its closest to the Sun?

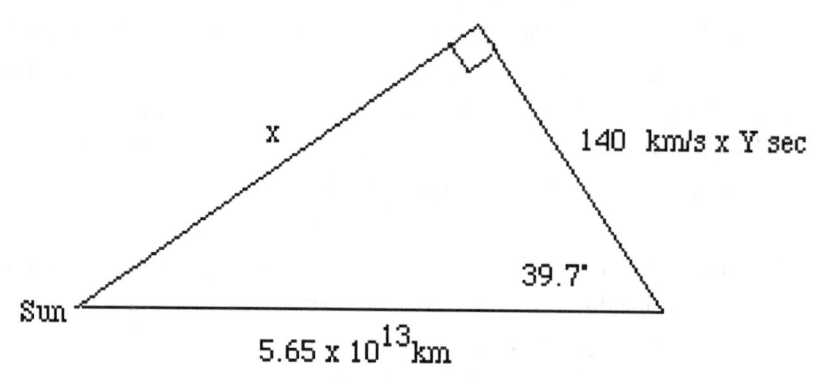

Closest approach will occur when the angle between its present space velocity vector and the line of sight vector is a right angle,

$$\cos(39.7°) = (140 \text{ km/s} \times Y \text{ sec}) / 5.65 \times 10^{13} \text{ km}$$
$$Y = 3.1 \times 10^{11} \text{ s} = (3.1 \times 10^{11} \text{ s}) / (3.16 \times 10^7 \text{ s/yr})$$
$$= 9,800 \text{ yr}$$

(f) Barnard's star currently has an apparent visual magnitude of 9.54. What will be its magnitude at closest approach?

We can use Equation 11-6 to write (time 1 = at closest approach, time 2 = today)

$$m_1 - m_2 = 5 (\log(d_1) - \log(d_2)) = 5 \log(d_1 / d_2)$$
$$m_1 - 9.54 = 5 \log(1.17 \text{ pc} / 1.83 \text{ pc}) = -0.97$$
$$m_1 = 8.57$$

Barnard's star will still be too faint to be seen with the unaided eye, though it will be barely visible in binoculars from a dark location.

19-15. Calculate a rotation curve for a mass distribution of constant density, adding up to $M(R_\odot)$ at the Sun's orbit. Compare this model to the actual rotation curve.

[*Note:* In this problem and in Appendix Table A7-1, $R_\odot$ is used to define the Sun's distance from the center of the Galaxy. Students may find this confusing, since $R_\odot$ is also used as the Sun's radius. We will use R_0, as in the chapter, for the Sun's distance

228

from the center of the Galaxy to avoid this confusion.]

Following Section 19-4A (see also Equation 14-6) where this problem is worked out, the velocity, V(R), of a mass with a circular orbit of radius R about the galaxy would have a velocity

$$V(R) = (G M(R) / R)^{1/2}$$

where M(R) is the mass of the galaxy internal to R.
For a constant density, the mass within R is given by (Section 19-4A)

$$M(R) = (4/3) \pi R^3 \rho$$

This assumes the Galaxy's mass distribution is spherically symmetric, which it isn't.
Combining the two equations

$$V(R) = [(4/3) \pi G R^3 \rho / R]^{1/2}$$
$$= [(4/3) \pi G R^2 \rho]^{1/2}$$
$$= [(4/3) \pi G \rho]^{1/2} R \propto R$$

This shows that the velocity is linearly proportional to the distance from the galactic center. This motion is the same as rigid body rotation (ω = constant), but without the galaxy being a solid-body! The rotation speed would increase with R, unlike the Galaxy's rotation curve in Figure 19-10 which shows rigid body rotation only in the center (inner half kpc).

If we use Figure 19-10 (or Section 19-4B) to find the rotational velocity at the Sun's distance from the center of the Galaxy, $V(R_0 = 8.5 \text{ kpc}) = 220$ km/s, we can determine the Galaxy's mass internal to the Sun's orbit

$$M(R_0) = V^2 R / G$$
$$= (2.20 \times 10^5 \text{ m/s})^2 \times (8.5 \times 3.086 \times 10^{19} \text{ m}) / (6.67 \times 10^{-11} \text{N m}^2/\text{kg}^2)$$
$$= 1.9 \times 10^{41} \text{ kg} = 9.6 \times 10^{10} M_\odot \approx 10^{11} M_\odot$$

Solving for the constant density ρ,

$$\rho = M(R_0) / [(4/3) \pi (R_0)^3]$$
$$= (1.9 \times 10^{41} \text{ kg}) / [(4/3) \pi (8.5 \times 3.086 \times 10^{19} \text{ m})^3]$$
$$= 2.51 \times 10^{-21} \text{ kg/m}^3$$

or for M in solar masses and R in kpc,

$$\rho = M(R_0) / [(4/3) \pi (R_0)^3]$$
$$= (9.6 \times 10^{10} M_\odot) / [(4/3) \pi (8.5 \text{ kpc})^3]$$
$$= 3.7 \times 10^7 M_\odot / \text{kpc}^3$$

So,

$$V(R) = [(4/3) \pi G \rho]^{1/2} R$$
$$= [(4/3) \pi (6.67 \times 10^{-11} \text{N m}^2/\text{kg}^2) (2.51 \times 10^{-21} \text{ kg/m}^3)]^{1/2} R$$

$$= (8.37 \times 10^{-16} \text{ m/s}) R_{[m]}$$

If we express V in km/s and R in kpc instead of m, where 1 kpc = 3.086 × 10^{19} m

$$V(R) = (8.37 \times 10^{-19} \text{ km/s}) (3.086 \times 10^{19}) R_{[kpc]}$$

$$= (2.58 \times 10^1 \text{ km/s}) R_{[kpc]}$$

$$= (25.8 \text{ km/s}) R_{[kpc]}$$

Checking, this expression gives $V(R_0)$ = 219 km/s ≈ 200 km/s.

Note that we could solved the problem without solving for the mass internal to the Sun's orbit and the density. Simply, we could have taken the expression

$$V(R) = [(4/3) \pi G \rho]^{1/2} R$$

from above, and found the constant $[(4/3) \pi G \rho]^{1/2}$ by setting V(R) = 220 km/s at R = 8.5 kpc,

$$[(4/3) \pi G \rho]^{1/2} = V(R_0)/R_0 = (220 \text{ km/s})/(8.5 \text{ kpc}) = 25.9 \text{ km/s kpc}$$

so

$$V(R) = (220 \text{ km/s}) (R/R_0) = (25.9 \text{ km/s}) R_{[kpc]}$$

19-16. For l = 45°, we observe that v_r = +30 km/s. What are the values of R and d?

Using one of the equations derived from the Oort formulas (Equation 19-16),

$$v_r = A \, d \sin(2l)$$

where the Oort constant A is given by (Equation 19-14) and Section 19-4B

$$A = -(R_0/2) (d\omega/dR)_{R0} \approx 14 \text{ km/s·kpc}$$

Combining these equations and using the values from the question,

$$v_r = (14 \text{ km/s·kpc}) \, d \sin(2l)$$

$$30 \text{ km/s} = (14 \text{ km/s·kpc}) \, d \, (\sin(2 \times 45°))$$

$$d = 2.14 \text{ kpc}$$

To find the galactocentric distance, use Equation 19-15

$$v_r = -2 A (R - R_0) \sin(l)$$

where R0 = 8.5 kpc (Section 19-4B).

$$30 \text{ km/s} = -2 (14 \text{ km/s·kpc}) (R - 8.5 \text{ kpc}) \sin(45°)$$

$$R = -1.5 + 8.5 \text{ kpc} = 7.0 \text{ kpc}$$

The star/cloud is 2.14 kpc from the Sun and 7.0 kpc from the center of the galaxy.

19-17. For what mass distribution would v(R) be constant?

Following Section 19-4A (see also Equation 14-6), the velocity, V(R), of a mass with a circular orbit of radius R about the galaxy would have a velocity

$$V(R) = (G M(R)/R)^{1/2}$$

where M(R) is the mass of the galaxy internal to R. Solving for the mass distribution and assuming that $V(R) \equiv V$ is a constant,

$$M(R) = R\,V(R)^2/G = R\,V^2/G$$

so

$$dM(R)/dr = V^2/G$$

The mass distribution for a symmetric can be described by the mass continuity equation (Equation 16-2)

$$dM(R)/dr = 4\pi R^2 \rho(R)$$

$$\rho(R) = (dM(R)/dr)/4\pi R^2$$
$$= (V^2/G)/4\pi R^2$$
$$\propto 1/R^2$$

So a mass distribution $\propto 1/R^2$ would give a constant rotation curve.

Chapter 20: The Evolution of Our Galaxy

20-1. The following sketch shows the distribution of the neutral hydrogen maxima in the spiral arms of our Galaxy; the position of our Sun is denoted by O. Draw the 21-cm line profiles you would expect to observe in the directions $l = 50°$, $110°$, and $230°$; label the points on these profiles corresponding to the spiral arms. (*Hint:* Do not make any detailed calculations of radial velocity; just make certain that the signs of the velocities and the relative positions of the peaks are correct.)

Refer to Figures 20-2, 19-8, and 19-9 in answering this question. Also review Section 19-4.

It is assumed that all the clouds have the same intrinsic intensity, and that orbits are circular. In all profiles, the relative intensities are determined by the distances to the clouds. If the clouds had different intrinsic intensities, this would alter the relative observed intensities from what is shown.

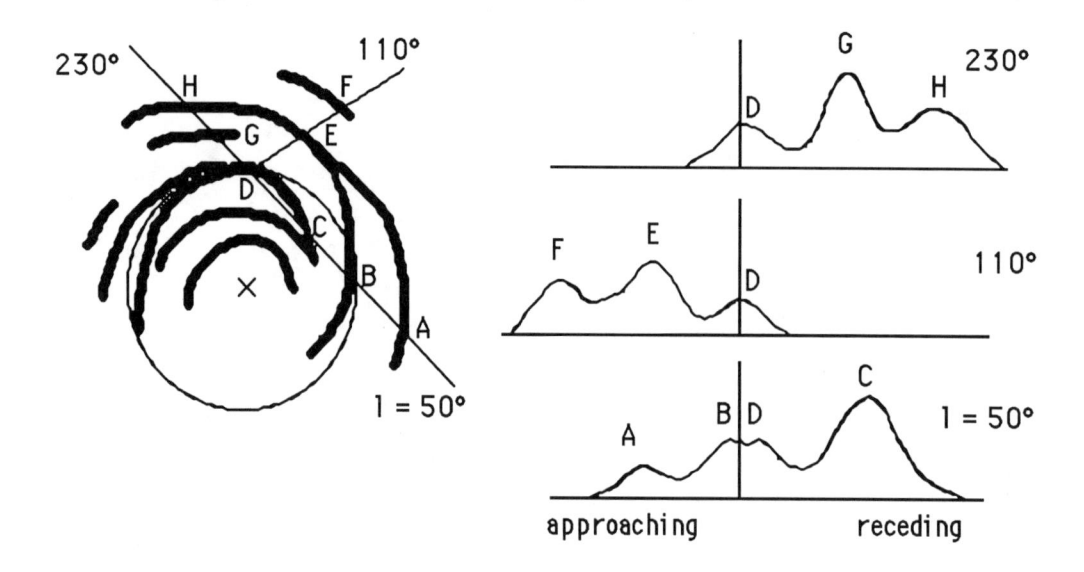

At $l = 50°$, the line profile will look similar to Figure 20-2: Cloud A is outside the solar circle, so it is weak and we are approaching it. Cloud C is near the tangent point, so it has a maximum recession radial velocity and is quite intense. Cloud B is on the solar circle so it has no net radial velocity and its intensity is between that of A and C. We are in cloud D, and see some emission from it (though not much, since we are embedded within this arm) - which blends with the profile from B.

At $l = 110°$, we see some emission from our location (D) with no net radial velocity. Since E and F are outside the solar circle in the second quadrant, we are approaching them, so we see them blueshifted. F is more blueshifted than E since it has a smaller orbital angular velocity, and is weaker since it is more distant.

At $l = 230°$, we again see some of our arm (D), with arms G and H being redshifted since we are receding from them. G is the stronger since it is the closer of the two.

20-2. Name three important physical characteristics that distinguish Population I stars from Population II stars. Explain these differences in terms of the evolution of our Galaxy.

Population I stars contain a larger metal abundance, are younger, and are distributed closer to the galactic plane than Population II stars (see Table 20-1).

The galaxy was initially more spherical in shape. As the galaxy evolved, matter became concentrated in the galactic disk and nuclear bulge as a result of the angular momentum. Hence, the older Population II stars -- formed during these early stages in the galaxy's history -- are found in the halo and uniformly (not concentrated) in the disk and bulge. Population I stars were formed more recently, and are found in the disk -- specifically in the spiral arm regions, site of current star formation. Initially the galactic material had a low metal abundance -- so Population II stars have low metal abundance. As metals were synthesized in stars and supernovae and ejected into space, the heavy element abundance increased in the interstellar medium. Stars formed from this enriched material will thus have a higher metal abundance -- Population I stars.

20-3. The objects that constitute our Galaxy are usually divided into five major population classes.
(a) What are these classes?

The five major population classes are (see Table 20-1 and also Section 14-3): Extreme Population I, Older Population I, Disk Population II, Intermediate Population II, and Halo Population II.

(b) Give an example of an object from each class.

Examples of each class include (see Table 20-1):

Extreme Population I: The interstellar medium, O and B stars, supergiants, T Tauri stars, young open clusters, Classical Cepheids, O-B associations, and HII regions.

Older Population I: The Sun, strong-line stars, A stars, Me dwarfs, Giants, and older open clusters.

Disk Population II: Weak-line stars, planetary nebulae, galactic bulge stars, novae, and short period RR Lyrae stars (P < 0.4 days).

Intermediate Population II: High velocity stars (z > 30 km / s), and long period variables (P < 250 days).

Halo Population II: Globular clusters, extreme metal poor stars (sub dwarfs), long period RR Lyrae stars (P > 0.4 days), and Population II Cepheids.

(c) Draw an edge-on view of our Galaxy, indicating the spatial distribution of each of the five classes. Label your diagram carefully.

The overall distribution of these population classes is given in the drawing. The scale heights of the regions are indicated, using values of |z| from Table 20-1. Note that the Extreme Population I stars lie in the spiral arms, the Old Population I stars lie in a patchy but not strictly spiral arm distribution; and that the density of stars varies from class to class.

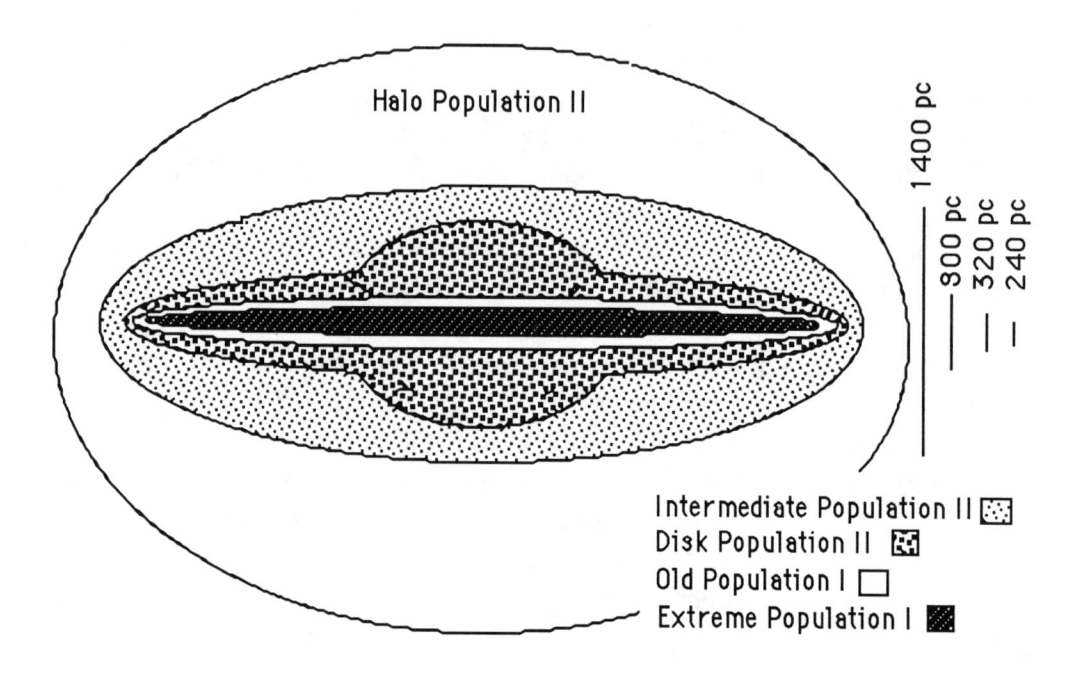

Halo Population II

1400 pc
300 pc
320 pc
240 pc

Intermediate Population II
Disk Population II
Old Population I
Extreme Population I

20-4. (a) What would be the apparent magnitude of a star like our Sun if it were at the distance of the galactic center (8.5 kpc) from the Sun?

Using $M_{bol\odot} = +4.75$ for our Sun and Equation 11-6 (see also Section 11-4C), we can estimate $m_{bol\odot}$ if it were at a distance of 8.5 kpc (the distance of the galactic center).

$$m - M = 5 \log(d_{pc}) - 5$$
$$m = 4.75 + 5 \log(8.5 \times 10^3) - 5 = 19.4$$

(b) The Milky Way contains 4×10^{11} stars. If we assume that all these stars are like our Sun ($M_{bol} = 4.75$), what is the absolute magnitude of the whole Galaxy? Compare this result with the apparent magnitude of the Sun.

[Note: Section 11-4C gives $M_{bol} = 4.75$ instead of $M_B = 4.7$ as stated in the problem.]

The absolute magnitude of the Milky Way is difficult to define strictly because it is not possible to be 10 pc from it -- it is too large. However, astronomers define the absolute magnitude to be that of an idealized object if all the energy were emitted from a small volume.

Assuming the Galaxy is 4×10^{11} times as luminous as the Sun, using Equation 11-5:
$$m - n = 2.5 \log (f_n / f_m)$$
so
$$M_{Sun} - M_{galaxy} = 2.5 \log (f_{galaxy} / f_{Sun})$$
$$M_{galaxy} = 4.75 - 2.5 \log (4 \times 10^{11}) = -24.3$$

The apparent magnitude of the Sun is -26.5, so if the Galaxy's total light output was placed at a distance of 10 pc, the Sun would appear 2.2 magnitudes (7.5 times) brighter than "the Galaxy."

20-5. When a particular globular cluster is at its farthest point from the galactic center (apogalacticon), its distance from the center is 10^4 pc. What is its period of galactic revolution? What assumptions must you make to arrive at a unique answer? Can you give any physical justification for your assumptions? (*Hint:* 1 pc = 2 x 10^5 AU. Assume that the mass of the Galaxy is 10^{12} $M_\odot$.)

We can use Kepler's Third law (Equation 1-24) to estimate the orbital period given the semi-major axis of the orbit. For orbital period in yr, semi-major axis in AU's, and mass in solar masses,

$$P^2 = a^3 / (M_{galaxy} + m_{cluster})$$

The Galaxy is much more massive than the globular cluster (10^{12} $M_\odot \gg 10^5$ $M_\odot$).

Since the eccentricity of the globular cluster's orbit is not given, let's assume two extreme scenarios -- that the orbit is nearly circular (a = 10^4 pc = 2 x 10^9 AU), and that the eccentricity is near 1 (a = 0.5 x 10^4 pc = 1 x 10^9 AU).

For semimajor axis a = 10^4 pc,

$$P^2 = (2 \times 10^9)^3 / (10^{12}) = 8.0 \times 10^{15}$$
$$P \approx 9 \times 10^7 \text{ yr} = 90 \text{ million yr}$$

For semimajor axis a = 0.5 x 10^4 pc,

$$P^2 = (1 \times 10^9)^3 / (10^{12}) = 1 \times 10^{15}$$
$$P \approx 3 \times 10^7 \text{ yr} = 30 \text{ million yr}$$

In this calculation, we have also assumed that the Galaxy's mass is concentrated symmetrically within the orbit of the cluster. Because this isn't true, the estimate of the period is low.

20-6. We can deduce the average time between stellar collisions by considering the figure in the text. If we have identical stars of radius R scattered through space with a mean number density N (stars per unit volume), then a star moving with speed V will sweep out the volume $\pi R^2 V$ per unit time. The average number of stars in this volume is $\pi R^2 VN$, so that in the time T = $1/\pi R^2 VN$, the star will collide with one other star (on the average)! The average distance between each collision is just L = VT = $1/\pi R^2 N$; this is called the *mean free path*. In each of the following situations, compute the mean collisional time and the mean free path for

(a) the solar neighborhood, where V = 20 km/s and N = $0.1/pc^3$; consider stars of radius R = $R_\odot$.

In the solar neighborhood,

$$N = 0.1 \text{ stars} / pc^3 = 0.1 \text{ stars} / (3.086 \times 10^{13} \text{ km})^3 = 3.4 \times 10^{-42} / km^3$$
$$R = R_\odot = 7.0 \times 10^5 \text{ km}$$

The time between collisions is

$$T = 1 / (\pi R^2 V N)$$

$$T = 1 / [\pi (7 \times 10^5 \text{ km})^2 (20 \text{ km/s}) (3.4 \times 10^{-42} / \text{km}^3)]$$
$$= 9.6 \times 10^{27} \text{ s} = 3.0 \times 10^{20} \text{ yr}$$

The age of the Universe is $1\text{-}2 \times 10^{10}$ yr. So collisions within the solar neighborhood are indeed rare! To estimate the time between collisions that may have occurred in our neighborhood of the galaxy, we'd have to divide the time for an individual star to collide by the number of stars in our neighborhood.

The average distance between collisions (mean free path) for a star is
$$L = VT = 1/\pi R^2 N$$
$$= (20 \text{ km/s}) (9.6 \times 10^{27} \text{ s}) = 1.9 \times 10^{29} \text{ km} = 6.2 \times 10^{15} \text{ pc}$$
This is a considerable distance - the size of the Universe is generally taken to be about 10^{24} km!

We shouldn't expect many collisions in this part of the Galaxy.

(b) a galactic nucleus, where V = 1000 km/s and where there are 10^9 stars (of radius $R = 10 R_\odot$) within a sphere of radius 5 pc. Comment briefly on your results.

In the galactic nucleus, $R = 10 R_\odot = 7.0 \times 10^6$ km, and the star density is
$$N = 10^9 / [(4/3) \pi (5 \text{ pc})^3] = 10^9 / [(4/3) \pi (1.54 \times 10^{14} \text{ km})^3]$$
$$= 6.5 \times 10^{-35} / \text{km}^3$$

So the mean collisional time is
$$T = 1 / [\pi (7 \times 10^6 \text{ km})^2 (1{,}000 \text{ km/s}) (6.5 \times 10^{-35} / \text{km}^3)]$$
$$= 1.0 \times 10^{17} \text{ s} = 3.2 \times 10^9 \text{ yr} = 3.9 \text{ billion yr}$$
This time is less than the age of the galaxy (and universe), so on the average stars will have suffered collisions in the galactic nucleus.

The mean free path is:
$$L = VT$$
$$= (1{,}000 \text{ km/s}) (1.0 \times 10^{17} \text{ s}) = 1.0 \times 10^{20} \text{ km} = 3.2 \times 10^6 \text{ pc}$$
$$= 3{,}200 \text{ kpc}$$
If the star was in orbit about the galactic nucleus, then it would take many revolutions about the nucleus before the star would collide with another. If the star was just passing through the galaxy, then the chance of collision is small since the mean free path is much larger than the galaxy.

20-7. The 21-cm line of HI gas has a rest frequency of 1.420406 GHz. A cloud at $b = 0°$, $l = 15°$ has an observed 21-cm emission line frequency of 1.420123 GHz. Use Figure 19-10 to determine, approximately, *two* possible distances to the cloud. Assume only circular motion about the galactic center.

The radial velocity of the cloud is given by the Doppler frequency shift (Equation 8-12):

$$v/c = (v_0 - v)/v$$
$$v/c = (1.420406 \text{ GHz} - 1.420123 \text{ GHz}) / (1.420123 \text{ GHz})$$
$$= 2.0 \times 10^{-4}$$
$$v = 60 \text{ km/s}$$

-- away from the observer, since the observed frequency (wavelength) is less than (greater than) the emitted frequency (wavelength) -- a redshift.

Using the notation from Figure 19-7, the distance of the cloud from the galactic center must be greater than $R_{min} = R_0 \sin(l) = 8.5 \text{ kpc} \sin(15°) = 2.2 \text{ kpc}$. The cloud can be on either the near or far side of the R_{min} orbit (which is the orbit tangential to our line of sight).

We can use Equation 19-7 to estimate the distance to the cloud (from Figure 19-10, $\Theta_0 \approx 220$ km/s):

$$v_r = \Theta \cos(\alpha) - \Theta_0 \sin(l)$$
$$60 \text{ km/s} = \Theta \cos(\alpha) - (220 \text{ km/s}) \sin(15°)$$
$$\Theta \cos(\alpha) \approx 117 \text{ km/s}$$

Now, from Figure 15-10, the rotational velocity in the inner part of the galaxy (2.2 kpc < R < 8.5 kpc) is between 200 km/s and 235 km/s -- let's estimate

$$\Theta \approx 210 \text{ km/s}$$

So
$$\cos(\alpha) \approx 0.56$$
$$\alpha \approx 56°$$

Using Figure 19-7 again, the law of sines (Equation 19-8) can be used to determine the distance to the nearer cloud, d:

$$\sin(l)/R = \sin(90°+\alpha)/R_0 = \sin[180°-(90°+\alpha)-(l)]/d$$
$$\sin(90°+56°)/8.5 \text{ kpc} = \sin(180°-(90°+56°)-15°)/d$$
$$0.56/8.5 \text{ kpc} = 0.33/d$$
$$d = 5.0 \text{ kpc} \quad \text{-- this is the distance to the nearer cloud}$$

The further cloud is found by substituting $-\alpha$ for α in the above solution:

$$\sin(90°-\alpha)/R_0 = \sin[180°-(90°-\alpha)-(l)]/d$$
$$\sin(90°-56°)/8.5 \text{ kpc} = \sin(180°-(90°-56°)-15°)/d$$
$$0.56/8.5 \text{ kpc} = 0.76/d$$
$$d = 11.5 \text{ kpc} \quad \text{-- this is the distance to the further cloud}$$

These estimates will differ if different values of Θ_0 or Θ are chosen. We could refine our estimates of d by taking the cloud's computed distances from the Earth, calculate (using the geometry in Figure 19-7) the distance of the cloud from the galactic center, R, re-estimate the value for Θ (corresponding to this distance R) from Figure 19-10, and re-calculate d according to the above procedure. This analysis could be repeated until the solution converges.

20-8. The following maximum velocities [relative to the Local Standard of Rest (LSR)] of HI gas are observed using the 21-cm emission along lines-of-sight corresponding to the given galactic longitudes: (i) 123 km/s, $l = 15°$; (ii) 95 km/s, $l = 30°$; (iii) 64 km/s, $l = 45°$; (iv) 29 km/s, $l = 60°$; (v) 7.5 km/s, $l = 75°$. Assume circular motions along the galactic center.

(a) Compute and draw a rough plot of the rotation curve of the inner part of the Galaxy ($R < R_o$) using these data.

Referring back to Chapter 19 on galactic rotation (Equation 19-22a and Figure 19-8), at a selected galactic longitude (l) the maximum radial velocity ($v_{r,max}$) is observed for orbits (at distance R_{min} from the galactic center) to which our line of sight is tangential.

$$\Theta(R_{min}) = v_{r,max} + \Theta_o \sin(l)$$
$$R_{min} = R_o \sin(l)$$

Θ_o (= 220 km/s) is the LSR's orbital velocity, Θ is the orbital velocity of the cloud being observed, and R_o (= 8.5 kpc) is the Sun's distance from the galactic center. Using these equations gives the following table of values.

	$v_{r,max}$	l	$\Theta(R_{min})$	R_{min}
i)	123 km/s	15°	180 km/s	2.2 kpc
ii)	95	30°	205	4.3
iii)	64	45°	220	6.0
iv)	29	60°	220	7.4
v)	7.5	75°	220	8.2

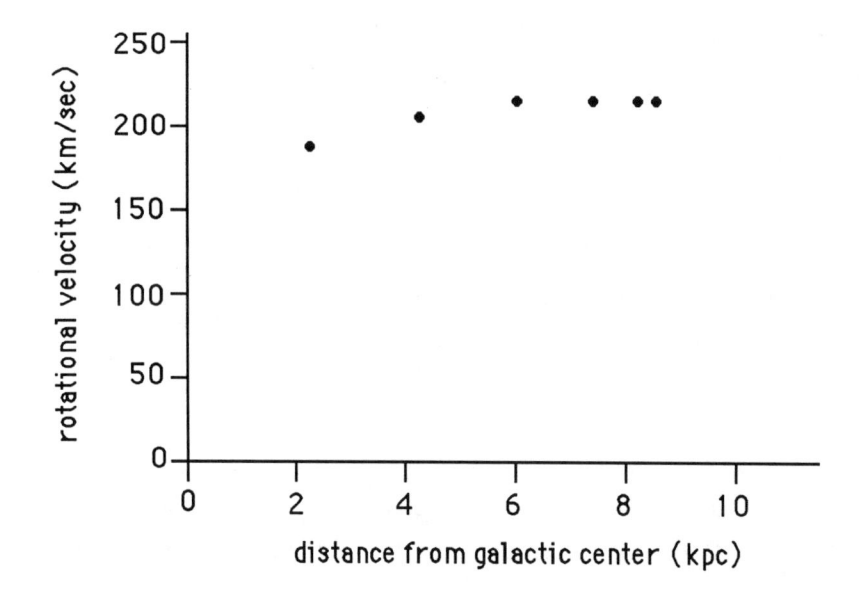

(b) Using the approximation that the mass distribution in the inner part of the Galaxy is spherically symmetric, determine the mass interior to the Sun's orbit, in solar masses.

For a spherical mass distribution internal to the Sun's orbit about the galactic center, the mass is given by (see Section 19-4A):

$$V = (GM/R)^{1/2}$$

or rewriting (with $V = \Theta(r)$),

$$M(r) = \Theta(r)^2 R / G$$

Converting to appropriate units of mass in solar masses, velocity in km/s, and distance in kpc,

$$(1.99 \times 10^{30} \text{ kg} / M_\odot)(M(r)) = (10^6 \text{ m}^2/\text{km}^2)\Theta(r)^2$$
$$\times (3.086 \times 10^{19} \text{ m}/\text{kpc}) R / (6.67 \times 10^{-11} \text{ N m}^2/\text{kg}^2)$$

$$M(r) = 2.3 \times 10^5 \, \Theta(r)^2 R$$

Calculating the mass internal to the six distances plotted in graph in part (a):

R [kpc]	2.2	4.3	6.0	7.4	8.2	8.5
M(r) [× $10^{10} M_\odot$]	1.6	4.2	6.7	8.2	9.1	9.5

(c) Using the same approximation, how does the mass density depend on R between about 6 and 8.5 kpc from the galactic center? (Give a rough proportionality.)

Between 6 and 8.5 kpc from the galactic center, the orbital velocity Θ is roughly constant, so, for spherical symmetry,

$$M(r) \propto R$$

or

$$M(r)/R^3 \propto R^{-2}$$

The mass density (mass per unit volume) decreases with the inverse square of the distance from the galactic center.

20-9. A bright radio source has several 21-cm absorption lines caused by neutral hydrogen clouds along the line-of-sight. Discuss qualitatively how observations of the Doppler shifts of the lines can be used to estimate a lower limit to the distance to the radio source. Under what circumstances would this limit *not* possess the ambiguity caused by two distances along the line-of-sight having the same radial velocity?

The absorption lines are caused by **HI** clouds, along the line of sight, between the radio source and the Earth. Using a model galactic rotation curve and Equation 19-16 for the radial velocity as a function of galactic latitude, distance to the cloud, and the Oort constant A:

$$v_r = A\, d\, \sin(2l)$$

The distance to the cloud can be calculated if the cloud's radial velocity is determined from the Doppler shift. However, in general there may be an ambiguity as to whether this lower limit corresponds to the nearer or further cloud with the same velocity (for example, cloud B or C in Figure 20-2). If the ambiguity can be resolved, the lower limit to the radio source will be the distance to the most distant intervening **HI** cloud along the line of sight.

There is one instance where there is no ambiguity as to the lower limit to the distance to the radio source -- if absorption lines are observed from clouds outside the solar circle (cloud D in Figure 20-2), then the radio source is beyond this cloud. There is no other cloud with the same radial velocity, so the distance ambiguity (that exists for velocities corresponding to clouds within the solar circle) does not exist.

20-10. Expanding gas rings are thought by some astronomers to be evidence of violent explosions near the center of the Galaxy.

(a) Calculate the kinetic energy of the 3-kpc arm, which is thought to contain about 10^8 $M_\odot$ of gas expanding at about 50 km/s.
The kinetic energy of the 3-kpc arm is (Equation 2-6):

$$KE = (1/2) \, m \, v^2 = (0.5) \, (10^8 \times 2 \times 10^{30} \text{ kg}) \, (5.0 \times 10^4 \text{ m/s})^2$$
$$= 2.5 \times 10^{47} \text{ J}$$

(b) Astronomers also believe that there is a ring of molecular gas of mass 10^7 $M_\odot$ expanding at 150 km/s at a distance of 200 pc from the galactic center. Calculate its kinetic energy.
The kinetic energy of the 200 pc gas ring is:

$$KE = (0.5) \, (10^7 \times 2 \times 10^{30} \text{ kg}) \, (1.50 \times 10^5 \text{ m/s})^2$$
$$= 2.25 \times 10^{47} \text{ J}$$

(c) Compare the energies already calculated with that of a single supernova explosion. Comment.
From Table 18-3, a Type **I** supernova has an energy of $\approx 10^{44}$ J, a Type **II** has energy $\approx 10^{43}$ J. To explain the expanding ring or arm of gas would require ≈ 2500 supernovae in each case (or about 5,000 supernovae total). This assumes that all the energy of a supernova is converted into kinetic energy of the ejected material.

20-11. Confirm the calculation in Section 20-3 for the free-fall time for the proto-Galaxy.
The free-fall time is given in Section 20-3B, Section 15-3A and Equation 15-10 [Note, Section 20-3B of the text incorrectly refers to this equation as 16-19]:

$$t_{ff} = (3 \pi / 32 \, G \, \rho_o)^{1/2} = 6.64 \times 10^4 / \rho_o^{1/2} \text{ s}$$

when ρ_o is in units of kg/m^3.
Assume that the current mass of the Galaxy is the same as the original mass and that the original radius of the Galaxy was twice the current radius:

$$M_o \approx 10^{12} \, M_\odot = 10^{12} \times (2 \times 10^{30} \text{ kg}) = 2 \times 10^{42} \text{ kg}$$
$$R_o \approx 2 \times 50 \text{ kpc} = 1 \times 10^5 \text{ pc} \approx 10^5 \times 3 \times 10^{16} \text{ m} = 3 \times 10^{21} \text{ m}$$

so the average density is

$$\rho_o = M_o / V_o = M_o / [(4 \pi / 3) \, R_o^3]$$

$$= (2 \times 10^{42} \text{ kg}) / [(4\pi/3)(3 \times 10^{21} \text{ m})^3]$$
$$= 2 \times 10^{-23} \text{ kg/m}^3$$

The free-fall time is therefore
$$t_{ff} = 6.64 \times 10^4 / (2 \times 10^{-2})^{1/2} \text{ s} = 1.6 \times 10^{16} \text{ s} \approx 5 \times 10^8 \text{ yr}$$

The free-fall time of the galaxy is approximately 500 million years. This agrees with the value at the beginning of Section 20-3.

20-12. Speculate on how one might indirectly observe cosmic-ray electrons in our Galaxy. Do the same for cosmic-ray protons. (*Hint:* Ask how such high-energy particles might produce radiation.) Such measurements indicate that cosmic rays are present throughout the Galaxy and even in the halo!

Charged particles moving in a magnetic field travel in a helical path, consequently emitting synchrotron radiation (since they are being accelerated) [see Figure 18-18]. Cosmic ray particles could therefore be indirectly observed by looking for synchrotron radiation. The galactic plane (and galactic nucleus) is a strong source of synchrotron radiation. Much of this radiation may be from galactic cosmic rays, while some radiation is produced by discrete sources (such as supernova remnants). The electrons, being lighter, would have higher velocities, and thus produce the bulk of the galactic synchrotron radiation.

20-13. The observed spiral pattern in our Galaxy is *not* produced by differential rotation "winding up" the arms. To demonstrate this, use the rotation curve in Figure 20-8 to estimate how long it would take for differential rotation to smooth out the distinct spiral pattern.

(a) What is the revolution period for a star at R = 5 kpc; for the Sun at 8.5 kpc; for a star at R = 20 kpc?

The circumference at galacto-radius R is
$$C = 2\pi R$$
which is traveled by a star traveling at velocity v in time t, so the time taken to revolve around the galactic center is
$$t = C/v = 2\pi R/v.$$
The rotation curve of the Galaxy (real data given in Figure 19-10; or model given in Figure 20-8) shows that the rotation speed is 220, 200 and 235 km/s at galactic radius 5, 8.5 and 20 kpc, respectively.

For a star at 5 kpc from the galactic center,
$$t = 2\pi (5.0 \times 10^3 \text{ pc} \times 3.086 \times 10^{16} \text{ m/pc}) / (2.20 \times 10^5 \text{ m/s})$$
$$= 4.41 \times 10^{15} \text{ s} = 1.40 \times 10^8 \text{ yr} = 140 \text{ million years}$$

For 8.5 kpc,
$$t = 2\pi (8.5 \times 10^3 \text{ pc} \times 3.086 \times 10^{16} \text{ m/pc}) / (2.20 \times 10^5 \text{ m/s})$$
$$= 7.49 \times 10^{15} \text{ s} = 2.37 \times 10^8 \text{ yr} \approx 240 \text{ million years}$$

For 20 kpc,

$$t = 2\pi (20.0 \times 10^3 \text{ pc} \times 3.086 \times 10^{16} \text{ m/pc}) / (2.35 \times 10^5 \text{ m/s})$$
$$= 1.65 \times 10^{16} \text{ s} = 5.22 \times 10^8 \text{ yr} \approx 520 \text{ million years}$$

(b) How long would it take for the star at R = 5 kpc to make one more revolution than the Sun around the center of the Galaxy? This section of the spiral arm would thus be "wound up one turn" from its original shape. [*Hint:* see Section 1-1 on the relationship between sidereal and synodic periods. Use a similar approach.]

A star at 5 kpc is an "inferior" object to the Sun in the terminology of Section 1.1:
$$1/S = 1/P - 1/E$$
where S is the "observed" time for the star to complete one orbit about the galactic center with respect to "the Universe" as seen from the Sun, P is the synodic (revolution) period of the star around the galactic center, and E is the synodic (revolution) period of the Sun about the galactic center.

For the "inferior" star at R = 5 kpc, expressing periods in millions of years,
$$1/S = 1/140 - 1/240 = 2.98 \times 10^{-3} = 1/336$$
$$S \approx 340 \text{ million years}$$
In ≈ 340 million years, the star at R = 5 kpc will complete one more revolution around the galactic center than will the Sun, so the spiral arm between R = 5 and 8.5 kpc would be wound one full turn.

(c) How long would it take for the Sun to make one more revolution than the star at R = 20 kpc around the center of the Galaxy?

For the "superior" star at R = 20 kpc, expressing periods in millions of years,
$$1/S = 1/E - 1/P$$
$$1/S = 1/240 - 1/520 = 2.24 \times 10^{-3} = 1/446$$
$$S \approx 450 \text{ million years}$$
In ≈ 450 million years, the star at R = 20 kpc will complete one less revolution around the galactic center than will the Sun, so the spiral arm between R = 8.5 and 20 kpc would be wound one full turn.

(d) Compare your results in (b) and (c) with the age of the Galaxy. Argue why the spiral pattern must be generated by some mechanism other than differential rotation.

The age of the Galaxy is approximately the age of the Universe, estimated to be between 10 and 20 billion years (see Section 22-3C). The inner (5 to 8.5 kpc) spiral arm will have wound up ≈ 30 to 60 times, the outer (8.5 to 20 kpc) arm will have would up ≈ 22 to 44 times during the age of the Galaxy.

The Sun's revolution period around the galactic center is ≈ 240 million years, so in less than two solar revolutions about the Galaxy the spiral arms between 5 and 8.5 kpc and between 8.5 and 20 kpc would have wound up. Clearly some mechanism other than differential rotation is necessary to explain the spiral arm pattern in our Galaxy and in other galaxies.

20-14. Using Figure 20-8, estimate the mass of the Galaxy in the (i) bulge; (ii) disk; (iii) dark halo; and (iv) total, interior to a galactic radius of (a) 3 kpc; (b) 8.5 kpc (the Sun's distance from the galactic center); (c) 20 kpc. Comment on the mass distribution in the Galaxy, particularly in regard to luminous and dark matter.
[Note, the problem in the text should refer to Figure 20-8, not 20-4.]

The mass, M_{gal}, internal to radius, R, from the galactic center can be estimated by using the revolution velocity, V, from Figure 20-8 and Kepler's third law (see Section 19-4A; Section 20-2D):

$$M_{gal} = R V^2 / G$$

For masses in solar masses, orbital radii in kpc, and velocities in km/s this becomes

$$M_{gal}[M_\odot] \, (1.99 \times 10^{30} \text{ kg}/M_\odot) = \{R[\text{kpc}] \times (10^3 \times 3.086 \times 10^{16} \text{ m/kpc})\}$$
$$\times \{V[\text{km/s}] \times (10^3 \text{ m /km})\}^2$$
$$/ \, (6.67 \times 10^{-11} \text{ N m}^2 / \text{kg}^2)$$

$$M_{gal}[M_\odot] = (2.32 \times 10^5) \times R[\text{kpc}] \times V[\text{km/s}]^2$$

The velocities, in km/s, measured from Figure 20-8, at different galacto-radii are:

galactic radius	bulge	disk	dark halo	total
3 kpc	140	80	100	190
8.5 kpc	90	135	150	220
20 kpc	60	130	180	235

The corresponding mass, in solar masses, at different galacto-radii are:

galactic radius	bulge	disk	dark halo	total
3 kpc	1.4 E 10	4.5 E 9	7.0 E 9	2.5 E 10
8.5 kpc	1.6 E 10	3.6 E 10	4.4 E 10	9.5 E 10
20 kpc	1.7 E 10	7.8 E 10	1.5 E 11	2.6 E 11

The bulge, as its name implies, lies primarily in the inner part of the galaxy; the disk is more uniformly distributed. Note that the dark halo contains a significant fraction (ranging from ≈ 30% at R = 3 kpc to ≈ 60% at R = 20 kpc) of the Galactic mass, particularly in the outer radii of the Galaxy because of its three dimensional structure. Much of this mass is not luminous.

20-15. Section 20-2B describes stars as being metal-poor if the metallicity [Fe/H] < -1, and metal-rich if [Fe/H] > -1. Calculate the relative abundance of N_{Fe}/N_H for the critical value [Fe/H] = -1. [*Hint:* See Table 10-2.]
[Note: The problem should refer to table 10-2, not 10-1.]

From Table 10-2, the solar abundance of iron to hydrogen is:

$$(N_{Fe} / N_H)_{solar} = 10^{7.67} / 10^{12} = 10^{-4.33} = 4.68 \times 10^{-5}$$

The relative abundance of iron to hydrogen is given by (Section 20-2):

$$[\text{Fe/H}] = \log (N_{Fe} / N_H) - \log (N_{Fe} / N_H)_{solar}$$

$$= \log[(N_{Fe}/N_H)/(N_{Fe}/N_H)_{solar}]$$

so the critical value of [Fe/H] = -1 corresponds to

$$\log(N_{Fe}/N_H) = [Fe/H] + \log(N_{Fe}/N_H)_{solar}$$
$$= -1 + \log(4.7 \times 10^{-5}) = -1 - 4.33 = -5.33$$
$$N_{Fe}/N_H = 4.68 \times 10^{-6}$$

This is 10% of the solar iron to hydrogen relative abundance.

20-16. (a) The nonthermal radio source Sgr A*, which may mark the Galaxy's core (Section 20-2C), is less than 0.1" in size. What is its corresponding linear size?

Using the small angle approximation (see Section 11-1A), where d is the linear size of an object at distance D, where the angle, θ, is in arcsec

$$\theta \approx \tan(\theta) = 206{,}265\, d/D$$
$$d = \theta D / 2.06 \times 10^5$$
$$= (0.1)(8.5 \times 10^3 \text{ pc})/(2.06 \times 10^5)$$
$$= 4.1 \times 10^{-3} \text{ pc} \approx 850 \text{ AU}$$

[Alternately, we can notice that if θ is in arcsec, d in AU, and D in pc, the equation is

$$\theta["] = d[AU]/D[pc]$$
$$d[AU] = \theta["] \times D[pc] = 0.1 \times 8{,}500 = 850 \text{ AU}$$

(b) From the timescale of variation in the gamma-ray luminosity of the galactic nucleus, astronomers estimate the size of the nuclear source to be less than 0.3 pc. What is the angular size of the gamma-ray emitting source?

$$\theta \approx \tan(\theta) = 206{,}265\, d/D$$
$$= (2.06 \times 10^5)(0.3 \text{ pc})/(8.5 \times 10^3 \text{ pc}) = 7.3"$$

20-17. Some spiral galaxies have rotation curves that are flat out to 100 kpc. If our Galaxy had such a curve, what would be its total mass be? (*Hint:* Try Kepler's third law in its binary star form to solve this problem.)

Note that this problem is similar to the Concept Application in Section 14-4.

Using Kepler's third law (Equation --24; Equation 12-1), an assuming $M_{galaxy} \gg M_{star}$ (with masses in solar masses), where a is the distance (in AU) of the star from the center of the galaxy and P is the rotational period (in years) of the galaxy at that distance,

$$M_{galaxy} + M_{star} = a^3/P^2$$
$$M_{galaxy} = a^3/P^2$$

The rotational period is given by

$$P = 2\pi a/v$$

where v is the rotational velocity of the galaxy, so

$$M_{galaxy} = a^3 / (2\pi a / v)^2$$
$$= a v^2 / 4\pi^2$$

The velocity as 220 km/s (the same as the velocity of the Sun around the center of the Galaxy; Figure 19-10; Section 19-4B; Section 14-4). Expressing this velocity in units of AU/year,

$$v = (220 \text{ km} / \text{s}) (3.16 \times 10^7 \text{ s/yr}) / (1.496 \times 10^8 \text{ km/AU})$$
$$= 46.5 \text{ AU} / \text{yr}$$

For a galaxy with this velocity at 100 kpc,

$$M_{galaxy} = (100 \text{ kpc} \times 2.06 \times 10^8 \text{ AU/kpc}) (46.5 \text{ AU} / \text{yr})^2 / 4\pi^2$$
$$= 1.1 \times 10^{12} M_\odot$$

The units work out since the masses are in solar masses, and right hand side is in units of AU^3/yr^2.

This is higher than the mass of a normal (typical) galaxy, since the size was taken to be 100 kpc radius -- larger than a typical galaxy.

For comparison, our Galaxy has a total mass of $7 \times 10^{11} M_\odot$ and a radius of 60 kpc (Appendix Table A7-1). Let's use the last equation above to see if we get this result, assuming the velocity of 220 km/s.

$$M_{Galaxy} = (60 \text{ kpc} \times 2.06 \times 10^8 \text{ AU/kpc}) (46.5 \text{ AU} / \text{yr})^2 / 4\pi^2$$
$$= 6.8 \times 10^{11} M_\odot \approx 7 \times 10^{11} M_\odot$$

This checks. Our technique is correct.

Chapter 21: Galaxies Beyond the Milky Way

21-1. Our Galaxy and the Andromeda Galaxy (M31) are by far the most massive members of the Local Group. If these two giant galaxies form a binary system and move about one another in circular orbits, then calculate

(a) the distance to the center of mass of the system from our Galaxy

(See Section 12-2A for background information on how to approach this problem, which is similar to that for binary star systems.)

The masses of galaxies in the Local Group are given in Table 23-1. [Note, Appendix Table A7-1 gives $M_{MW} = 7 \times 10^{11}\ M_\odot$.]

$M_{M31} = 3 \times 10^{11}\ M_\odot$; $M_{MW} = 4 \times 10^{11}\ M_\odot$; $D_{MW\text{-}M31} = 690$ kpc.

The location of the center of mass is defined by (Equation 12-4):

$$D_{CM\text{-}M31}\ M_{M31} = D_{CM\text{-}MW}\ M_{MW}$$

$$(D_{MW\text{-}M31} - D_{CM\text{-}MW})\ M_{M31} = D_{CM\text{-}MW}\ M_{MW}$$

$$(690\text{ kpc} - D_{CM\text{-}MW})\ 3 \times 10^{11} = (D_{CM\text{-}MW})\ 4 \times 10^{11}$$

$$2070 - 3\ D_{CM\text{-}MW} = 4\ D_{CM\text{-}MW}$$

$$D_{CM\text{-}MW} \approx 300\text{ kpc}$$

The Milky Way is ≈ 300 kpc away from the M31-Milky Way center of mass.

(b) the orbital period

The orbital period is given by Kepler's Third Law (Equation 1-24 or 12-1), for periods in yr, masses in $M_\odot$, and semi-major axis (distance between the galaxies) in AU.

$$(M_{M31} + M_{MW})\ P^2 = a^3$$

$$P^2 = [690 \times (2.06 \times 10^8)]^3 / (3 \times 10^{11} + 4 \times 10^{11}) = 4.1 \times 10^{21}\text{ yr}^2$$

$$P = 6.4 \times 10^{10}\text{ yr} \quad \text{-- this is greater than the age of the Universe.}$$

Perform the same computations for the orbit of the pair M32 and M31.

(a') Repeating the same for the M31 - M32 pair:

From Table 23-1, $M_{M32} = 2 \times 10^9\ M_\odot$, and

from Figure 23-1, $D_{M31\text{-}M32} \approx 30$ kpc

The center of mass is given by:

$$D_{CM\text{-}M31}\ M_{M31} = D_{CM\text{-}M32}\ M_{M32}$$

$$(D_{M32\text{-}M31} - D_{CM\text{-}M32})\ M_{M31} = D_{CM\text{-}M32}\ M_{M32}$$

$$(30\text{ kpc} - D_{CM\text{-}M32})\ 3 \times 10^{11} = (D_{CM\text{-}M32})\ 2 \times 10^9$$

$$9000 - 300\ D_{CM\text{-}M32} = 2\ D_{CM\text{-}M32}$$

$$D_{CM\text{-}M32} = 29.8\text{ kpc}$$

The galaxy M32 is 29.8 kpc away from the M31-M32 center of mass, and M31 is 0.2 kpc from the center of mass of this system.

(b') The orbital period is:

$(M_{M31} + M_{M32}) P^2 = a^3$

$P^2 = [30 \times (2.06 \times 10^8)]^3 / (3 \times 10^{11} + 2 \times 10^9) = 7.82 \times 10^{17}$

$P = 8.8 \times 10^8$ yr ≈ 0.9 billion yr

so these galaxies have orbited several times in the age of the Universe.

21-2. Use the rotation curves in Figure 21-9 to calculate the masses of NGC 7664 and NGC 4378.
From Figure 21-9, the rotational velocity of NGC 7664 is 200 km/s at a radius of 25 kpc; NGC 4378 is 300 km/s at 22 kpc. The mass internal to these points in the galaxies is given by Kepler's Third Law (Equation 1-24 or 12-1):

$M = a^3 / P^2$

where M is mass in solar masses, a is distance (semi-major axis) in AU, and P is the orbital period in yr. For circular orbits, the period is given by:

$P = 2 \pi a / v$

Determining P for these two galaxies:

$P_{NGC\ 7664} = 2 \pi (25 \times 3.1 \times 10^{16}$ km$) / (200$ km$/$s$)$

$= 2.43 \times 10^{16}$ s $= 7.7 \times 10^8$ yrs

$P_{NGC\ 4378} = 2 \pi (22 \times 3.1 \times 10^{16}$ km$) / (300$ km$/$s$)$

$= 1.43 \times 10^{16}$ s $= 4.5 \times 10^8$ yrs

And the masses are thus,

$M_{NGC\ 7664} = (25 \times 2.06 \times 10^8)^3 / (7.7 \times 10^8)^2$

$= 2.3 \times 10^{11} M_\odot$

$M_{NGC\ 4378} = (22 \times 2.06 \times 10^8)^3 / (4.5 \times 10^8)^2$

$= 4.6 \times 10^{11} M_\odot$

Alternately, we can combine the two equations (see Section 20-2D),

$M_{gal} = R V^2 / G$

where (see Problem 20-14 for derivation) for masses in solar masses, radii in kpc, and velocities in km/s,

$M_{gal}[M_\odot] = (2.32 \times 10^5) \times R[kpc] \times V[km/s]^2$

which get the same results as above

$M_{NGC\ 7664} = (2.32 \times 10^5) \times 25 \times 200^2 = 2.3 \times 10^{11} M_\odot$

$M_{NGC\ 4378} = (2.32 \times 10^5) \times 22 \times 300^2 = 4.6 \times 10^{11} M_\odot$

21-3. The Sun orbits around the Galaxy at a radius of about 8.5 kpc. In a mere 110 million years from now, the Sun will be on the other side of the Galaxy. The nearby galaxies will appear to shift in the sky relative to the more distant background galaxies — a galactic parallax. What then would be the number of parsecs in a "galsec," the distance of a galaxy whose galactic parallax equals 1.0"?

Using the Sun's galacto-radius for "super" parallax measurements, the "baseline" for parallax measurements increases from 1 AU to 8.5 kpc = 1.75×10^9 AU (this assumes we use the radius of the galactic orbit for the definition of galactic parallax, as we use the Earth's orbital radius about the Sun for the parallax). By the argument of similar triangles, a "galsec" would be 1.75×10^9 pc = 5.7 billion light yr -- thus we could easily determine parallax distances to distant galaxies and quasars! Unfortunately, the timescale over which measurements need to be made (110 million yr) would not make this a good graduate thesis project.

21-4. Based on the photographs and text in this chapter, plot I(r) versus r for a hypothetical (a) E galaxy and (b) S0 galaxy. What kind of assumptions must you make? How might an Sa galaxy differ from your second plot? Should you use logarithms for either or both axes?

From Section 21-1A of the text, for elliptical galaxies and the spheroidal (bulge) components of S0 galaxies the intensity function is given by

$$\log[I(r)] \propto r^{-1/4}.$$

For the disk component of an S0 galaxy

$$I'(r) = I'_o e^{-\alpha r}$$

where I'_o is a constant. Note, the disk and bulge components have different, independent profiles. These trends in the surface brightness are seen in the photographs.

(a) The easiest way to make these graphs is to use a spreadsheet or graphics program, though the calculations can be done easily with a calculator. To make the graphs below, we started by making a column of intensities ranging from 100, 90, 80, ..., 10, 9, 8, ... The second column is the Log of the intensity; third column is $(\log I)^{-4}$ which is the radius r. We can then plot, as desired, I versus r, or Log (I) versus r, or as shown in the next graph, Log(I) versus Log(r).

For an elliptical galaxy, or the spheroidal component of an S0 galaxy, a graph of the logarithm of the intensity versus a logarithm of the radius will be a straight line with slope -1/4.

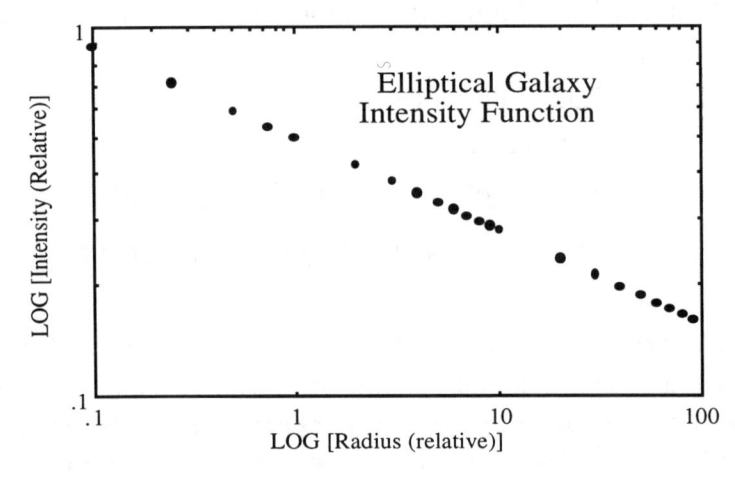

(b) For the disk component of an S0 galaxy, we took the radius column from part (a), and calculated $50 e^{-r}$, where we assume $\alpha = 1$ (which may not be true for the disk components of real galaxies, but which is adequate here) and that the disk component has intensity $\approx 1/2$ that of the spherical bulge component at small radii. By adjusting the coefficient I_0, and the index α, one can make galaxies with different bulge-disk relative brightnesses and different disk profiles. We show a linear and logarithmic intensity function graph. Note that the spherical component in these graphs is the same as the elliptical model component in part (a).

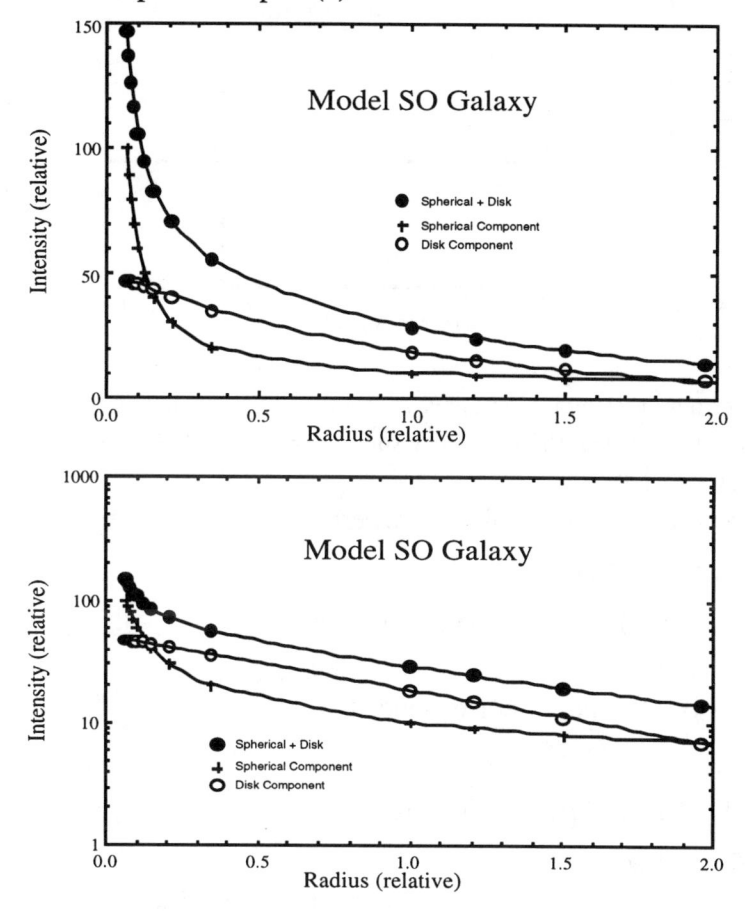

21-5. Plot the angular diameter versus distance (in Mpc) for a galaxy with linear diameter of 30,000 pc. At what distance does the galaxy subtend an angle of d = 4 arcsec?

Using the small angle approximation, the angular size, θ, in arcsec of a galaxy of linear size, d, at a distance D will be

$$\theta_{arcsec} \approx \tan(\theta) = 206{,}265 \, d / D$$
$$= (2.063 \times 10^5)(3.0 \times 10^5 \text{ pc}) / (D_{Mpc} \times 10^6) = 6.19 \times 10^4 / D_{Mpc}.$$

[We will see in section 25-2, that at great distances we have to make corrections for the curvature of the Universe, but for now this Euclidian result will suffice.]

A log-log graph best shows the angular size — distance relationship, since the angular size is proportional to the reciprocal of the distance. Graphically, one finds that a galaxy will appear to be 4" in size at a distance about 1500 Mpc; or we can solve the equation to show

$$D = 6.19 \times 10^4 \text{ Mpc} / \theta_{arcsec} = 6.19 \times 10^4 \text{ Mpc} / 4 \approx 1{,}550 \text{ Mpc}.$$

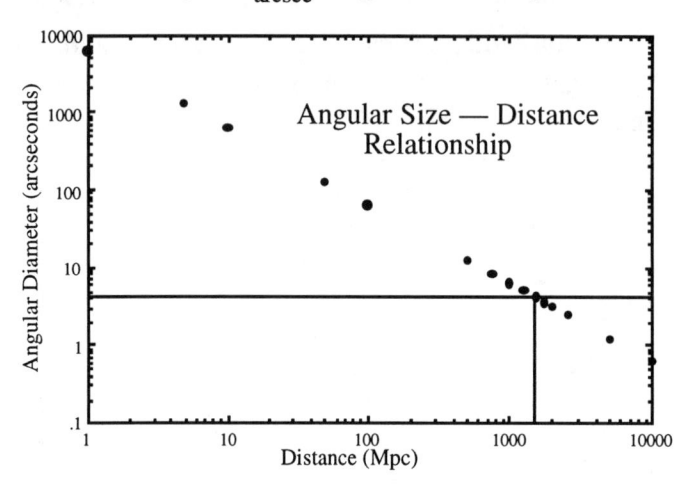

Angular Size — Distance Relationship

21-6. Using the data from Problem 5, discuss our ability to classify accurately the morphologies of galaxies at distances of 10, 100, and 1000 Mpc. Assume 1-arcsec resolution data. What types are easy to classify? Which are hard?

Using the small angle approximation, where d is the linear size of an object at distance D and θ is the angle in arcsec

$$\theta_{arcsec} \approx \tan(\theta) = 206{,}265 \, d / D$$

For a distance of 10 Mpc, 1" corresponds to

$$d = \theta_{arcsec} D / (2.06 \times 10^5)$$
$$= (1)(10 \times 10^6 \text{ pc}) / (2.06 \times 10^5)$$
$$\approx 5 \times 10^1 \text{ pc} = 50 \text{ pc}$$

while 1" would correspond to distances of 500 pc (= 0.5 kpc) and 5000 pc (= 5 kpc) at distances of 100 Mpc and 1000 Mpc respectively.

For comparison, the angular size of a typical 30 kpc galaxy would be

$$\theta = (2.06 \times 10^5) \, d / D$$
$$= (2.06 \times 10^5)(3 \times 10^4 \text{ pc}) / (10 \times 10^6 \text{ pc})$$
$$= 620"$$

while the angular size would be 62" and 6.2" for galaxies at distances of 100 Mpc and 1000 Mpc, respectively.

With 1" resolution, morphological classification would be easy at distances of 10 Mpc and 100 Mpc. Beyond 1000 Mpc classification would be increasingly difficult. The easiest galaxies to classify would be the giant galaxies (spiral and elliptical). Face on

21-7. In light of a collapsing gas cloud model of galaxy formation, discuss the observations that halo population stars (Population **II**) have low metal content and that disk population (Population **I**) objects with higher metal abundance define thinner disks.

In the collapsing gas cloud model the galaxy was initially spherical covering roughly the extent of the current halo. As the galaxy evolved the gas cloud collapsed into the current disk. Hence the halo stars were formed first (from the primordial material) while the disk stars are continuing to form. The older halo stars are therefore low metal abundance Population **II** stars since the original gas had low metal abundance while the more recently formed disk stars are higher metal abundance Population **I** stars. Because the disk is flattening with time, the disk in which stars of given metal abundance are found becomes thinner as the metal abundance increases.

21-8. Qualitatively discuss why we know that large amounts of dust do not account for the high M/L ratios of galaxies found by the dynamical methods of Section 21-5C.

Large amounts of dust cannot account for the high M/L ratios of galaxies because we would be able to observe the dust indirectly. The dust would absorb light, decreasing the luminosities in the visible relative to other wavelengths. But the dust would re-radiate the energy at infrared wavelengths appropriate to the dust temperature — allowing us to detect the dust. In addition, the dust would redden the light from stars, the reddening which could easily be measured.

21-9. How many brown dwarfs with $M \approx 10^{-3} M_\odot$ would be needed to give our Galaxy an M/L ratio of 10?

The Milky Way has a mass out to 18-20 kpc estimated dynamically to be $\approx 3.4 \times 10^{11}$ $M_\odot$ (see Section 19-4C or Section 20-2D). [Note: Table 23-1 and Appendix Table A7-1 give total galaxy mass of 4×10^{11} and 7×10^{11} $M_\odot$, respectively, for different Galaxy sizes.] If all the stars were, on the average, like the Sun the M/L ratio would be ≈ 1 $M_\odot/L_\odot$ and the luminosity would thus be 3.4×10^{11} $L_\odot$. To get a M/L ratio ≈ 10 would require that 10% of the mass is in Sun-like stars ($L \approx 3.4 \times 10^{10}$ $L_\odot$) and 90% of the mass ($\approx 3.06 \times 10^{11}$ $M_\odot$) is in dark material which contributes no (little) luminosity.

If this dark matter were brown dwarfs then the total number of needed to account for 90% of the galactic mass would be

3.06×10^{11} $M_\odot$ / 10^{-3} $M_\odot$ $\approx 3 \times 10^{14}$ brown dwarfs.

This is thousands of times more brown dwarfs than luminous stars. If there were this many brown dwarfs throughout the galaxy to account for the non-luminous matter indicated by the rotation curve, then we should have found this population by now.

21-10. Assume that the Hubble telescope has recently observed a galaxy that has been given the name Westphal 1. If this galaxy is in the direction $l = 20°$, $b = 30°$, has an observed diameter of 6" x 4", is of type Sc, has a redshift $cz = 12,5000$ km/s, and has an observed magnitude of $B = 16.5$, what is its absolute magnitude M_B?

Follow the outline for determining luminosities of extragalactic objects given in Section 21-1B. This problem is similar to Problem 21-14. Students may want to compare the answers, since the galaxy in that problem and this problem are similar.

We first correct for dust obscuration within the Milky Way Galaxy, for b > 0:

$A_B = 0.19 \times [1 + \cos(b) \times$
$\{0.1948 \cos(l) + 0.0725 \sin(l) + 0.1168 \cos(2l)$
$- 0.0921 \sin(2l) + 0.1147 \cos(3l) + 0.0784 \sin(3l)$
$+ 0.0479 \cos(4l) + 0.0847 \sin(4l)\}]$
$\times [\csc\{b + 0.25° - 1.7° \sin(l) - 1.0° \cos(3l)\}]$

Solving, for $l = 20°$ and $b = 30°$,

$A_B = 0.19 \times [1 + 0.455 \cos(30°)] \times [1 / \sin(29.17°)]$
$= 0.19 \times 1.394 \times 2.052$
$= 0.544$

The second correction is for extinction internal to the galaxy in question, which for spiral galaxies is

$A_B(i) = 0.70 \log(\sec(i))$.

To find the inclination, i, we assume that the galaxy is circular when viewed face-on, but elliptical in shape when viewed from an angle. Assume that at 0° inclination the galaxy size would be 6" by 6", so the observed angular size of 4" by 6" is due to the inclination.

$\cos(i)$ = observed diameter / real diameter = 4" / 6" = 2/3
$i = \arccos(2/3) = 48.2°$

and

$\sec(i) = 1 / \cos(i) = 3/2 = 1.5$

so

$A_B(i) = 0.70 \log(1.5) = 0.123$

The third correction is the K-correction, which is due to the Doppler shift resulting in the observed wavelength range not being the same as the emitted wavelength range. This correction is dependent on the galaxy type / morphology, and for Sc galaxies is

$10^4 K_B(cz) = [0.075 - 0.010 \times \text{(galaxy morphology unit - Sb)}] cz$

We assume that the text means that an Sc galaxy is 1 morphology unit different from an Sb, so this expression becomes

$K_B(cz) = [0.075 - 0.010 \times (1)] (12,500) \times 10^{-4}$
$= 0.081$

Applying these three corrections, remembering to subtract (rather than add) magnitudes to make the corrected apparent magnitude to be brighter than the observed magnitude, gives:

$$m_B = B_{observed} - A_B - A_B(i) - K_B(cz)$$
$$= 16.5 - 0.544 - 0.123 - 0.081$$
$$\approx 15.8$$

The corrected apparent blue magnitude is 15.8.

Now we can use Equation 11-7 to derive the corrected absolute magnitude,

$$M = m + 5 - 5\log(d_{pc}).$$

To convert the observed redshift to distance, we need information in Section 22-2, which the student has not yet seen! The relationship between redshift, cz, and distance is given by the Hubble Law (Equation 22-3c)

$$d = cz / H$$

where the Hubble constant H will be assumed to be 50 km / s / Mpc, as in Chapter 22 (note, a value of 75 km / s / Mpc may be more realistic).

$$d = (12{,}500 \text{ km/s}) / (50 \text{ km/s / Mpc}) = 250 \text{ Mpc} = 2.5 \times 10^8 \text{ pc}$$

The absolute magnitude is thus

$$M_B = 15.8 + 5 - 5\log(2.5 \times 10^8)$$
$$= -21.2$$

the absolute magnitude of the galaxy Westphal 1 is $M_B = -21.2$. Notice that this is similar to the absolute magnitude of the Milky Way Galaxy (also an Sc, or Sb, galaxy) given as M = -21 in Section 21-1B or Table 23-1.

21-11. The Pinwheel Galaxy (M 33, also called the Triangulum Galaxy) is a nearby (D = 690 kpc) nearly face-on Sc-type spiral galaxy.

(a) How many parsecs in M 33 correspond to an angular size of 1" as seen from the Earth?

Using the small angle approximation, where d is the linear size of an object at distance D and θ is the angle in arcsec

$$\theta_{arcsec} \approx \tan(\theta) = 206{,}265 \, d / D$$
$$d = \theta_{arcsec} D / (2.06 \times 10^5)$$
$$= (1)(6.9 \times 10^5 \text{ pc}) / (2.06 \times 10^5)$$
$$= 3.3 \text{ pc}$$

A distance of 3.3 parsec in the galaxy M 33 would appear on Earth to subtend 1".

(b) If the rotation curve is like the Sc galaxy NGC 7664 (Figure 21-9), what is the revolution period for a star 10 kpc from the galactic center of M 33?

Reading a velocity of 190 km/s at R = 10 kpc for galaxy NGC 7664 in Figure 21-9, the rotation period for the galaxy M 33 would be

$$t = 2\pi R / v = 2\pi [(1.0 \times 10^4 \text{ pc}) \times (3.086 \times 10^{13} \text{ km/pc})] / (190 \text{ km/s})$$
$$= 1.02 \times 10^{16} \text{ s} = 3.2 \times 10^8 \text{ yr} = 320 \text{ million years}$$

(c) What is the velocity of the star in part (b) in parsecs/year?
Converting a velocity in km/s to parsecs/year:
$$190 \text{ km/s} = 1.9 \times 10^5 \text{ m} / 1 \text{ s}$$
$$= [(1.9 \times 10^5 \text{ m}) / (3.086 \times 10^{16} \text{ m/pc})] / [(1 \text{ s}) / (3.16 \times 10^7 \text{ s/yr})]$$
$$= 1.9 \times 10^{-4} \text{ pc / yr}$$

(d) How far in arcsec would the star appear to move in 100 years?
In 100 years, the star would move a distance in the galaxy of
$$(1.9 \times 10^{-4} \text{ pc / yr}) \times 100 \text{ yr} = 1.9 \times 10^{-2} \text{ pc}$$

From part (a), this corresponds to an observed angular motion in 100 years of
$$(1.9 \times 10^{-2} \text{ pc}) / (3.3 \text{ pc / arcsec}) \approx 6 \times 10^{-3} \text{ arcsec} = 0.006"$$

(e) If astronomers can measure positions with an accuracy of 0.01", comment on how long it would take before the proper motion of the star in M 33 could be readily detected.
To detect a motion of 0.01" would take
$$0.01" / (0.006" \text{ per 100 years}) \approx 170 \text{ years}.$$
This period of time would be needed to "just barely" see the motion of the star. Several hundred years would have to pass before a real good measurement could be made — and by then better measurement techniques would probably be possible.

21-12. You observe a star in the field of an elliptical galaxy and want to determine whether the star is a permanent member of the galaxy, an object escaping from the galaxy, or a foreground star in our own Galaxy. You observe the spectrum of the star and find that the Hα line (λ_o = 656.3 nm) has an observed wavelength of 656.9 nm. The galaxy has a mass of 10^{12} $M_\odot$ and a radius of 100 kpc. What can you conclude?

Using the Doppler formula (Equation 8-13 or 19-1), we can find the radial velocity of the star:
$$v_r = (\Delta\lambda / \lambda_o) c = [(656.9 \text{ nm} - 656.3 \text{ nm}) / 656.3 \text{ nm}] \times (3.0 \times 10^5 \text{ km/s})$$
$$= 2.74 \times 10^2 \text{ km/s} = 274 \text{ km/s}$$

Since the observed wavelength is longer than the emitted wavelength, the star is moving away from us with this velocity. Note that this is the radial velocity only of the star. Because the star may also have a tangential component to its motion, its space velocity most likely is larger (see Section 19-1).

The velocity dispersion for a galaxy is (Equation 21-2):
$$<v^2> = 0.4 \, G M / r_h$$
where r_h is the radius enclosing half the galaxy mass.
$$<v^2> = 0.4 \, (6.67 \times 10^{-11} \text{ N m}^2 / \text{kg}^2) \, (10^{12} \, M_\odot \times 1.99 \times 10^{30} \text{ kg}/M_\odot)$$
$$/ \, (100 \text{ kpc} \times 3.086 \times 10^{19} \text{ m/kpc})$$

$$= 1.72 \times 10^{10} \text{ m}^2/\text{s}^2$$
$$\langle |v| \rangle = 1.3 \times 10^5 \text{ m/s} = 1.3 \times 10^2 \text{ km/s} = 130 \text{ km/s}$$

The observed radial velocity for the star is ≈ twice the velocity dispersion (so the space velocity is even more than twice the velocity dispersion) of the galaxy. The interpretation of the nature of the star is thus uncertain, so additional information is needed. From the velocity information alone, one cannot determine whether the star is bound to the galaxy or escaping the galaxy — some stars which are bound to the galaxy will have velocities greater than the mean velocity.

One piece of information which would be useful in determining whether the star is in our Galaxy or a member of the other galaxy is the radial velocity of the galaxy. What we really need to consider is the velocity of the star relative to the velocity of the galaxy. So far we have discussed only the star's radial velocity relative to us. The problem is more complicated than one would suspect from the limited information given. Additional information in the problem would make it easier to analyze the nature of the star.

21-13. Several possibilities (dust, comets or asteroids, brown dwarfs, black holes) are suggested in the chapter as being unlikely to be the source of the dark matter in galaxies. Explain why each of these classes of objects is unlikely to be the dark matter.

Dust - Dust is not truly "dark." It is detectable by the reddening and polarization effects which it has on background light. In concentrated clouds it tends to produce a dark clump or heat up and glow in the infrared.

Comets or asteroids - If the dark mass is 5 to 30 times the total mass of observed stars in the Galaxy, a typical solar mass star (such as the Sun) would need to have 5 to 30 solar masses of comets and/or asteroids associated with it, an unreasonable amount from what we know about our solar system. Comets and asteroids are not likely to be found in large amounts between the stars as we think they are by-products of solar system formation.

Brown dwarfs - See Problem 21-9 for a numerical estimate of the number of brown dwarfs required. Way too many brown dwarfs would be needed — we should have detected some by now if they were that plentiful, since some would be in our local neighborhood.

Black holes - Black holes in binary systems should produce x-ray emission from their accretion disks. Unless black holes are completely isolated, which is not likely since the majority of stars are in binary systems, we would expect to detect them from their emission.

21-14. The galaxy NGC 5055 is classified as Sbc II-III. It has an apparent B magnitude of 9.30 and is located at galactic longitude $l = 106°$, latitude $b = 75°$. Its redshift is 550 km / s, and its observed axis ratio is 0.60. For Sbc galaxies, the edge-on axis ratio is 0.13. What would this object's absolute face-on corrected B magnitude be using the corrections discussed in Section 21-1B? Would the K correction be of any practical concern?

Follow the outline for determining luminosities of extragalactic objects given in Section 21-1B. This problem is similar to Problem 21-10.

We first correct for dust obscuration within the Milky Way Galaxy, for b > 0:
$$A_B = 0.19\,(1 + S_N \cos(b))\,|C|$$
$$A_B = 0.19 \times [1 + \cos(b) \times$$
$$\{0.1948 \cos(l) + 0.0725 \sin(l) + 0.1168 \cos(2l)$$
$$- 0.0921 \sin(2l) + 0.1147 \cos(3l) + 0.0784 \sin(3l)$$
$$+ 0.0479 \cos(4l) + 0.0847 \sin(4l)\}\,]$$
$$\times [\csc\{b + 0.25° - 1.7° \sin(l) - 1.0° \cos(3l)\}]$$

Solving, for $l = 106°$, latitude $b = 75°$,
$$A_B = 0.19 \times [1 + 0.09566 \cos(75°)] \times [1 / \sin(72.87°)]$$
$$= 0.19 \times 1.0248 \times 1.0464$$
$$= 0.204$$

The second correction is for extinction internal to the galaxy, NGC 5055, which for spiral galaxies is
$$A_B(i) = 0.70 \log(\sec(i)).$$
To find the inclination, i, we use the observed axial ratio, b/a, of 0.60 and the edge-on axial ratio, α, for this type of spiral of 0.13.
$$\cos^2(i) = [(b/a)^2 - \alpha] / (1 - \alpha^2)$$
$$= [(0.60)^2 - 0.13] / [1 - (0.13)^2]$$
$$= 0.234$$
$$\cos(i) = 0.484$$
$$i = 61.1°$$
Thus,
$$A_B(i) = 0.70 \log[\sec(61.1°)]$$
$$= 0.70 \log(2.069°) = 0.221$$

The third correction is the K-correction, which is due to the Doppler shift resulting in the observed wavelength range not being the same as the emitted wavelength range. This correction is dependent on the galaxy type / morphology, and for Sbc galaxies is
$$10^4\,K_B(cz) = [0.075 - 0.010 \times (\text{galaxy morphology unit - Sb})]\,cz$$
We assume that the text means that an Sbc galaxy is 0.5 morphology unit different from an Sb, so this expression becomes
$$K_B(cz) = [0.075 - 0.010 \times (0.5)]\,(550) \times 10^{-4}$$
$$= 0.00385 \approx 0.004$$
The K correction is insignificant, being nearly two orders of magnitude less than each of the other two corrections. This is not a surprise since the galaxy is very close to us, so there is very little redshifting of the emitted light out of the rest filter band.

Applying these three corrections, remembering to subtract (rather than add) magnitudes to make the corrected apparent magnitude to be brighter than the observed magnitude,

gives:
$$m_B = B_{observed} - A_B - A_B(i) - K_B(cz)$$
$$= 9.30 - 0.204 - 0.221 - 0.004$$
$$\approx 8.87$$
The corrected apparent blue magnitude is 8.87.

Now we can use Equation 11-7 to derive the corrected absolute magnitude,
$$M = m + 5 - 5 \log(d_{pc}) .$$

To convert the observed redshift to distance, we need information in Section 22-2, which the student has not yet seen! The relationship between redshift, cz, and distance is given by the Hubble Law (Equation 22-3c)
$$d = cz / H$$
where the Hubble constant H will be assumed to be 50 km / s / Mpc, as in Chapter 22 (note, a value of 75 km / s / Mpc may be more realistic).
$$d = (500 \text{ km/s}) / (50 \text{ km/s / Mpc}) = 10 \text{ Mpc} = 1.0 \times 10^7 \text{ pc}$$
One must be careful in interpreting all the observed velocity as being cosmological. The redshift is small enough that a significant fraction of it may be due to its motion within its galaxy cluster. But we will assume the redshift is cosmological.

The absolute magnitude is thus
$$M_B = 8.87 + 5 - 5 \log(1.0 \times 10^7)$$
$$= -21.1$$
The absolute magnitude of the Sbc II-III galaxy NGC 5055 is $M_B = -21.1$.

Notice that this is similar to the absolute magnitude of the Milky Way Galaxy (also an Sc, or Sb, galaxy) given as M = -21 in Section 21-1B or Table 23-1; and nearly the same as the galaxy Westphal 1 ($M_B = -21.2$) in Problem 21-10.

Chapter 22: Hubble's Law and the Distance Scale

22-1. An approximate distance for the separation between our Galaxy and the Magellanic Clouds is 50 kpc. What will be the observed apparent magnitude for the following stars in the Magellanic Clouds:

(a) an RR Lyrae variable with a period of 0.5 day

A 0.5 day period RR Lyrae star has $M_V = +0.5$ (from Table 18-1), so at a distance of 50 kpc its apparent magnitude is given by Equation 11-6:

$$m = M + 5 \log(d_{pc}) - 5$$
$$= 0.5 + 5 \log(5.0 \times 10^4) - 5 = 0.5 + 23.5 - 5$$
$$= 19$$

(b) a classical Cepheid with a pulsation period of 100 days

For a Classical (Population **I**) Cepheid with P = 100 day (Figure 18-3):

$$L \approx 5 \times 10^4 \, L_\odot$$

so from Equations 11-5 and 11-6:

$$M_\odot - M = 2.5 \log (L / L_\odot)$$
$$M = M_\odot - 2.5 \log [(5 \times 10^4 \, L_\odot) / L_\odot]$$
$$= 4.8 - 11.7 \approx -7$$
$$m = -7.0 + 23.5 - 5 = 11.5$$

(c) a Population **II** Cepheid with a ten-day period

Similarly, for a Population **II** Cepheid with P = 10 days,

$$L \approx 5 \times 10^2 \, L_\odot$$
$$M = M_\odot - 2.5 \log [(5 \times 10^2 \, L_\odot) / L_\odot]$$
$$= 4.8 - 6.7 \approx -2$$
$$m = -2 + 23.5 - 5 = 16.5$$

It is thus clear why Cepheid variable stars are such useful distance indicators -- they are visible with even modest sized telescopes to great distances.

22-2. What methods of distance determination are most useful for finding the distance to
Refer to Table 22-1 and Figure 22-3 for a summary of distance indicators.

(a) the Pleiades
By main sequence fitting (specifically compared to the Hyades whose distance can be determined by the moving cluster method).

(b) a globular cluster in our Galaxy
By the Period-Luminosity relationship for RR Lyrae and Population **II** Cepheids.

(c) the Large Magellanic Cloud
The distance to this galaxy in the Local Group can be determined from the period-luminosity relationship for Population **I** (Classical) Cepheids, brightest stars, and the

size of **HII** regions (we need to know whether **HII** regions in irregular galaxies differ from those in spiral galaxies).

(d) M 31, the Andromeda galaxy

By the period luminosity relation for Population **I** (Classical) Cepheids, brightest stars, apparent size of **HII** regions, and novae - supernovae.

(e) the Virgo cluster of galaxies

Distance to this cluster, a member of the local supercluster, is determined by supernovae, globular clusters, brightest stars, brightest galaxies in the cluster.

(f) the Hercules cluster of galaxies

By brightest galaxies in the cluster, redshift and the Hubble Law.

(g) our Sun

By solar system triangulation, radar distancing of planets (e.g. Venus).

(h) the nucleus of our Galaxy

By distribution of globular clusters in the sky, galactic rotation curve.

22-3. The Ca **II** K stellar absorption line has a rest wavelength of 393.3 nm. In a particular galaxy, the Ca **II** K line is observed at a wavelength of 410.0 nm. What is the distance to the galaxy, assuming

(a) H = 50 km/s · Mpc

We first compute the recessional velocity of the galaxy (Equation 8-13 or 22-2):

$$v/c = z \equiv \Delta\lambda/\lambda_o = (\lambda - \lambda_o)/\lambda_o$$
$$= (410 \text{ nm} - 393.3 \text{ nm}) / 393.3 \text{ nm}$$
$$= 4.25 \times 10^{-2}$$
$$v = 1.27 \times 10^4 \text{ km/s}$$

Since z « 0.8, we do not have to worry about cosmological effects (Equation 22-3d), so we can use Hubble's Law (Equation 22-3b) to determine the distance to the galaxy:

d = v / H

For H = 50 km / s / Mpc,

$$d = (1.27 \times 10^4 \text{ km/s}) / (50 \text{ km/s/Mpc})$$
$$\approx 250 \text{ Mpc}$$

(b) H = 100 km / s / Mpc

$$d = (1.27 \times 10^4 \text{ km/s}) / (100 \text{ km/s/Mpc})$$
$$\approx 125 \text{ Mpc}$$

22-4. How would the derived age of the Universe (Equation 22-4) change if

(a) the expansion is decelerating (H decreasing with time)

[For a further discussion of this material, see Sections 25-1B and 25-2B and Figure 25-2.]

The age of the Universe estimated from Equation 22-4 :
$$t = 1/H = d/V$$
assumes that the Hubble constant has remained unchanged.

If the expansion is decelerating (i.e. H is decreasing with time) H was greater in the early stages of the Universe, so much of the expansion occurred in a shorter time than the present H would imply. The Universe should be younger than calculated from the current value of H. The "size" (scale) of the Universe is greater than we'd expect from the current value of H (or V) due to this early rapid expansion -- thus $t = d / V$ is greater than the age of the Universe.

(b) the expansion is accelerating (H increasing with time)
Following the argument above, if the expansion is accelerating then the true age of the Universe is greater than is calculated by a naive application of the current Hubble constant.

(c) What observations could be made to determine if either of these possibilities is correct?
A long term (multi-million year) project would be that suggested in Problem 21-3 -- use the galactic rotation to determine distances to quasars and distant galaxies. Use these distances and the recessional velocities to determine whether the Hubble Constant is changing. Or, if we are impatient, observe these same quasars and distant galaxies, trying to accurately determine their distances. Currently we have no good method (other than the Hubble Law, which we are attempting to test!) for determining distances to these most distant objects.

We could try to determine the average density of the Universe (that is, the luminous plus non-luminous mass per unit volume). As we will see in Chapter 25, the average density can tell us whether the Universe is open, closed, or flat and the curvature of the Universe.

22-5. Most of the nearest galaxies do not obey Hubble's law. Explain why.
Most nearby galaxies do not obey the Hubble law because their random motions overwhelm the velocity attributable to the expansion of the Universe. These "random" velocities are due the galaxy motions within the cluster of galaxies or motions of the clusters of galaxies within the supercluster. Only when we look at galaxies that are sufficiently distant is the expansion velocity large enough to be much greater than the "local" velocities of galaxies.

22-6. Devise a lecture demonstration of Hubble's law using some easily obtainable elastic materials.
One way to demonstrate Hubble's law is to blow up a rather large balloon on which galaxies are painted. If one wants to be quantitative, one can measure the distances between "galaxies" and compute the "recessional velocities."

An alternative is to attach tennis balls (paper cups, or whatever) representing galaxies to an elastic cord (such as a long bungee cord) representing the spacetime of the Universe. Stretch the cord to show the expansion of the Universe. Let go to show the collapse of a closed Universe. [The collapse demo is not recommended if you have a student volunteer rather than an inanimate object holding the other end of the cord!]

In deference to the traditional raisin bread analogy, you could bring in some raisin bread and a pan of dough, or even bake some in class. This is not as easy to visualize, but eating the results is fun.

An example of what not to do, consider a suggestion made in a publication for elementary school teachers. They recommend placing a large, clean handkerchief on a table. Place some marbles in the center. Disrupt the marbles so they roll to the edge of the handkerchief. Challenge your class to figure out what's wrong with this suggested demonstration. (See Problem 25-11.)

22-7. Plot Equation 22-3d for redshifts up to $z = 5$.

For a flat Universe, the distance to objects as a function of the observed redshift, cz, is (Equation 22-3d)

$$d = cz(1 + z/2) / H(1 + z)^2$$

Using $H = 50$ km/s/Mpc,

$$d = [(3.0 \times 10^5 \text{ km/s}) / (50 \text{ km/s/Mpc})] \, z(1 + z/2) / (1 + z)^2$$
$$= (6000 \text{ Mpc}) \, z(1 + z/2) / (1 + z)^2$$

And, using $H = 100$ km/s/Mpc,

$$d = (3000 \text{ Mpc}) \, z(1 + z/2) / (1 + z)^2$$

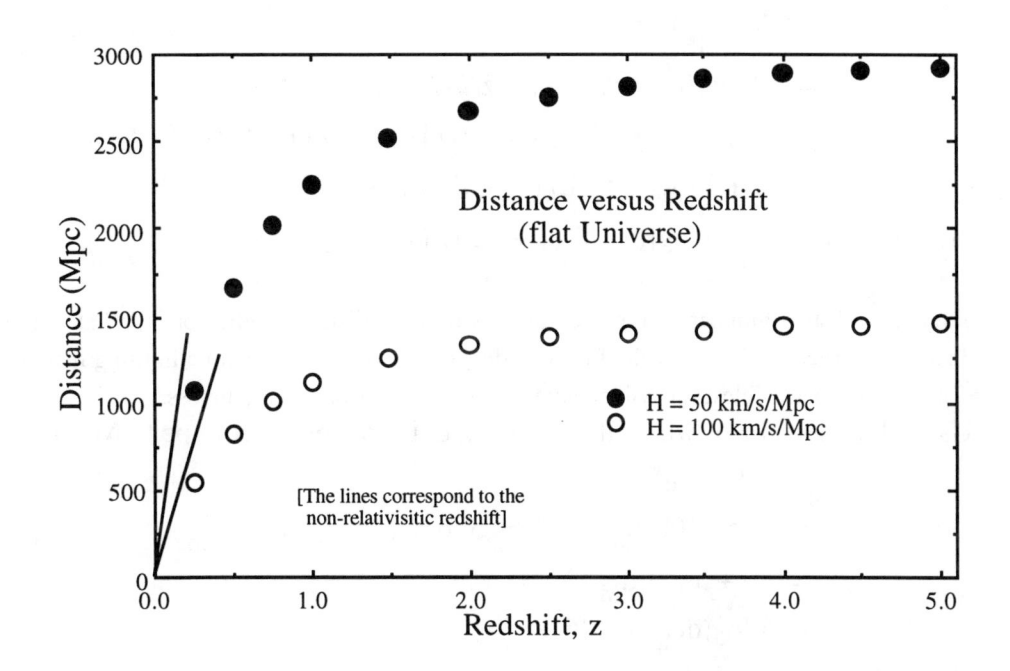

$z =$	0.25	0.5	0.75	1.0	1.5	2.0	3.0	4.0	5.0
d (H=50 km/s/Mpc)	1080	1667	2020	2250	2520	2667	2813	2880	2917
d (H=100 km/s/Mpc)	540	833	1010	1125	1260	1333	1406	1440	1458

Distance versus Redshift (flat Universe)

● $H = 50$ km/s/Mpc
○ $H = 100$ km/s/Mpc

[The lines correspond to the non-relativisitic redshift]

Notice how the distance increases slowly with redshift for large redshifts. For redshifts « 1, the distance - redshift relationship is d ≈ cz / H, as shown by the lines in the graph.

22-8. We often use the equation m - M = 5 log(d) - 5 to obtain absolute magnitudes for distant galaxies. The distance d is usually estimated by Hubble's law. Parametrize this equation using h from Section 22-2B.

From Section 22-2B, H = 100 h km/s·Mpc

which can be converted to an equation with pc rather than Mpc (= 10^6 pc)

$$H = 10^{-4} \text{ h km/s·pc}$$

Assuming that the galaxy is close enough to use Equation 22-3c,

$$d_{pc} = cz/H = 10^4 cz/h = 10^4 v/h$$

which is like Equation 22-5.

Substituting into the magnitude equation, where cz ≡ v is in units of km/s,
$$m - M = 5 \log(d_{pc}) - 5$$
$$= 5 \log(10^4 cz/h) - 5 = 5 \log(10^4) + 5 \log(cz/h) - 5$$
$$= 15 + 5 \log(cz/h)$$
or in terms of v,
$$m - M = 15 + 5 \log(v/h)$$

If the galaxy is distant enough that we must use Equation 22-3d,
$$d_{pc} = cz(1 + z/2)/H(1 + z)^2$$
$$= 10^4 cz(1 + z/2)/h(1 + z)^2 = 10^4 v(1 + z/2)/h(1 + z)^2$$

Substituting into the magnitude equation, where cz ≡ v is in units of km/s,
$$m - M = 5 \log(d_{pc}) - 5$$
$$= 5 \log[10^4 (cz/h)(1 + z/2)/(1 + z)^2] - 5$$
$$= 5 \log(10^4) + 5 \log[(cz/h)(1 + z/2)/(1 + z)^2] - 5$$
$$= 15 + 5 \log[(cz/h)(1 + z/2)/(1 + z)^2]$$
or, $m - M = 15 + 5 \log[(v/h)(1 + z/2)/(1 + z)^2]$

22-9. Table 22-1 lists a number of types of objects used as distance indicators. What is the maximum distance of detectability for each of these indicators for a telescope with a limiting magnitude of +25? (The Hubble Space Telescope will be able to detect objects of this brightness.)
Using Equation 11-6 (where the distance, d, is converted from pc to Mpc)
$$m - M = 5 \log(d_{pc}) - 5$$
$$= 5 \log(10^6 d_{Mpc}) - 5 = 5 \log(10^6) + 5 \log(d_{Mpc}) - 5$$
$$= 30 + 5 \log(d_{Mpc}) - 5$$
$$= 5 \log(d_{Mpc}) + 25$$

For a limiting magnitude m = +25, the distance (in Mpc) to which we can observe an object is

$$5 \log(d_{Mpc}) = m - M - 25$$
$$\log(d_{Mpc}) = (25 - M - 25)/5 = -M/5$$
$$d_{Mpc} = 10^{-M/5} \quad \text{-- how convenient}$$

For the objects in Table 22-1, $d_{Mpc} = 10^{-M/5}$ corresponds to a distance of:

object	M_V(max)	distance
A-M main sequence	>0	1 Mpc
O-B star	-6	16 Mpc
supergiant	-7	25 Mpc
RR Lyrae	0.5	0.8 Mpc
Classical Cepheid	-6	16 Mpc
W Virginis	-3	4 Mpc
globular cluster	-9	63 Mpc
nova	-8	40 Mpc
HII region	-9	63 Mpc
supernova	-20	10,000 Mpc
brightest galaxies	-21	16,000 Mpc

Compare these distances to the distance of (Chapter 23) the Andromeda Galaxy in our Local Group (0.69 Mpc), the Virgo Cluster of Galaxies (15.7 Mpc), the Coma Cluster of Galaxies ($\approx$ 90 Mpc).

22-10. Show that for small values of z, Equation 22-3d reduces to Equation 22-3c.

For small values of z, the binomial expansion (Appendix A9-6),

$$(1 \pm x)^n = 1 \pm nx + (1/2)(n)(n-1)x^2 \pm (1/6)(n)(n-1)(n-2)x^3 + ...$$

can be used to approximate $(1+z)^{-2}$

$$(1+z)^{-2} = 1 + (-2)z + (1/2)(-2)(-3)z^2 + (1/6)(-2)(-3)(-4)z^3 + ...$$
$$= 1 - 2z + 3z^2 - 4z^3 + ...$$

where for z << 1, the cubed term can be ignored (note, for z = 0.1, the z^3 term is 4% that of the z term).

Substituting this expression for z in equation 22-3d:

$$d = cz(1 + z/2) / H(1+z)^2$$
$$= (cz/H)(1 + z/2)(1 - 2z + 3z^2)$$
$$= (cz/H)(1 - 2z + 3z^2 + z/2 - z^2 + (3/2)z^3)$$
$$= (cz/H)(1 - (3/2)z + 2z^2 + (3/2)z^3)$$
$$= (cz/H) - (3/2)(c/H)z^2 + 2(c/H)z^3 + ...$$

The last two terms are small for z << 1, reducing the equation to Equation 22-3c

$$d = cz/H$$

22-11. For relatively nearby galaxies, the hydrogen Balmer-alpha line can be redshifted out of the visible part of the spectrum.

(a) For what redshifts would the Balmer-alpha line of hydrogen be shifted out of the visible into the infrared portion ($\lambda > 720$ nm) of the electromagnetic spectrum?

Use the Doppler shift formula (Equation 22-1), where λ is the observed wavelength and λ_o is the emitted wavelength = 656.3 nm (see section 8-1C):

$$z = \Delta\lambda/\lambda_o = (\lambda - \lambda_o)/\lambda_o$$
$$= (720 \text{ nm} - 656.3 \text{ nm}) / 656.3 \text{ nm}$$
$$= 0.097$$

For redshifts $z \geq 0.097$, the Balmer-alpha line will be redshifted to wavelengths longer than 720 nm, into the infrared portion of the spectrum.

(b) To what distances does this correspond? Express your answer in terms of the Hubble parameter h. Since $z < 0.8$, we can use Equation 22-3c to determine the distance corresponding to $z \geq 0.097$:

$$d = cz/H$$

and in terms of the dimensionless parameter h (Section 22-2B),

$$d = (3 \times 10^5 \text{ km/s}) \, z / (100 \, h \text{ km/s·Mpc}) = (3000 \, z / h) \text{ Mpc}$$

(c) To what distance do these correspond for a Hubble constant $H = 100$ km/s·Mpc; for $H = 50$ km/s·Mpc?

For $H = 50$ km/s·Mpc ($h = 0.5$),
$$d = [(3000)(0.097)/(0.5)] \text{ Mpc}$$
$$= 582 \text{ Mpc}$$

For $H = 100$ km/s·Mpc,
$$d = [(3000)(0.097)/(1.0)] \text{ Mpc}$$
$$= 291 \text{ Mpc}$$

Thus galaxies further away than 582 Mpc, for $H = 50$ km/s·Mpc (or 291 Mpc for $H = 100$ km/s·Mpc) would have the Hydrogen Balmer-alpha line shifted out of the visible into the infrared portion of the spectrum.

[Note that all the galaxy clusters listed in Table 23-2 have redshifts < 0.097 (the most distant one, the Corona Borealis cluster, has $cz = 21{,}600$ km/s → $z = 0.072$) which means the hydrogen Balmer-alpha line would be in the visible portion of the spectrum.]

22-12. For relatively distant galaxies (and quasars — see Chapter 24), the hydrogen Lyman-alpha line can be redshifted into the visible part of the spectrum.

(a) For what redshifts would the Lyman-alpha line of hydrogen be shifted out of the ultraviolet into the visible portion ($\lambda \approx 390$ nm to 720 nm) of the electromagnetic spectrum?

To see the range of z for which the Lyman-alpha line will be shifted into the visible range (390 nm to 720 nm), we need to consider both ends of the optical window. Use the Doppler shift formula (Equation 22-1), where λ is the observed wavelength and λ_o is the emitted wavelength = 121.6 nm (see section 8-1C):

$$z = \Delta\lambda / \lambda_o = (\lambda - \lambda_o) / \lambda_o$$
$$z_{lower} = (390 \text{ nm} - 121.6 \text{ nm}) / 121.6 \text{ nm}$$
$$= 2.21$$
$$z_{upper} = (720 \text{ nm} - 121.6 \text{ nm}) / 121.6 \text{ nm}$$
$$= 4.92$$

For redshifts $2.21 \leq z \leq 4.92$, the ultraviolet Lyman-alpha line will be redshifted to visible wavelengths between 390 nm and 720 nm.

(b) For a flat Universe, to what range in distance do these redshifts correspond? Express your answer in terms of the Hubble parameter h.

Since $z > 0.8$, we cannot use Equation 22-3c but must instead use Equation 22-3d to determine the distance corresponding to $z = 2.21$ and 4.92:

$$d = (c z / H) (1 + z/2) / (1 + z)^2$$

and in terms of the dimensionless parameter h (Section 22-2B),

$$d = [(3 \times 10^5 \text{ km/s}) \, z / (100 \, h \text{ km/s} \cdot \text{Mpc})] (1 + z/2) / (1 + z)^2$$
$$= (3000 \, z / h) [(1 + z/2) / (1 + z)^2] \text{ Mpc}$$

(c) To what distance do these correspond for a Hubble constant $H = 100$ km/s·Mpc; for $H = 50$ km/s·Mpc? Compare these distances to the age of the Universe.

For $H = 100$ km/s·Mpc (h = 1.0),
the lower z limit, $z \geq 2.21$, corresponds to a distance

$$d = (3000 \, z / h) [(1 + z/2) / (1 + z)^2] \text{ Mpc}$$
$$= [(3000)(2.21)/(1.0)][(1 + 2.21/2)/(1 + 2.21)^2] \text{ Mpc}$$
$$= 1{,}360 \text{ Mpc}$$
$$= (1.36 \times 10^9 \text{ pc}) \times (3.26 \text{ ly/pc})$$
$$= 4.4 \times 10^9 \text{ light years} = 4.4 \text{ billion light years}$$

and the upper z limit, $z \leq 4.92$, corresponds to a distance

$$d = [(3000)(4.92)/(1.0)][(1 + 4.92/2)/(1 + 4.92)^2] \text{ Mpc}$$
$$= 1{,}460 \text{ Mpc}$$
$$= (1.46 \times 10^9 \text{ pc}) \times (3.26 \text{ ly/pc})$$
$$= 4.8 \times 10^9 \text{ light years} = 4.8 \text{ billion light years}$$

Therefore, for $H = 100$ km/s·Mpc (h = 1.0), the Lyman-alpha line will be redshifted into the visible portion (390 nm to 720 nm) of the spectrum for objects at a distance of

$$2.21 \leq z \leq 4.92$$
$$1{,}360 \text{ Mpc} \leq d \leq 1{,}460 \text{ Mpc}$$
$$4.4 \text{ billion light years} \leq d \leq 4.8 \text{ billion light years}$$

which compares with the age of the Universe of ≈ 10 billion years.

For H = 50 km/s·Mpc (h = 1.0),
the lower z limit, z ≥ 2.21, corresponds to a distance
$$d = (3000\, z/h)\, [(1 + z/2)/(1+z)^2]\text{ Mpc}$$
$$= [(3000)(2.21)/(0.5)]\, [(1 + 2.21/2)/(1+ 2.21)^2]\text{ Mpc}$$
$$= 2{,}710 \text{ Mpc}$$
$$= (2.71 \times 10^9 \text{ pc}) \times (3.26 \text{ ly/pc})$$
$$= 8.8 \times 10^9 \text{ light years} = 8.8 \text{ billion light years}$$
and the upper z limit, z ≤ 4.92, corresponds to a distance
$$d = [(3000)(4.92)/(0.5)]\, [(1 + 4.92/2)/(1+ 4.92)^2]\text{ Mpc}$$
$$= 2{,}910 \text{ Mpc}$$
$$= (2.91 \times 10^9 \text{ pc}) \times (3.26 \text{ ly/pc})$$
$$= 9.5 \times 10^9 \text{ light years} = 9.5 \text{ billion light years}$$

Therefore, for H = 50 km/s·Mpc (h = 0.5), the Lyman-alpha line will be redshifted into the visible portion (390 nm to 720 nm) of the spectrum for objects at a distance of
$$2.21 \leq z \leq 4.92$$
$$2{,}710 \text{ Mpc} \leq d \leq 2{,}910 \text{ Mpc}$$
$$8.8 \text{ billion light years} \leq d \leq 9.5 \text{ billion light years}$$
which compares with the age of the Universe of ≈ 20 billion years.

For either value of H (or h) the galaxies with distance about 45 - 48% the age of the Universe will be redshifted into the visible portion of the spectrum.

22-13. For a classical Cepheid, estimate the uncertainties in the distances derived by Equation 22-7 if the slope of the P-L relationship were uncertain by 3%.

The Cepheid period-luminosity relation was first discussed in Section 18-2C (see also Figure 18-3 and Equation 18-2). The relation was represented, in modified form, in Section 22-3B and Equation 22-7,
$$\langle M_V \rangle = -3.53 \log_{10}(P) + 2.13\, (\langle B_0 \rangle - \langle V_0 \rangle) - 2.13$$
where the brackets "< >" indicate the average flux converted to magnitude. The 2.13 x ($\langle B_0 \rangle - \langle V_0 \rangle$) term is the color correction term. The 2.13 factor in this term is the slope of the P-L relation. If this value was uncertain to 3%, then the slope would be 2.13 ± 0.064. The absolute magnitude of the Cepheid would thus be uncertain by about 0.064 magnitude.

The distance is found from the absolute magnitude by rearranging the distance modulus (Equation 11-6)
$$\log(d) = (m - M + 5)/5$$
$$d = 10^{(m - M + 5)/5}$$

If the absolute magnitude is in error by 0.064,

$$d' = 10^{(m - (M + 0.064) + 5)/5} = 10^{[(m - M + 5)/5] + (0.064/5)}$$
$$= 10^{[(m - M + 5)/5]} \, 10^{(0.064/5)}$$
$$= d \, 10^{0.0128} = 1.03 \, d$$

This is a 3% error in the distance.

Thus, a 3% error in the slope of the Cepheid variable period-luminosity relation corresponds to an error of 0.064 magnitude error in the absolute magnitude, which corresponds to a 3% error in the distance of the Cepheid.

22-14. If a nova were seen in the Large Magellanic Cloud with V = 10.8 and the maximum rate of decline over the first two magnitudes was found to be 0.1 mag/day, what would be your estimate of the distance to the LMC?

Using Equation 22-10 relating a nova's maximum absolute magnitude, M_V^{max}, and mean rate of initial decline, m, in units of magnitude per day over the first two magnitudes of decline,

$$M_V^{max} = -9.96 - 2.31 \log(m)$$

We assume that the maximum rate of decline stated in the problem is really the mean rate of decline during the rapid decline just after maximum (see Figure 18-12 for a representative nova light curve). Thus,

$$M_V^{max} = -9.96 - 2.31 \log(0.1) = -7.65$$

Using the distance modulus (Equation 11-6)
$$m - M = 5 \log(d) - 5$$
$$5 \log(d) = 10.8 - (-7.65) + 5 = 23.45$$
$$\log(d) = 4.69$$
$$d = 4.9 \times 10^4 \text{ pc} \approx 50 \text{ kpc}$$

This distance would place the nova at the distance of the Large Magellanic Cloud (see Table 23-1, which gives a distance of 50 kpc).

Chapter 23: Large-Scale Structure in the Universe

23-1. Calculate the crossing time for the Hercules super-cluster and compare it with the age of the Unniverse ($\approx 15 \times 10^9$ years).

[Note: This problem uses some information on velocity dispersions contained in Section 23-1F of the 3rd edition, which has been deleteed from the 4th edition.]

From Figure 23-12, the Hercules supercluster extends from redshift $\approx$ 9,500 to 13,500 km/s in the radial direction. Using H_o = 75 km / s / Mpc, this corresponds to an extent of (Equation 22-3b):

$$d = v / H = (4{,}000 \text{ km/s}) / (75 \text{ km/s/Mpc}) \approx 50 \text{ Mpc}$$

If the velocity distribution is characteristic of the typical velocity dispersions, with a range in rich clusters being 400 - 1500 km/s *(from Section 23-1F of the 3rd edition)*, corresponding to long and short crossing times, respectively:

$$\tau_{c\text{-long}} = d / v$$
$$= (50 \times 10^6 \text{ pc}) (3.1 \times 10^{13} \text{ km/pc}) / (400 \text{ km/s})$$
$$= 3.9 \times 10^{18} \text{ s} \approx 1.2 \times 10^{11} \text{ yrs}$$

$$\tau_{c\text{-short}} = (50 \times 10^6 \text{ pc}) (3.1 \times 10^{13} \text{ km/pc}) / (1500 \text{ km/s})$$
$$= 1.0 \times 10^{18} \text{ s} \approx 3.3 \times 10^{10} \text{ yrs}$$

both which are greater than the age of the Universe ($\approx 1.5 \times 10^{10}$ yrs). Thus, clusters within this supercluster cannot have traversed the supercluster. This also implies that the supercluster cannot have formed recently -- it must have been present during the origin of the Universe.

23-2. (a) Demonstrate that a low-density hydrogen gas at 10^7 K produces X-rays. What is the peak wavelength of emission?

Using the Wien displacement law (Equation 8-39), a gas at 10^7 K will produce thermal emission which peaks at a wavelength of:

$$\lambda_{max} = (2.9 \times 10^{-3} \text{ m K}) / T$$
$$= (2.9 \times 10^{-3} \text{ m K}) / 10^7 \text{ K} \approx 3 \times 10^{-11} \text{ m} = 0.03 \text{ nm} = 0.3 \text{ Å}$$

From Table 8-1, we see that this is in the short wavelength portion of the X-ray spectral range. The gas will emit over the entire X-ray region of the electromagnetic spectrum (according to the Planck Law, Equation 8-35; see also Figure 8-14), with some γ-ray emission.

(b) Compare the mass of the hot gas in a cluster with the mass contained in galaxies.

Models for X-ray observations of galaxy clusters (Section 23-4) indicate that typical clusters contain hot gas in an "x-ray core" with a density of $n \approx 10^3$ ions/m³ in a volume ranging from 0.05 to 1.5 Mpc. Consider a core with R $\approx$ 0.5 Mpc, or volume

$$V = (4/3) \pi R^3 = (4/3) \pi [(0.5 \times 10^6 \text{ pc}) \times (3.1 \times 10^{16} \text{ m/pc})]^3$$

$$= 1.6 \times 10^{67} \text{ m}^3$$

so the number of ions is

$$N = nV = (10^3 / \text{m}^3) \times (1.6 \times 10^{67} \text{ m}^3) = 1.6 \times 10^{70}$$

If this gas is hydrogen, then the mass of the gas is

$$M = N m_H = (1.6 \times 10^{70}) \times (1.67 \times 10^{-27} \text{ kg}) = 2.6 \times 10^{43} \text{ kg}$$
$$= 1.3 \times 10^{13} \text{ M}_\odot$$

For comparison, the Milky Way Galaxy has a mass of $\approx 4 \times 10^{11}$ M$_\odot$ (Table 23-1) [or 7×10^{11} M$_\odot$ from Appendix Table A7-1]. The hot cluster gas is thus about 30 times more massive than our spiral galaxy -- quite a considerable mass.

23-3. We state in the chapter that intergalactic space must have a dust density of less than 4×10^{-30} kg/m^3. Assume that ten times this amount really exists. What would the intergalactic extinction be, in magnitudes per megaparsec? How would that affect our ability to see the Virgo cluster?

Assuming that the intergalactic dust density is $\rho \approx 4 \times 10^{-29}$ kg/m^3 we can estimate the intergalactic extinction by comparison with the interstellar dust densities (Section 15-2G).

$$\rho_{\text{gas interstellar}} \approx 3 \times 10^5 \text{ atoms}/\text{m}^3 \approx 5 \times 10^{-22} \text{ kg}/\text{m}^3$$

where most of the gas is hydrogen

Since dust comprises about 1% of the mass (see Section 15-1) of the interstellar medium,

$$\rho_{\text{dust interstellar}} \approx 0.01 \, \rho_{\text{gas interstellar}} \approx 5 \times 10^{-24} \text{ kg}/\text{m}^3$$

so

$$\rho_{\text{dust intergalactic}} / \rho_{\text{interstellar}} \approx 4 \times 10^{-29} / 5 \times 10^{-24} \approx 8 \times 10^{-6}$$

Assume that the absorption coefficient of intergalactic dust is the same as interstellar dust (assumes that the grains are similar in structure). The absorption would be:

$$K_{v \text{ intergalactic}} / K_{v \text{ interstellar}} \approx 8 \times 10^{-6}$$

$$K_{v \text{ intergalactic}} \approx 8 \times 10^{-6} \, K_{v \text{ interstellar}}$$

$$\approx (8 \times 10^{-6}) \times (1 \text{ magnitude} / \text{kpc})$$

$$\approx 8 \times 10^{-6} \text{ magnitude} / \text{kpc}$$

$$\approx 8 \times 10^{-3} \text{ magnitude} / \text{Mpc}$$

The Virgo cluster is at a distance of 15.7 Mpc, so the extinction is 0.13 magnitudes. This is a small amount of extinction, but it could be detected (by carefully looking at the color indexes of galaxies). The effect on our ability to see the Virgo cluster is minimal.

Note that if we use the upper limit to the intergalactic dust density ($\rho = 4 \times 10^{-30}$ kg/m^3), then the intergalactic extinction for the Virgo cluster would be ≈ 0.01 magnitudes -- insignificant.

23-4. How close does a spiral galaxy like the Milky Way have to get to a cD galaxy to be tidally disrupted? The Roche instability limit is given by (Equation 3-9):

$$d = 2.44 \, (\rho_M / \rho_m)^{1/3} \, R$$

where ρ_M is the average density of the cD galaxy, ρ_m is the average density of the Milky Way, and R is the radius of the cD galaxy.

The Milky Way (including the halo) is roughly spherical with $R \approx 20$ kpc and $M \approx 4 \times 10^{11} \, M_\odot$ (Section 20-2D; Table 23-1), so

$$\rho_m = (4 \times 10^{11} \, M_\odot) \, (2.0 \times 10^{30} \, kg/M_\odot)$$
$$/ \{(4/3) \, \pi \, [(2.0 \times 10^4 \, pc) \times (3.1 \times 10^{16} \, m/pc)]^3\}$$
$$= 8.0 \times 10^{41} \, kg \, / \, 1.0 \times 10^{63} \, m^3$$
$$= 8.0 \times 10^{-22} \, kg/m^3$$

For the cD galaxy, Section 23-1E states that cD galaxies can have halos of up to 1 Mpc but that typical cDs have diameters of 200 kpc. Let's take $R \approx 100$ kpc, and $M \approx 10^{13} \, M_\odot$, so

$$\rho_m = (1.0 \times 10^{13} \, M_\odot) \, (2.0 \times 10^{30} \, kg/M_\odot)$$
$$/ \{(4/3) \, \pi \, [(1.0 \times 10^5 \, pc) \times (3.1 \times 10^{16} \, m/pc)]^3\}$$
$$= 2.0 \times 10^{43} \, kg \, / \, 1.2 \times 10^{65} \, m^3$$
$$= 1.7 \times 10^{-22} \, kg/m^3$$

So the closest the Milky Way galaxy can get to a typical cD galaxy without being disrupted is

$$d = 2.44 \times (1.7 \times 10^{-22} / 8.0 \times 10^{-22})^{1/3} \, R_{cD}$$
$$= 1.5 \, R_{cD}$$
$$\approx 150 \, kpc$$

Given the assumptions we made about the size and mass of the galaxies, we find that the Milky Way would have to be very near the cD galaxy to be disrupted. For a given mass of the cD galaxy, the radius we adopt does not influence the distance at which the Milky Way is disrupted (since $\rho_M^{1/3} \, R$ = constant). Our selection of the radius determines only whether the Milky Way disrupts "outside" or "inside" the cD galaxy.

23-5. Estimate the free-fall time for a cluster of galaxies and compare your result with the age of the Universe ($\approx 15 \times 10^9$ years). What do you conclude?

Following the derivation for the free-fall time for a cloud (Equation 15-10):

$$\tau_{ff} = (3\pi / 32 \, G \, \rho_o)^{1/2} = (6.64 \times 10^4) / \rho_o^{1/2} \, s \quad \text{(where } \rho_o \text{ in kg/m}^3\text{)}$$

We must estimate the mass density of a cluster of galaxies. From Table 23-1, the Local Group has a mass of about $7.5 \times 10^{11} \, M_\odot$ -- the sum of the masses of the individual galaxies. The characteristic length of the Local Group is about the distance between the

Milky Way and M31 (≈ 700 kpc). Assuming an initially spherical distribution gives a volume of about 1.4×10^9 kpc^3, so

$$\rho_o = (7.5 \times 10^{11} M_\odot) / (1.4 \times 10^9 \text{ kpc}^3)$$
$$= [(7.5 \times 10^{11} M_\odot) \times (2 \times 10^{30} \text{ kg}/M_\odot)]$$
$$/ [(1.4 \times 10^9 \text{ kpc}^3) \times (2.9 \times 10^{58} \text{ m}^3/\text{kpc}^3)]$$
$$= 1.5 \times 10^{42} \text{ kg} / 4.1 \times 10^{67} \text{ m}^3$$
$$= 3.7 \times 10^{-26} \text{ kg}/\text{m}^3$$

so the free-fall time is

$$\tau_{ff} = (6.64 \times 10^4) / (3.7 \times 10^{-26})^{1/2} \text{ s}$$
$$= 3.5 \times 10^{17} \text{ s}$$
$$= 1.1 \times 10^{10} \text{ yr} \quad \text{-- comparable to the age of the Universe}$$

23-6. An approximation to the luminosity function of galaxies in rich clusters is $\Phi(L) = \Phi^*(L/L^*)^{-5/4}$ for $L<L^*$ and $\Phi(L) = 0$ for $L>L^*$, where Φ^* is 0.005/Mpc^3 for H = 50 km/s·Mpc, $L^* \approx (10^{10})L_\odot$, and $L_{min} \approx (10^3)L_\odot$. The number of galaxies per cubic megaparsec is

$$N_{gal} = \int_{Lmin}^{Lmax} \Phi(L) \, dL$$

and the total luminosity per cubic megaparsec is

$$l_{gal} = \int_{Lmin}^{Lmax} \Phi(L) \, L \, dL$$

(a) Show that most galaxies have luminosities near the low end of the range.

The approximation to the luminosity function of galaxies is plotted in Figure 23-5. By inspecting the plot, it is apparent that there are more galaxies near the low end of the luminosity range. One can calculate the number density of galaxies by integrating the numeric fit to the luminosity function: $\Phi(L) = \Phi^*(L/L^*)^{-5/4}$ for $L < L^*$. Since the number of galaxies drops off exponentially at higher luminosities, these galaxies can be ignored in the count of the total number of galaxies: $\Phi(L) = 0$ for $L > L^*$. Integrating, we get the total number of galaxies:

$$N_{gal} = \int_{Lmin}^{Lmax} \Phi(L) \, dL$$
$$= \int_{Lmin}^{Lmax} \Phi^* (L/L^*)^{-5/4} \, d(L/L^*)$$
$$= \Phi^* [1/(-1/4)] \{ (L/L^*)^{-1/4} \Big|_{Lmin}^{Lmax} \}$$
$$= (-4) \Phi^* [(L_{max}/L^*)^{-1/4} - (L_{min}/L^*)^{-1/4}]$$
$$= (-4) \Phi^* [1 - (L_{min}/L^*)^{-1/4}]$$

where $L_{max} = L^*$ in the last step.

Substituting $\Phi^* = 0.005 / \text{Mpc}^3$, $L^* = 10^{10} L_\odot$, $L_{min} = 10^3 L_\odot$

271

$$N_{gal} = (-4)(0.005)[(1) - (10^3/10^{10})^{-1/4}] / Mpc^3$$
$$= (-2.0 \times 10^{-2}) \times (-5.523 \times 10^1) / Mpc^3$$
$$= 1.1 / Mpc^3$$

Now let's look at the number of galaxies in luminosity intervals:
$$N_{gal} = (-4) \Phi^* [(L_{upper}/L^*)^{-1/4} - (L_{lower}/L^*)^{-1/4}].$$
This calculation can be done for the intervals below:

$L_{lower}[L_\odot]$	10^3	10^4	10^5	10^6	10^7	10^8	10^9
$L_{upper}[L_\odot]$	10^4	10^5	10^6	10^7	10^8	10^9	10^{10}
$N_{gal}(L_l - L_u)$	0.49	0.28	0.16	0.09	0.05	0.03	0.02

Inspection of this table reveals that most of the galaxies have luminosities near the low end of the range.

(b) Show that most of the luminosity of a cluster is produced in the higher mass galaxies.

Similarly, the luminosity per cubic Megaparsec is given by:
$$l_{gal} = \int_{Lmin}^{Lmax} \Phi(L) \, L \, dL$$
$$= \int_{Lmin}^{Lmax} \Phi^* (L/L^*)^{-5/4} \, L \, d(L/L^*)$$
$$= \int_{Lmin}^{Lmax} \Phi^* (L/L^*)^{-5/4} (L/L^*)(L^*) \, d(L/L^*)$$
$$= \int_{Lmin}^{Lmax} \Phi^* L^* (L/L^*)^{-1/4} \, d(L/L^*)$$
$$= \Phi^* L^* [1/(3/4)] \{ (L/L^*)^{3/4} \Big|_{Lmin}^{Lmax} \}$$
$$= (4/3) \Phi^* L^* [(L_{max}/L^*)^{3/4} - (L_{min}/L^*)^{3/4}]$$
$$= (4/3) \Phi^* L^* [1 - (L_{min}/L^*)^{3/4}]$$
where $L_{max} = L^*$ in the last step.

Substituting $\Phi^* = 0.005 / Mpc^3$, $L^* = 10^{10} L_\odot$, $L_{min} = 10^3 L_\odot$
$$l_{gal} = (4/3)(0.005)(10^{10} L_\odot)[(1) - (10^3/10^{10})^{3/4}] / Mpc^3$$
$$= (6.67 \times 10^7 L_\odot) \times (1.000 \times 10^0) / Mpc^3$$
$$= 6.67 \times 10^7 L_\odot / Mpc^3$$

Now let's look at the total luminosity of galaxies in luminosity intervals:
$$l_{gal} = (4/3) \Phi^* L^* [(L_{upper}/L^*)^{3/4} - (L_{lower}/L^*)^{3/4}].$$
This calculation can be done for the intervals below:

$L_{lower}[L_\odot]$	10^3	10^4	10^5	10^6	10^7	10^8	10^9
$L_{upper}[L_\odot]$	10^4	10^5	10^6	10^7	10^8	10^9	10^{10}
$l_{gal}(L_l - L_u)$	1.7×10^3	9.8×10^3	5.5×10^4	3.1×10^5	1.7×10^6	9.8×10^6	5.5×10^7

Inspection of this table reveals that most of the luminosity of a cluster is produced by galaxies in the higher end of the range.

23-7. Only the radial velocity is observable for galaxies that are members of clusters. By integrating to obtain the mean value, determine how v^2 is related to v_r^2.

Consider only the velocity of galaxies within a cluster (ignore the radial velocity due to the Hubble expansion). The velocity vectors of the galaxies within the cluster will have random orientations to our line of sight. If the angle between our line of sight and the galaxy velocity, v, is θ, then the observed radial velocity is $v_r = v \cos(\theta)$, and the transverse velocity is $v_t = v \sin(\theta)$.

The mean value of the radial velocity squared is:

$$v_r^2 = \int v_r^2 \, dv_r \,/\, \int dv_r$$
$$= \int v^2 \cos^2(\theta) \, d(v \cos(\theta)) \,/\, \int d(v \cos(\theta))$$

The integration limits are 0 to π,

$$v_r^2 = \{(1/3) [v \cos(\theta)]^3\}\big|_0^\pi \,/\, [v \cos(\theta)]\big|_0^\pi$$
$$= (1/3) ((v)^3 (-1 - 1)) / ((v) (-1 -1)) = v^2 / 3$$

so $v^2 = 3 v_r^2$

23-8. Discuss how the existence of superclusters affects the Hubble diagram. What are the dangers in deriving the Hubble constant using galaxies within the Local (Virgo) Supercluster?

The Hubble diagram (law) is a plot of the radial velocity versus distance of galaxies. For distant galaxies, the radial velocity is due to the expansion of the Universe. The local supercluster is about 30 Mpc in diameter with the Milky Way (and the Local Group) at one edge. For H_o = 50 km/s/Mpc, a galaxy at the other end of the supercluster would have a cosmological redshift of 1,500 km/s. However, our galaxy's orbital motion about the supercluster is $\approx$ 500 km/s (Section 23-2B). If this motion is typical of galaxies within our supercluster, significant deviations in the Hubble diagram will occur for nearby galaxies - leading to an inaccurate value of the Hubble Constant.

For distant galaxies, the motion of the Milky Way in the local supercluster will result in an "error" in velocity of up to 500 km/s (or scatter in the Hubble diagram). In addition, the motion of distant galaxies in their superclusters would also contribute to scatter in the Hubble diagram (resulting in an error in determination of H). The effect would diminish at greater distances, where $v_{Hubble} > v_{random}$.

23-9. Discuss what happens to the kinetic energy of a galaxy that is cannibalized by a central cD galaxy. Relate this to the extended haloes of cD galaxies.

As a galaxy, which is moving relative to the cD galaxy, is eaten by the central cD galaxy in a cluster, the kinetic energy of the galaxy is transfered into random stellar motions within the galaxies. When the galaxy is "absorbed" into the cD, this increase in stellar velocities results in an expansion of the cD galaxy -- resulting in a large halos of stars around the dense bright nuclear region.

23-10. Discuss the evidence that M/L increases as the characteristic scale of the system increases. What does this imply about the distribution of dark matter?

According to Section 23-5, the mass to light ratio, M/L, for clusters is often as high as $\approx 300:1$ to $\approx 500:1$. This compares to the typical M/L ratio for a double galaxy of $\approx 100:1$ and for a single galaxy $\approx 50:1$ or less. On a smaller scale, M/L is typically ≈ 1 for a star and ≈ 0 for a planet. On the largest scales it might be amusing to know M/L of the entire Universe. Anyway, this all suggests that there is considerable dark matter between stars in a galaxy and considerably more dark matter between galaxies in a cluster, and presumably between clusters in a supercluster. Is there also dark matter in the voids?

23-11. Imagine that an astronomer in the Virgo Cluster of galaxies (D = 15.7 Mpc) wanted to observe our Local Group cluster of galaxies.

(a) What angular size would the Local Group subtend in the sky?

Using the small angle approximation, for θ in radians, where d is the linear size of an object at distance D,

$$\theta \approx \tan(\theta) = d/D$$

and assuming the Local Group subtends approximately 1 Mpc,

$$\theta \approx 1 \text{ Mpc} / 15.7 \text{ Mpc} = 6.4 \times 10^{-2} \text{ radians}$$
$$= 3.6°$$

(b) How large would the Milky Way Galaxy appear?

The Milky Way Galaxy is approximately 40 kpc in diameter, so it would subtend

$$\theta \approx 0.04 \text{ Mpc} / 15.7 \text{ Mpc} = 2.5 \times 10^{-3} \text{ radians}$$
$$= 0.15° \approx 9 \text{ arcmin}$$

(c) What would the apparent visual magnitude of the Milky Way Galaxy? Compare this with our naked-eye limiting magnitude of m = +6.

Using Equation 11-6 and an absolute magnitude for the Galaxy (Table 23-1) M = -21,

$$m = M + 5 \log(d_{pc}) - 5$$
$$= -21 + 5 \log(15.7 \times 10^6) - 5$$
$$= 10.0$$

This is 4 magnitudes fainter than the naked-eye limiting magnitude, so the Milky Way would not be visible with the naked-eye (nor, actually, with binoculars) from the Virgo Cluster.

(d) Assuming that our observer could build a telescope to detect m = +20 objects, could she detect all the Local Group galaxies listed in Table 23-1?

The limiting absolute magnitude (Equation 11-7) visible with this telescope would be

$$M = m + 5 - 5 \log(d_{pc})$$
$$= 20 + 5 - 5 \log(15.7 \times 10^6) = -11.0$$

Galaxies brighter than M = -11 would be visible (such as the Milky Way, Andromeda Galaxy (M31), the irregular galaxies, and some of the dwarf ellipticals) but several fainter dwarf ellipticals listed in table 23-1 (such as Leo II, Ursa Minor and Draco) would not be visible. An observer in the Virgo cluster with this telescope would have a reasonably good idea what our Local Group cluster is like.

(e) Based upon your answer to (d), comment on how observational selection effects may influence our understanding of distant galaxy clusters.

The hypothetical observer would not detect the lowest luminosity galaxies for the Local Group, which is a nearby cluster to the Virgo cluster. In more distant clusters the observer would not see most (any?) of the dwarf ellipticals and many irregular galaxies, which comprise the greatest number of galaxies in a cluster.

23-12. Observations of the maximum light from novae and supernovae in other galaxies are important as distance indicators (see Figure 22-3 and Table 22-1).

[Note the text incorrectly says Figure 23-3, instead of Figure 22-3.]

(a) Could a nova or supernova be observed in a galaxy in the Virgo Cluster (D = 15.7 Mpc) using a modest size Earth-based telescope with limiting magnitude m = +23?

The limiting absolute magnitude (Equation 11-7) visible with this telescope would be

$$M = m + 5 - 5 \log(d_{pc})$$
$$= 23 + 5 - 5 \log(15.7 \times 10^6)$$
$$= -7.0$$

Objects with absolute magnitude brighter than M = -7 would be visible in the Virgo cluster with this telescope.

From Table 22-1, supernovae have maximum luminosity $M \approx -16$ to -20, so they would be easily seen; novae have maximum luminosity $M \approx -8$ so they would be barely visible at maximum light but not prior to or after maximum.

(b) Could a nova or supernova be observed in a galaxy in the Corona Borealis cluster (asssume a cosmological redshift: cz = 21,600 km/s) using this telescope? Assume H = 50 km/s·Mpc.

Since z = 0.072 « 1, we can use the non-relativistic Doppler shift (Equation 22-3c):

$$d = cz / H$$
$$= (21{,}600 \text{ km/s}) / (50 \text{ km/s} \cdot \text{Mpc})$$
$$= 432 \text{ Mpc}$$

The limiting absolute magnitude (Equation 11-7) visible with this telescope would be

$$M = m + 5 - 5 \log(d_{pc})$$
$$= 23 + 5 - 5 \log(432 \times 10^6)$$
$$= -15.2$$

Objects with absolute magnitude brighter than M = -15 would be visible in the Corona Borealis cluster with this telescope. From Table 22-1, supernovae have maximum luminosity $M \approx -16$ to -20, so they would be seen (the lower luminosity supernovae may be difficult to see); novae have maximum luminosity $M \approx -8$ so they would not be visible.

Inspection of Table 22-1 indicates that, of the items listed, only the supernovae would be visible in the Corona Borealis cluster. This is why supernovae are important distance indicators in our study of the scale of the Universe, since no other stellar type objects are visible at these great distances.

23-13. Using Figure 23-11, estimate the linear extent of the void region near cz = 5500 km/s (assume H = 50 km/s·Mpc). Compare this size to the size of the Local Group, to the size of the Virgo cluster, and to the distance between the Local Group and the Virgo cluster.

The void stretches from cz ≈ 5000 km/s to ≈ 6400 km/s, corresponding to a redshift z of (6400 km/s) / (300,000 km/s) = 0.021 — so we can use the non-relativisitc Doppler shift, Equation 22-3c. Assuming H = 50 km/s·Mpc,

d_{inner} = c z / H

= (5000 km / s) / (50 km / s · Mpc)

= 100 Mpc

d_{outer} = (6400 km / s) / (50 km / s · Mpc)

= 128 Mpc

The void is thus 128 Mpc - 100 Mpc = 28 Mpc in extent.

Comparing this to the size of the Local group cluster of galaxies of 1 Mpc and the Virgo cluster of galaxies of 3 Mpc, we see that the void is much larger than either cluster. The Virgo cluster is 15.7 Mpc from the Local Group (we belong to the Virgo Supercluster which includes Virgo and the Local Group) — the void is nearly twice as large as this distance. In fact, simple inspection of Figure 23-11 reveals that the void is comparable in scale to the largest superclusters of galaxies. An important cosmological question is "Is there anything in the galaxy voids?"

Chapter 24: Active Galaxies and Quasars

24-1. The special theory of relativity says that no material object can move faster than the speed of light. The classical Doppler formula says that when the redshift z is greater than unity, we have v > c (which is impossible).

(a) Referring to Chapter 8, derive the *exact relativistic relation* between v and z. Express your result in the form v = f(z).

The redshift, z, is defined by Equation 22-1:

$$z \equiv (\Delta\lambda/\lambda_o) = (\lambda - \lambda_o)/\lambda_o$$

where λ_o is the rest wavelength and λ is the observed wavelength.

The classical Doppler formula is (Equation 22-2; Equation 8-13)

$$z = v/c$$

The relativistic Doppler formula is (Equation 8-14a):

$$\lambda = \lambda_o [(1 + v/c)/(1 - v/c)]^{1/2}$$

so we get Equation 24-1:

$$z \equiv \Delta\lambda/\lambda_o = (\lambda - \lambda_o)/\lambda_o = (\lambda/\lambda_o) - 1 = [(1 + v/c)/(1 - v/c)]^{1/2} - 1$$

Solving for v/c:

$$(z + 1)^2 = (1 + v/c)/(1 - v/c)$$
$$(1 - v/c)(z + 1)^2 = (1 + v/c)$$
$$(z + 1)^2 - (v/c)(z + 1)^2 = 1 + (v/c)$$
$$(z + 1)^2 - 1 = (v/c) + (v/c)(z + 1)^2 = (v/c)[(z + 1)^2 + 1]$$
$$v/c = [(z + 1)^2 - 1]/[(z + 1)^2 + 1]$$

(b) Make a table with three columns: column one for z, where you enter five exemplary values of z from 0 to 3.0; the second column headed $(v/c)_{cla}$, where you compute the classical *nonrelativistic* result for your five values of z; and the last column headed $(v/c)_{rel}$, where you use your formula from part (a).

z	$(v/c)_{classical}$	$(v/c)_{relativistic}$
0.0	0.0	0.00
0.5	0.5	0.38
1.0	1.0	0.60
1.5	1.5	0.72
2.0	2.0	0.80
2.5	2.5	0.85
3.0	3.0	0.88

24-2. Assume that the optical galaxy associated with the radio source Centaurus A is the same size as our Galaxy and draw a scale diagram of the radio and optical emission regions of Centaurus A. Clearly indicate the dimensions and the relative positions of the various components.

Centaurus A has radio features on several size scales - many much larger than the optical extent of the galaxy. From the inner region outward: there is a variable compact (point-like) source coincident with the center (core) of the galaxy, this component being dominant at millimeter wavelengths; a thin, single-sided non-uniform radio jet (Figure 24-11) is seen extending about 1 kpc from this core radio source; larger scale features in the single-sided jet are seen at radio and x-ray wavelengths, consisting of several blobs; this jet leads into one of the two symmetrically placed (about the core) lobes of radio emission about 10 kpc in size, stretching several galactic diameters from the optical galaxy (Figure 24-10) -- these jet and lobe features are strong and most easily mapped at centimeter wavelengths; on a larger scale two large lobes of radio emission of diameters 200 and 400 kpc lie in a rough extension of the inner lobes, these outer lobes being dominant at decimeter wavelengths, extending about 10 galactic diameters from the optical galaxy. For scale, the Milky Way Galaxy is ≈ 40 kpc in diameter.

24-3. The radio galaxy Cygnus A has an observed radio *flux density* of 2.18 x 10^{-23} W/m²·Hz at a frequency of 10^3 MHz. (Note that the unit of bandwidth $\Delta \nu$ is 1 Hz.) The observed redshift of the galaxy is $\Delta \lambda / \lambda_o = z = 0.170$.

(a) If the radiation is received at 10^3 MHz, at what (rest) frequency was it emitted by Cygnus A?
We can use the classical Doppler formula. From Equation 8-12:

$$\nu = \nu_o / (1 + v/c)$$

where ν is the observed frequency and ν_o is the emitted frequency, we derive

$$\nu / \nu_o = 1 / (1 + v/c)$$
$$\nu_o / \nu = 1 + v/c$$
$$z \equiv v/c = (\nu_o - \nu) / \nu ,$$

and from Equation 8-13,

$$\Delta \lambda / \lambda_o = (\lambda - \lambda_o) / \lambda_o = v/c \equiv z .$$

So $0.170 = (\nu_o - \nu) / \nu = (\nu_o / \nu) - 1$

$\nu_o = 1.170 \nu = 1.170$ x 10^3 MHz $= 1.17$ x 10^3 MHz   --- rest frequency

(b) What is the distance to Cygnus A? (Use a Hubble constant of $H_o = 50$ km/s·Mpc.)
Using the Hubble Law (Equation 22-2 and 22-3a) or Equation 22-3c:
$v = zc = Hd \implies d = cz/H$
the distance to Cygnus A is:
$$d = (3.0 \times 10^5 \text{ km/s})(0.170) / (50 \text{ km/s} \cdot \text{Mpc}) = 1.0 \times 10^3 \text{ Mpc}$$

(c) What is the *radio luminosity* (W/Hz) of this radio source at 10^3 MHz?
Given the radio flux density of 2.18 x 10^{-23} W/m²·Hz and a distance of
$$d = (1.0 \times 10^3 \times 10^6 \text{ pc})(3.086 \times 10^{16} \text{ m/pc}) = 3.09 \times 10^{25} \text{ m}$$
the radio luminosity at 10^3 MHz (received) is
$$L = F 4 \pi d^2$$

$$= (2.18 \times 10^{-23} \text{ W/m}^2/\text{Hz}) \, 4\pi \, (3.09 \times 10^{25} \text{ m})^2$$
$$= 2.6 \times 10^{29} \text{ W/Hz}$$

(d) To find the total radio luminosity of Cygnus A, we must multiply the result of part (c) by the *bandwidth* $\Delta\nu$ of our detector. Assume $\Delta\nu = 10^4$ Hz and compute the energy radiated per second at radio frequencies.

If the bandwidth, $\Delta\nu$, of our detector is 10^4 Hz, then the energy radiated per second at these radio frequencies is:

$$L_{\text{"total"}} = L \, \Delta\nu = (2.6 \times 10^{29} \text{ W/Hz}) \times (10^4 \text{ Hz}) = 2.6 \times 10^{33} \text{ W}$$

Note that this bandpass includes only a small portion of the radio spectrum, so this is not really the total radio luminosity. To determine the total radio luminosity we would have to integrate over the entire radio spectrum of the source. Note, this luminosity is $\approx (2/3) \times 10^7 \, L_\odot$.

(e) What is the minimum mass of hydrogen (in solar masses) that must be converted to helium during each second to provide this luminosity?

Referring to Section 16-1D, the conversion of four ^{1}H nuclei into one ^{4}He nucleus releases 4.3×10^{-12} J. To account for the luminosity in part (d) there must be

$$(2.6 \times 10^{33} \text{ J/s}) / (4.3 \times 10^{-12} \text{ J}) = 6.0 \times 10^{44} \text{ reactions/s}.$$

This reaction rate corresponds to a mass of hydrogen converted to helium of:

$$(6.0 \times 10^{44} \text{ reactions/s}) \times (4 \text{ H atoms/reaction}) \times (1.67 \times 10^{-27} \text{ kg/H})$$
$$= 4.0 \times 10^{18} \text{ kg/s} = 2.0 \times 10^{-12} \, M_\odot/\text{s} = 6.3 \times 10^{-5} \, M_\odot/\text{yr}.$$

Thus, if conversion (fusion) of hydrogen into helium is the source of energy in Cygnus A, $2.0 \times 10^{-12} \, M_\odot$ of hydrogen must be consumed per second.

(f) If Cygnus A continues to radiate at this rate for 10^8 years, how many solar masses of hydrogen must be converted to helium? Express this result in terms of the mass of our Galaxy ($\approx 10^{12} \, M_\odot$).

If this rate of energy production is to continue for 10^8 yr, then $6.3 \times 10^3 \, M_\odot$ of hydrogen would be consumed. This is $\approx 6.3 \times 10^{-9}$ the mass of the Milky Way Galaxy.

24-4. 3C 9 is a quasi-stellar object that has a redshift of 2.0 and an apparent visual magnitude of 18.2. Answer the following questions using the cosmological interpretation of the redshift.
(a) What is the speed of recession?
Assuming that the redshift is cosmological, we need to use the relativistic Doppler shift (Equation 24-1; Equation 8-14a):

$$z \equiv \Delta\lambda/\lambda_o = (\lambda - \lambda_o)/\lambda_o = (\lambda/\lambda_o) - 1 = [(1 + v/c)/(1 - v/c)]^{1/2} - 1$$

to derive (see Problem 24-1):

$$v/c = [(z+1)^2 - 1]/[(z+1)^2 + 1]$$

so for z = 2, v / c = 0.8, or v = 2.4 x 10^5 km / s.

(b) What is the distance to 3C 9?

Using a value of the Hubble constant, H = 50 km / s · Mpc, and the Hubble law for z > 1, the distance to 3C 9 is (Equation 22-3d):

$$d = (cz/H)(1+z/2)/(1+z)^2$$
$$= [(3 \times 10^5 \text{ km/s})(2.0)/(50 \text{ km/s} \cdot \text{Mpc})]((1+2/2)/(1+2)^2)$$
$$= 2.7 \times 10^3 \text{ Mpc}$$

If we used a value H = 100 km / s · Mpc, the distance would be d = 1.3 x 10^3 Mpc.

(c) What is the intrinsic luminosity relative to that of our Galaxy?

From Equation 11-20b, the luminosity relative to the Sun is:

$$\log(L_{3C\,9}/L_\odot) = 1.9 - 0.4\, M_{bol(3C\,9)}$$

Given m = +18.2 for 3C 9, and using d = 2.7 x 10^3 Mpc, the absolute magnitude is (Equation 11-7):

$$M = m + 5 - 5\log(d_{pc})$$
$$= 18.2 + 5 - 5\log(2.7 \times 10^9) = -24.0$$

Assuming no bolometric correction,

$$\log(L_{3C\,9}/L_\odot) = 1.9 - 0.4(-24.0) = 11.5$$
$$L_{3C\,9} = 10^{11.5} L_\odot = 3.2 \times 10^{11} L_\odot$$

The luminosity of the Milky Way is $L_{MW} \approx 2.5 \times 10^{10} L_\odot$ (Section 21-1B), so the luminosity of 3C 9 is

$$L_{3C\,9} = 3.2 \times 10^{11} L_\odot / 2.5 \times 10^{10} L_\odot \approx 13\, L_{MW}$$

If we had selected a distance of d = 1.3 x 10^3 Mpc, corresponding to H = 100 km / s · Mpc, the luminosity of 3C 9 would be

$$M = -22.4$$
$$L_{3C\,9} = 6.9 \times 10^{10} L_\odot \approx 2.8\, L_{MW}$$

(d) What is the maximum size of the emitting region if 3C 9 exhibits luminosity variations on a time scale of two months?

If variations occur in a characteristic time scale of 2 months, then the emitting region must be less than 2 light months (0.05 pc) in size. This is the amazing aspect of quasars -- they emit more energy than a typical spiral galaxy, yet they are much smaller than the average distance between stars in our galaxy -- a tremendous luminosity emitted from a small volume.

24-5. (a) Quasar 3C 273 has a redshift of 0.16. What is its distance?

Since the redshift is much less than 1, let's first use the classical Doppler shift formula (Equation 22-2; Section 24-4A):

$$z = v/c$$
$$v = zc = 0.16 \times (3 \times 10^5 \text{ km/s}) = 4.8 \times 10^4 \text{ km/s}$$

and using the Hubble law (Equation 22-3c):

$$d = v/H = zc/H$$

If $H = 50$ km/s·Mpc, then the distance to 3C 273 is

$$d = (0.16)(3 \times 10^5 \text{ km/s})/(50 \text{ km/s} \cdot \text{Mpc}) = 960 \text{ Mpc}$$

If $H = 100$ km/s·Mpc, the $d = 480$ Mpc.

If we'd used the relativistic Doppler formula for redshift (Equation 24-1; see Problem 24-1 or 24-4):

$$v/c = [(z+1)^2 - 1]/[(z+1)^2 + 1]$$
$$= [(0.16+1)^2 - 1]/[(0.16+1)^2 + 1] = 0.147 = 4.4 \times 10^4 \text{ km/s}$$

(this is a velocity difference of ≈9% from the non-relativisitic formula)

corresponding to a Hubble distance (Equation 22-3d)

$$d = (cz/H)(1+z/2)/(1+z)^2$$
$$= [(3 \times 10^5 \text{ km/s})(0.16)/(50 \text{ km/s} \cdot \text{Mpc})]((1+0.16/2)/(1+0.16)^2$$
$$= 7.7 \times 10^2 \text{ Mpc} = 770 \text{ Mpc} \quad \text{--- for } H = 50 \text{ km/s} \cdot \text{Mpc}$$

or $d = 385$ Mpc for $H = 100$ km/s·Mpc

Assuming that the student will use the non-relativistic Doppler shift, the rest of the problem is done using this approximation.

(b) The V magnitude of 3C 273 is 12.8. What is its flux density at V band? Its luminosity at V band?

[We assume in these calculations the optical spectrum of 3C 273 is like that of the Sun - which is *not* true - for lack of sufficient information. We lack information on the bolometric correction for 3C 273.]

From Equation 11-20b, the luminosity relative to the Sun is:

$$\log(L_{3C\ 273}/L_\odot) = 1.9 - 0.4 M_{bol(3C\ 273)}$$

Given $m = +12.8$ for 3C 273, and using $d = 960$ Mpc $= 9.6 \times 10^8$ pc (for $H = 50$ km/s·Mpc), the absolute V band magnitude is (Equation 11-7):

$$M_V = m_V + 5 - 5\log(d_{pc})$$
$$= 12.8 + 5 - 5\log(9.6 \times 10^8) = -27.1$$

Assuming no bolometric correction,

$$\log(L_{3C\ 273}/L_\odot) = 1.9 - 0.4(-27.1) = 12.7$$
$$L_{3C\ 273} = 10^{12.7} L_\odot = 5.0 \times 10^{12} L_\odot = 2.0 \times 10^{39} \text{ W}$$

[For comparison, the luminosity of the Milky Way is $L_{MW} \approx 2.5 \times 10^{10} L_\odot$ (Section 21-1B), so the luminosity of 3C 273 is

$$L_{3C\ 273} = 5.0 \times 10^{12} L_\odot / 2.5 \times 10^{10} L_\odot = 200\ L_{MW}\ ]$$

If we had selected a distance of d= 480 Mpc, corresponding to H = 100 km / s · Mpc, the luminosity of 3C 273 would be

$$M = -25.6$$

$$L_{3C\ 273} = 10^{12.1} L_\odot = 1.3 \times 10^{12} L_\odot \approx 50\ L_{MW} = 0.5 \times 10^{39}\ W$$

The flux is related to the luminosity of 3C273 in the V band

$$f_V = L_V / 4\pi d^2$$
$$= (2.0 \times 10^{39}\ W) / 4\pi (960 \times 10^6 \times 3.09 \times 10^{16}\ m)^2$$
$$= 1.8 \times 10^{-13}\ W/m^2 \quad \text{(for d = 960 Mpc, H = 50 km / s · Mpc)}$$

If H = 100 km / s · Mpc, then $f_V = 4.5 \times 10^{-14}\ W/m^2$.

The flux density, S_V, is the energy per unit time (W = J/s) per unit area (/m²), or flux, per unit frequency (/Hz) received on the Earth from an object. For the visual bandpass, $\lambda \approx 550$ nm ($\nu \approx 5.5 \times 10^{14}$ Hz), and $\Delta\lambda \approx 100$ nm ($\Delta\nu \approx 10^{14}$ Hz) (see Figure 11-3), so the visual flux density, S_V, is:

$$S_{V\ 3C273} = (1.8 \times 10^{-13}\ W/m^2) / (10^{14}\ Hz)$$
$$= 1.8 \times 10^{-27}\ W/m^2/Hz = 1.8 \times 10^{-1}\ Jy = 180\ mJy$$

where we have used a unit used by astronomers called the jansky, where 1 Jy = 10^{-26} W / m² / Hz; 1 mJy ≡ 1 milli-Jy.

(c) The fuzz around 3C 273 has a diameter of 15". What is its linear size?
Using the small angle approximation, the angular size in arcsec of the fuzz is:

$$\theta_{["]} = 206265\ r/d\ ,\text{ where r is the linear size, and d is the distance.}$$

For d = 960 Mpc (for H = 50 km / s · Mpc),

$$r = (15) \times (9.6 \times 10^8\ pc) / (2.06 \times 10^5) = 7.0 \times 10^4\ pc = 70\ kpc.$$

For d = 480 Mpc (H = 100 km / s · Mpc), r = 35 kpc.
In either case, this warm fuzzy around 3C 273 is about the size of a galaxy -- perhaps it is the galaxy in which the quasar 3C 273 is imbedded!

(d) The absolute magnitude of the fuzz is -25. What is its luminosity?
Note the fuzz is 2.1 magnitudes fainter than the quasar (calculated in part (b)). The luminosity of the fuzzy, assuming no bolometric correction, is (Equation 11-20b):

$$\log(L_{fuzzy}/L_\odot) = 1.9 - 0.4\ M_{bol\ fuzzy} = 1.9 - 0.4\ (-25) = 11.9$$

$$L_{fuzzy} = 7.9 \times 10^{11}\ L_\odot$$

This is about the luminosity of a bright galaxy, confirming our suspicion that the fuzzy was a galaxy.

24-6. The quasar PKS 1402+044 has a redshift of 3.2. What is its distance? Note that z > 1!
Using Equation 24-1 (see also the result of Problem 24-1), since z > 1 we need to use the relativistic Doppler shift:

$$v = \{[(z+1)^2 - 1] / [(z+1)^2 + 1]\} c$$
$$= \{[(3.2+1)^2 - 1] / (3.2+1)^2 + 1]\} c$$
$$= 0.89 c = 2.7 \times 10^5 \text{ km/s}$$

corresponding to a Hubble distance (Equation 22-3d):

$$d = (cz/H)(1 + z/2) / (1 + z)^2$$
$$d = [(3 \times 10^5 \text{ km/s})(3.2) / (50 \text{ km/s} \cdot \text{Mpc})]((1 + 3.2/2) / (1 + 3.2)^2$$
$$= 2.8 \times 10^3 \text{ Mpc} = 280 \text{ Mpc} \quad \text{--- for } H = 50 \text{ km/s} \cdot \text{Mpc}$$

If $H = 100 \text{ km/s} \cdot \text{Mpc}$, $d = 1.4 \times 10^3 \text{ Mpc}$

24-7. Observations of the radio jet in Cen A indicate that the spectral index is about 0.5. The 20-cm flux density of the strongest blob of the jet is 2.3 Jy (janskies, 1 Jy = 10^{-26} W/m^2·Hz). If this emission is synchrotron radiation, what should the flux density be at 2.2 µm?
The spectral index is defined by (Section 24-1B):

$$F(\nu) = F_o \nu^{-\alpha}$$

so $F(\nu_1)/F(\nu_2) = (\nu_1/\nu_2)^{-\alpha} = (\lambda_2/\lambda_1)^{-\alpha} = (\lambda_1/\lambda_2)^{\alpha} = F(\lambda_1)/F(\lambda_2)$

$F(2.2 \text{ µm}) / F(20 \text{ cm}) = (2.0 \times 10^{-1} \text{ m} / 2.2 \times 10^{-6} \text{ m})^{-0.5} = 3.32 \times 10^{-3}$

$F(2.2 \text{ µm}) = 3.32 \times 10^{-3} (2.3 \text{ Jy}) = 7.6 \times 10^{-3} \text{ Jy} = 7.6 \text{ mJy}$

(1 mJy ≡ 1 milli-Jansky)

24-8. (a) How close must a star pass a black hole of mass 10^6 M$_\odot$ to be disrupted tidally?

For a star to be tidally disrupted by a 10^6 M$_\odot$ black hole, it must stray to within the Roche limit of the black hole. The Roche limit is given by Equation 3-9:

$$d = 2.44 (\rho_M / \rho_m)^{1/3} R$$

where ρ_M is the density of the black hole
    ρ_m is the density of the star
    R is the radius of the black hole.

For ρ_m use $\rho_\odot = 1400 \text{ kg/m}^3$

For ρ_M assume the black hole fills the Schwarzschild radius ($R_s = 3 \times 10^6$ km for a 10^6 M$_\odot$ black hole):

$$\rho_M = M/V = M / [(4/3) \pi R^3]$$

$$= [(10^6 \, M_\odot) (2.0 \times 10^{30} \, kg)] / [(4/3) \pi (3 \times 10^9 \, m)^3]$$
$$= 2.0 \times 10^{36} \, kg / 1.1 \times 10^{29} \, m^3$$
$$= 1.8 \times 10^7 \, kg/m^3$$

So $d = 2.44 \, (1.8 \times 10^7 / 1.4 \times 10^3)^{1/3} \, R_s$
$$= 5.7 \times 10^1 \, R_s = 57 \, R_s$$
$$= (5.7 \times 10^1)(3.0 \times 10^6 \, km) = 1.7 \times 10^8 \, km$$
$$= 250 \, R_\odot \approx 1.1 \, AU$$

If the star approaches within 57 Schwarzschild radii, ≈ 1.1 AU, of the $10^6 \, M_\odot$ black hole it will be tidally disrupted.

(b) What *yearly* rate of matter infall is needed to power a quasar at 10^{39} W if a black hole of mass $10^6 \, M_\odot$ lies in its core?

In Section 24-5A, it was shown that the change in the potential energy when matter (mass M) falls from infinity to the Schwarzschild radius of a black hole of:

$R = 3 \, km \times M/M_\odot$

is $PE = -GMM/R$

$10^{39} \, W = -(6.67 \times 10^{-11} \, N \, m^2/kg^2)(10^6 \, M_\odot \times 2.0 \times 10^{30} \, kg/M_\odot) \, M$
$/ (3.0 \times 10^3 \, m \times 10^6)$

$M = (10^{39} \, W) / (4.45 \times 10^{16} \, J/kg)$
$= 2.25 \times 10^{22} \, kg/s = 1.1 \times 10^{-8} \, M_\odot/s = 0.36 \, M_\odot/yr$

This assumes that the conversion from potential energy to light is 100% efficient. (See Concept Application in Section 24-5A.)

24-9. In Cen A, the radio jet and the innermost lobe are separated by a distance of 4 arcmin.
(a) What is their physical separation?
The radio galaxy Centaurus A is at a distance (d) of 4 Mpc (Section 24-3C). Using the small angle approximation, the physical separation (r) of the radio jet and innermost lobe, given an angular separation (θ) of 4 arcmin is given by

$\theta_{[rad]} = r/d$
$\theta_{[']} = 3.44 \times 10^3 \, r/d$
$r = 4 \times (4 \times 10^6 \, pc) / (3.44 \times 10^3) = 4.7 \times 10^3 \, pc = 4.5 \, kpc$

(b) How long would it take relativistic electrons to travel from the jet to the lobe?
Relativistic electrons travel close to the speed of light, so they would travel from the jet to the lobe -- a distance of

$r = (4.7 \times 10^3 \, pc) \times (3.26 \, l.y./pc) = 15,000 \, l.y$ --- in 15,000 yr.

24-10. The quasar 1059+730 has a redshift of 0.089.

(a) What is the distance to the quasar if the redshift is cosmological? What is the range, given the uncertainty in H_o?

Since the redshift is $\ll 1$, let's try the non-relativistic Doppler shift (Equations 22-1 and 22-2; see also Section 24-4A):

$z \equiv \Delta\lambda / \lambda = v/c$

and Hubble's Law (Equation 22-3b and 22-3c):

$d = v/H = zc/H$

If $H = 50$ km/s · Mpc,

$d = (0.089)(3.0 \times 10^5 \text{ km/s}) / (50 \text{ km/s · Mpc}) = 534$ Mpc

If $H = 100$ km/s · Mpc, $d = 267$ Mpc.

The relativistic Doppler shift (Equation 24-1; Problem 24-1):

$v = \{[(z+1)^2 - 1] / [(z+1)^2 + 1]\} c$

gives a distance of

$d = \{[(z+1)^2 - 1] / [(z+1)^2 + 1]\} c / H$.

If $H = 50$ km/s · Mpc,

$d = \{[(0.089+1)^2 - 1] / [(0.089+1)^2 + 1]\}(3.0 \times 10^5 \text{ km/s}) / (50 \text{ km/s · Mpc})$
$= 510$ Mpc

If $H = 100$ km/s · Mpc, then $d = 255$ Mpc.

Note that the uncertainty in the Hubble constant is much more important than whether we select to use the relativistic or non-relativistic Doppler shift formula.

(b) The fuzz around the quasar has an angular size of 9" x 16". What is its physical size? How does this compare with the size of a typical spiral galaxy?

The fuzz around the quasar is at a distance, d, of 534 Mpc (using $H = 50$ km/s · Mpc and the non-relativisitic Doppler shift). Using the small angle approximation, the physical size (r) of the fuzz, given an angular size (θ) of 9 x 16 arcsec is:

$\theta_{[rad]} = r/d$

$\theta_{["]} = 2.06 \times 10^5 \, r/d$

$r_1 = 9 \times (5.34 \times 10^8 \text{ pc}) / (2.06 \times 10^5) = 2.3 \times 10^4$ pc $= 23$ kpc

$r_2 = 16 \times (5.34 \times 10^8 \text{ pc}) / (2.06 \times 10^5) = 4.1 \times 10^4$ pc $= 41$ kpc

So, if $H = 50$ km/s · Mpc, the physical size of the fuzz is 23 x 41 kpc.

If $H = 100$ km/s · Mpc, the physical size of the fuzz is 12 x 21 kpc.

The Milky Way has a diameter of about 30 kpc, so the fuzz is of a size comparable to a galaxy -- assuming that the quasar redshift is cosmological.

(c) The supernova observed in 1059+730 had an apparent V magnitude of 19.6. Use this value to estimate the distance to the quasar. Compare this value with your results from (a).

A Type I supernova typically has M ≈ -20 (Table 18-3), so the distance can be found from Equation 11-6:

$\log(d_{pc}) = (m - M + 5) / 5 = (19.6 - (-20) + 5) / 5 = 8.92$

$d = 8.3 \times 10^8$ pc $= 8.3 \times 10^2$ Mpc $= 830$ Mpc

This is slightly greater than the distance estimated from the redshift in part (a). If the absolute magnitude of the supernova was -18 (typical of a Type II supernova), then the distance would be 331 Mpc, more in agreement with part (a). Without further information on the classification of the supernova, it can be said that the distance derived from the supernova is consistent with that derived from the Doppler shift and Hubble law.

24-11. The argument is often made that the dimensions of an object with variable brightness cannot exceed the speed of light times the time scale of the variations. Critically examine this argument by considering
(a) two nonspherical geometries

Consider first a slightly curved or flat screen (partial shell) located in front of an "exciting" source. By some unspecified mechanism, the source incites the screen to vary. Since the light travel time from the exciting source to the shell is about the same for all points on the shell, the shell will vary simultaneously in its own rest frame. There will be some time delay between variations in different parts of the shell as seen at the Earth, since the light travel time from points on the shell to the Earth will be different. However, this delay will be less than the light travel time from one side of the shell to the other.

Second, consider a cylinder, nearly end on to the observer. If a shock wave, or some other plane parallel "signal", travels down the cylinder inciting the material in the cylinder to vary, we will see the entire face of the material vary simultaneously.

(b) special relativistic effects (note that observed time intervals are inversely proportional to observed frequencies)

As discussed in Section 24-5B, if a jet emits blobs of material close to the line of sight of the observer, special relativistic effects will enhance the apparent flux density and also result in apparent time dilation. Variations in source intensity would appear to occur in a shorter time interval than is occurring in the source's rest frame, and the total energy produced would be less than that estimated by the observer -- partially resolving the energy problem in quasars.

24-12. The quasar NRAO 140 (z = 1.26) has a compact radio component that moves at an angular speed of 0.15 milli-arcsec / year. What is the apparent velocity of the component as a percentage of c? (Multiply your answer by 1 + z to correct for relativistic time dilation. Use a Hubble constant of 50 km/s·Mpc.)

From the redshift of NRAO140 (z = 1.26) we can compute its distance, using the relativistic Doppler shift. Using the results of Problem 24-1 (see also Equation 24-1) the cosmological recessional velocity of the quasar is:

$v / c = [(z + 1)^2 - 1] / [(z + 1)^2 + 1]$

$$= [(1.26 + 1)^2 - 1] / [(1.26 + 1)^2 + 1]$$
$$= 0.673$$
$$v = 2.0 \times 10^5 \text{ km/s}$$

corresponding to a Hubble distance (Equation 22-3d), for H = 50 km/s · Mpc:
$$d = (cz/H)(1 + z/2)/(1+z)^2$$
$$d = [(3 \times 10^5 \text{ km/s})(1.26)/(50 \text{ km/s} \cdot \text{Mpc})]((1 + 1.26/2)/(1 + 1.26)^2)$$
$$= 2.4 \times 10^3 \text{ Mpc} = 2,400 \text{ Mpc} = 7.4 \times 10^{25} \text{ m}$$

Converting the observed angular motion (0.15 milli-arcsec / yr) to velocity,
$$\theta = (0.15 \times 10^{-3} \text{ "/yr}) \times (1 \text{ rad} / 2.06 \times 10^5 \text{ "}) \times (1 \text{ yr} / 3.16 \times 10^7 \text{ s})$$
$$= 2.3 \times 10^{-17} \text{ rad/s}$$

Using the small angle formula for the velocity
$$v = d \tan(\theta) \approx d\,\theta = (7.4 \times 10^{25} \text{ m})(2.3 \times 10^{-17} /\text{s})$$
$$= 1.7 \times 10^9 \text{ m/s} = 5.7 c$$

Correcting for relativistic time dilation,
$$v_{corr} = 5.7 c (1 + z) = 5.7 c (1 + 1.26) = 12.9 c$$

The component appears to move with a velocity of ≈ 13 times the speed of light -- superluminal motion.

24-13. A quasar has an emission line, identified as Ly-α of hydrogen (λ_o = 121.6 nm), that is observed at 581.2 nm. Calculate the redshift and distance to the quasar for a Hubble constant of 50 km/s · Mpc. How fast is the quasar moving away from us?

The redshift is given by (Equation 22-1):
$$z \equiv \Delta\lambda / \lambda_o = (\lambda - \lambda_o)/\lambda_o$$
$$= (581.2 \text{ nm} - 121.6 \text{ nm}) / 121.6 \text{ nm}$$
$$= 3.78$$

Using the relativistic Doppler shift (see Problem 24-1, derived from Equation 24-1) to determine the recessional velocity:
$$v = [(z+1)^2 - 1] / [(z+1)^2 + 1] c$$
$$= [(3.8 + 1)^2 - 1] / [(3.8 + 1)^2 + 1] c$$
$$= 0.92 c = 2.75 \times 10^5 \text{ km/s}$$

corresponding to a Hubble distance (Equation 22-3d), for H = 50 km/s · Mpc:
$$d = (cz/H)(1 + z/2)/(1+z)^2$$
$$d = [(3 \times 10^5 \text{ km/s})(3.78)/(50 \text{ km/s} \cdot \text{Mpc})]((1 + 3.78/2)/(1 + 3.78)^2)$$
$$= 2.87 \times 10^3 \text{ Mpc} = 2,870 \text{ Mpc}$$

24-14. At what what redshift is the Ly-α line brought into a visible light detector that is sensitive to photons of wavelength greater than 370 nm?
[See also Problem 22-12a, where the visible limit is given as 390 nm.]

Use the Doppler shift formula (Equation 22-1 or Section 24-4A), where λ is the observed wavelength and λ_o is the emitted wavelength = 121.6 nm (see Section 8-2C):

$$z \equiv \Delta\lambda/\lambda = (\lambda - \lambda_o)/\lambda_o$$
$$= (370 \text{ nm} - 121.6 \text{ nm}) / 121.6 \text{ nm}$$
$$= 2.04$$

Objects with redshift 2.04 or greater will have the Lyman-α line shifted to wavelengths longer than 370 nm.

24-15. Discuss the evidence that Seyfert 1, Seyfert 2, LINERs, radio galaxies, BL Lac objects, and quasars are all related and should be lumped together into one class called AGNs. What kinds of physical phenomena or conditions make us detect these objects as *different* classes when we observe them?

All these classes of objects share similarities in terms of luminosity and range of distances but exhibit a much wider range in parameters than might be expected of a unified class of objects. It has been proposed that the viewing angle (the angle between our line of sight and the disk of the underlying galaxy and/or accretion disk surrounding a supermassive (black hole?) object at the center of the host galaxy) causes us to see the objects with different structural, dynamical, and compositional parameters. For example, BL Lacs and highly variable quasars may be oriented such that a jet emanating from the core is pointed more or less toward the observer, so that we are seeing down into the inner parts of the supposed accretion disk while less active quasars and radio galaxies have this jet oriented more in the plane of the sky. Similar orientation effects cause us to distinguish between the Seyfert 1 and 2 galaxies, by hiding gas behind the disk when viewed edge-on. Some evolutionary effects must also be considered, where the quasar phenomenon may have been more prevalent in the early days of the Universe. Of concern also is the environment in which the object finds itself. Recent interest has centered around interactions between passing galaxies as being responsible for the initiation of quasar like activity.

24-16. How fast would a dust cloud have to be moving to occult a typical Seyfert galaxy BLR in the time scale of one year? Assume that the cloud is in circular orbit around the galaxy nucleus.

From Section 24-3A, the broad-line region (BLR) has an upper limit to its size of 10^{14} m. So for a dust cloud to occult the BLR it would have to move roughly 10^{14} m in its orbit around the galaxy nucleus. To do this in a year, the cloud must move at a velocity

$$v = d/t$$
$$= 10^{14} \text{ m} / 3.16 \times 10^7 \text{ s} \approx 3 \times 10^6 \text{ m/s} \approx 0.01 \, c$$

If the cloud is far from the BLR, a straight line distance of 10^{14} m is a good approximation to the distance traveled in a circular path. If the cloud is just outside the BLR, the distance traveled in its circular orbit is half the circumference

$$d' = (1/2) \pi d$$

or roughly 1.5 times the linear estimate.

24-17. To appreciate the observational difficulties associated with trying to detect a galaxy's stellar emission around a distant quasar, calculate the apparent magnitude (ignore the k-correction) and angular size that a large, luminous galaxy (M = -21; R = 50 kpc) would have at a redshift of (Use H = 50 km / s · Mpc):

(a) $z = 0.1$

For the nearby galaxy, we can use the non-relativistic Hubble law (Equation 22-3c) to get its distance:

$$d = cz/H$$
$$= (3 \times 10^5 \text{ km/s})(0.1) / 50 \text{ km/s} \cdot \text{Mpc}$$
$$= 6 \times 10^2 \text{ Mpc} = 600 \text{ Mpc}$$

The apparent magnitude can be calculated from Equation 11-6:

$$m = M + 5 \log(d_{pc}) - 5$$
$$= -21 + 5 \log(600 \times 10^6) - 5$$
$$= 17.9$$

This would be visible with modern telescopes. (In fact, a 0.5 m class telescope equipped with a CCD could easily detect an object of this brightness.)

The angular size, in arcsec, can be calculated using the small angle approximation, where r is the linear size and d the distance to the galaxy:

$$\theta_{[\text{''}]} \approx \tan(\theta) = 206{,}265 \, r/d$$
$$= (2.06 \times 10^5)(50 \times 10^3 \text{ pc}) / (600 \times 10^6 \text{ pc})$$
$$\approx 17 \text{ arcsec}$$

Again, this angular size object should be visible in modern telescopes.

(b) $z = 1.0$

For the distant galaxy, we must use the relativistic Hubble law (Equation 22-3d) to get its distance:

$$d = (cz/H)(1 + z/2)/(1 + z)^2$$
$$= [(3 \times 10^5 \text{ km/s})(1.0) / 50 \text{ km/s} \cdot \text{Mpc}](1 + 1/2)/(1 + 1)^2$$
$$= (2.25 \times 10^3 \text{ Mpc}) \approx 2{,}250 \text{ Mpc}$$

The apparent magnitude can be calculated from Equation 11-6:

$$m = M + 5 \log(d_{pc}) - 5$$
$$= -21 + 5 \log(2250 \times 10^6) - 5$$
$$= 20.8$$

This would be visible, though more difficult, with modern telescopes.

The angular size, in arcsec, can be calculated using the small angle approximation, where r is the linear size and d the distance to the galaxy:

$$\theta_{[\text{''}]} \approx \tan(\theta) = 206{,}265 \, r/d$$
$$= (2.06 \times 10^5)(50 \times 10^3 \text{ pc}) / (2250 \times 10^6 \text{ pc})$$
$$\approx 4.6 \text{ arcsec}$$

Again, this angular size object should be visible in modern telescopes as a smudge. However if the quasar is much brighter than the underlying galaxy it would be difficult to detect this small an angular feature surrounding an object hundreds of times brighter.

Note that that the galaxy magnitude is the total integrated magnitude, so the surface brightness of the galaxy would be less than 20.8 magnitudes per unit area (such as per square arcsec) in the sky, making the detection even more difficult than may be suggested by the results of the calculations above. The fact that galaxies surrounding quasars have been detected for only the nearest objects supports this argument.

24-18. Using Figure 24-15, determine the redshift of the quasar 3C 273. How well does your measurement compare with the accepted value of z = 0.158?

Let's take the H-β emission line, which has a rest (emitted) wavelength of 486.1 nm (Table 24-1). [We could have calculated the wavelength of the H-β emission line using Equation 8-25:

$1/\lambda_{ab} = R(1/n_b^2 - 1/n_a^2)$
$= (10.96774 \, \mu m^{-1})(1/2^2 - 1/4^2) = 2.056 \, \mu m^{-1}$
$\lambda_{ab} = 0.4863 \, mm = 486.3 \, nm$]

To determine the observed wavelength, first determine the scale of the spectrum then measure the distance from a wavelength marker to the observed H-β line, using the scale to determine the corresponding wavelength. Measuring, the distance between the markers 5016 Å (=501.6 nm) and 6030 Å (= 603.0 nm) is 3.34 mm. So the scale of the spectrum in the figure is

(603.0 nm - 501.6 nm) / 3.34 mm = 30.4 nm / mm

The H-β line is 2.07 mm to the right of the 5016 Å wavelength marker, so the observed wavelength is

501.6 nm + (30.4 nm / mm) x (2.07 mm) = 564 nm

corresponding to a redshift (Equation 22-1,

$z \equiv \Delta\lambda/\lambda = (\lambda - \lambda_o)/\lambda_o$
= (564 nm - 486.1 nm) / 486.1 nm
= -0.160

This is close to (to within ≈ 1%) the accepted value of 0.158, which is exceptional considering the measurement errors in measuring such a small figure!

24-19. Which of the emission lines listed in Table 24-1 would be observable in the visible portion of the electromagnetic spectrum (from 390 nm to 720 nm) for a quasar with the following redshifts?
[See also Problem 22-12, which investigates the redshift by considering at what redshifts certain spectral lines are redshifted into the visible spectrum.]
To see what wavelengths are shifted into the visible portion of the spectrum, use the equation redshift formula (Equation 22-1; see Section 8-1A), setting the observed wavelength to λ and solving for λ_o:

$z \equiv (\lambda - \lambda_o)/\lambda_o$
$z\lambda_o = \lambda - \lambda_o$

$$(1+z)\lambda_o = \lambda$$
$$\lambda_o = \lambda/(1+z)$$

(a) z = 0.1
For the lower wavelength limit,
$$\lambda_o = 390 \text{ nm}/(1+0.1) = 345 \text{ nm}$$
and for the upper wavelength limit,
$$\lambda_o = 720 \text{ nm}/(1+0.1) = 655 \text{ nm}$$
So for an object with redshift z = 0.1, objects with rest wavelength in the range
$$345 \text{ nm} \leq \lambda_o \leq 655 \text{ nm}$$
would be seen in the visible, corresponding to the 8 lines between [OII] 372.7 and (barely) [NII] 654.8. The H-α line would be just outside the long wavelength end of this range.

(b) z = 1.0
For the lower wavelength limit,
$$\lambda_o = 390 \text{ nm}/(1+1) = 195 \text{ nm}$$
and for the upper wavelength limit,
$$\lambda_o = 720 \text{ nm}/(1+1) = 360 \text{ nm}$$
So for an object with redshift z = 1, objects with rest wavelength in the range
$$195 \text{ nm} \leq \lambda_o \leq 360 \text{ nm}$$
would be seen in the visible, corresponding to the 1 line [MgII] 279.8. The [CIII] 190.9 line would be just outside the short wavelength end of this range.

(c) z = 4.0
For the lower wavelength limit,
$$\lambda_o = 390 \text{ nm}/(1+4) = 78 \text{ nm}$$
and for the upper wavelength limit,
$$\lambda_o = 720 \text{ nm}/(1+4) = 140 \text{ nm}$$
So for an object with redshift z = 4, objects with rest wavelength in the range
$$78 \text{ nm} \leq \lambda_o \leq 140 \text{ nm}$$
would be seen in the visible, corresponding to the 2 lines HI (Ly-α) 121.6 and NV 124.0.

24-20. (a) Calculate the Schwarzschild radius of a 10^7 $M_\odot$ black hole.

The Schwarzschild radius of a black hole is given by Equations 17-7a and 17-7b:
$$R = 3 (M/M_\odot) \text{ km}$$
where M is the mass of the black hole in solar masses.

For a $10^7 M_\odot$ black hole,

$$R = 3 \times 10^7 \text{ km}$$
$$\approx 43 R_\odot \approx 0.2 \text{ AU}$$

Such a massive object in a small, relatively speaking, volume.

(b) What is the average density inside the Schwarzschild radius?
The average density is

$$\rho = M/V = M/[(4/3) \pi R^3]$$
$$= (10^7 M_\odot \times 1.99 \times 10^{30} \text{ kg}/M_\odot)/[(4/3) \pi (3 \times 10^{10} \text{ m})^3]$$
$$= 1.8 \times 10^5 \text{ kg}/\text{m}^3$$
$$\approx 180 \text{ times the density of water}$$

Note that a $10^8 M_\odot$ solar mass black hole would have an average density comparable to terrestrial rocks and water!

$$\rho = (10^8 M_\odot \times 1.99 \times 10^{30} \text{ kg}/M_\odot)/[(4/3) \pi (3 \times 10^{11} \text{ m})^3]$$
$$= 1.8 \times 10^3 \text{ kg}/\text{m}^3 \approx 1.8 \text{ times the density of water}$$

Note, since for a black hole, the Schwarzschild radius

$$R \propto M$$

we have

$$\rho \propto M/R^3 \propto M/M^3 \propto M^{-2}$$

The more massive the black hole, the (much) lower the density.

Chapter 25: Cosmology: The Big Bang and Beyond

25-1. What is the approximate volume of our Galaxy (express your answer in cubic kiloparsecs)? By what scale factor must the dimensions of our Universe shrink if there is to be no empty space between the galaxies? Is this stage of cosmic expansion a reasonable time for galaxy formation?

[Note, in this edition there is not mention of the number of galaxies in the observable Universe, which we take from the previous edition to be $\approx 10^9$ galaxies. The average density could also be estimated by considering the number of galaxies in a galaxy cluster, perhaps, but this would overestimate the average galaxy density since it would not take into account the voids.]

Including the halo we can approximate our Galaxy as a sphere with a radius r = 50 kpc (Table 14-1; note Appendix Table A7-1 gives 60 kpc). The volume is thus

$$V = (4/3)\pi r^3 = (4/3)\pi (50 \text{ kpc})^3 = 5.2 \times 10^5 \text{ kpc}^3$$

Assuming (see note above) that there are $\approx 10^9$ galaxies in the observable Universe (a sphere of radius 15 billion light yr $\approx 5 \times 10^9$ pc), the average density of galaxies is:

$$n_{gal} \approx 10^9 / [(4/3)\pi (5 \times 10^9 \text{ pc})^3]$$

$$\approx 2 \times 10^{-21} \text{ galaxies / pc}^3 = 2 \times 10^{-12} \text{ galaxies / kpc}^3$$

$$1 / n_{gal} = 5 \times 10^{11} \text{ kpc}^3 / \text{ galaxy}$$

The volume "filling factor" of a typical galaxy is thus

$$V / (1 / n_{gal}) = 5.2 \times 10^5 / (5 \times 10^{11}) \approx 10^{-6}.$$

For the volume to contract this amount, the linear size of the Universe would be $\approx (10^{-6})^{1/3} = 10^{-2}$ the current value. If the expansion of the Universe is constant in time, this corresponds to a time after formation of the Universe of

$$(10^{-2}) \times (15 \text{ billion yr}) \approx 150 \text{ million yr.}$$

This is not an unreasonable time for galaxy formation to occur in the Universe (it is well after the decoupling time).

25-2. Planck's law for the intensity of blackbody radiation (Chapter 8) is

$$I_\lambda = (2hc^2 / \lambda^5)(e^{hc/\lambda kT} - 1)^{-1}$$

As the Universe expands with a scale factor (radius) R(t), the intensity varies as $I_\lambda \propto R^{-5}$ while the wavelength goes as $\lambda \propto R$.

(a) Show that $T \propto R^{-1}$ if the blackbody formula is to remain valid.

Planck's Law for blackbody radiation (Equation 8-35b):

$$I_\lambda = (2hc^2 / \lambda^5)(e^{hc/\lambda kT} - 1)^{-1}$$

can be written as a proportionality of the form:

$$I_\lambda \propto \lambda^{-5}(e^{a/\lambda T} - 1)^{-1} \qquad \text{where a is a constant.}$$

If we assume that the intensity varies as $I_\lambda \propto R^{-5}$ and the wavelength $\lambda \propto R$, then

$$R^{-5} \propto R^{-5}(e^{a/RT} - 1)^{-1}$$

$$e^{a/RT} - 1 \propto 1$$

$$e^{a/RT} \propto b \qquad \text{where b is a constant}$$

$$a/RT \propto \ln(b)$$

$$RT \propto d \qquad \text{where d is a constant}$$

$$T \propto R^{-1}$$

So the temperature of the Universe is inversely proportional to the radius.

(b) At what wavelength does the blackbody curve reach a maximum for the observed 2.7 K background radiation?

Using Wien's Law (Equation 8-39; Section 25-3), for T = 2.7 K:

$$\lambda_{max} = 2.898 \times 10^{-3} \text{ m K} / T$$

$$\lambda_{max} = 2.898 \times 10^{-3} \text{ m K} / 2.7 \text{ K}$$

$$= 1.1 \times 10^{-3} \text{ m} = 1.1 \text{ mm}$$

The cosmic blackbody radiation peaks in the microwave (radio) region of the electromagnetic spectrum ($\lambda_{max} = 1.1$ mm), in agreement with Figure 25-3A.

25-3. If the Hubble constant is observed to be $H_0 = 50 \pm 5$ km/s·Mpc, what are the permissible ranges for the Hubble time ($t_0 \approx H_0^{-1}$), the size of the Universe ($r_{hor} \approx cH_0^{-1}$), and the critical mass density ($\rho_0 \propto H_0^2$)?

For a Hubble constant of $H_0 = 50 \pm 5$ km/s · Mpc, the error is 10% so

$$H_0 = 50 \text{ km/s} / [(3.09 \times 10^{13} \text{ km/pc}) \times (10^6 \text{ pc/Mpc})]$$

$$= 1.6 \times 10^{-18} \text{ s}^{-1} \text{ (with a 10\% error)}$$

The Hubble time is

$$t_0 \approx H_0^{-1} = 6.2 \times 10^{17} \text{ s} = 2 \times 10^{10} \text{ yr}$$

$$= 20 \text{ billion yr}$$

with an error of 10%, or

$$t_0 \approx 20 \pm 2 \text{ billion yr}$$

The size of the Universe is

$$r_{hor} \approx cH_0^{-1} = (3.0 \times 10^8 \text{ m/s}) / (1.6 \times 10^{-18} \text{ s}^{-1})$$

$$\approx 1.9 \times 10^{26} \text{ m} \approx 6 \times 10^9 \text{ pc}$$

$$= 6,000 \text{ Mpc} \approx 20 \text{ billion light yr}$$

with an error of 10%, or

$$r_{hor} \approx 6{,}000 \pm 600 \text{ Mpc}$$

The critical mass density (Equation 25-22):
$$\rho_o \propto H_o^2$$

The critical mass density of the Universe (Section 25-2C) for $H_o = 50$ km/s · Mpc is
$$\rho_o \approx 5 \times 10^{-27} \text{ kg/m}^3$$

Since the critical density varies as the square of the Hubble constant, a 10% error in H_o results in a ≈ 20% error in ρ_o, so
$$\rho_o \approx (5 \pm 1) \times 10^{-27} \text{ kg/m}^3 \text{ .}$$

25-4. The following diagram (in the text) illustrates the famous expanding-balloon analogy for our Universe. All space is represented by the surface of the spherical balloon, and clusters of galaxies are represented by spots painted on this surface. The radius of the balloon corresponds to R(t) — the radius of the Universe.

(a) As the balloon expands, the spots remain at constant angular separations (θ) from one another. Let the balloon expand at a constant rate and verify that
$$\Delta s/\Delta t = (1/R)(\Delta R/\Delta t)\, s$$
where s is the separation between any two spots on the surface and Δs/Δt is the speed of recession of one spot from another. (Note that this is Hubble's law.)

θ, R, and s are related by
$$\theta = s/R$$
$$s = \theta R$$
where θ remains constant during the expansion. Taking the time derivative, for θ constant,
$$ds/dt = \theta \, dR/dt$$
or
$$\Delta s/\Delta t = \theta \, \Delta R/\Delta t$$

Substituting $\theta = s/R$
$$\Delta s/\Delta t = (s/R)\,\Delta R/\Delta t = (1/R)(\Delta R/\Delta t)\,s \qquad \text{QED}$$

where $(1/R)(\Delta R/\Delta t)$ corresponds to Hubble's constant.

(b) The photons from distant galaxies may be represented by ants crawling along the balloon's surface at speed 1. Show that, for uniform cosmic expansion (ΔR/Δt = constant), there is a distance s from beyond which these ants cannot ever reach our Galaxy (this distance is called the *horizon*).

To reach us from a distance s, the ant's speed (v = 1) must be greater than the recessional speed produced by the expansion
$$v = 1 > \Delta s/\Delta t$$
so
$$v = 1 > (1/R)(\Delta R/\Delta t)\, s$$

$s < v R (\Delta R/\Delta t)^{-1} = R (\Delta R/\Delta t)^{-1} = R / (\Delta R/\Delta t)$

If there is an upper limit to the velocity of the ant (v = 1), and if the cosmic expansion is uniform $(\Delta R/\Delta t)$ = constant, then there is a distance s = $R / (\Delta R/\Delta t)$, from beyond which the ant can never reach our Galaxy.

Taking from part (a) that $(1 / R) (\Delta R/\Delta t)$ corresponds to the Hubble constant H_o, then the distance beyond which we cannot see, r_{hor}, corresponds to the distance, s, when the velocity v = c:

$r_{hor} \approx c / H_o$.

(c) Discuss the effects that take place if the balloon's expansion is decelerated [the increase of R(t) is slowed down].

If the balloon is decelerating ($\Delta R/\Delta t$ is decreasing), then the "Hubble constant" is decreasing, and the distance to the horizon increases with time.

25-5. Consider an expanding gaseous sphere of uniform mass density ρ, total mass M, and radius R(t). A gas particle at the surface of this sphere will move radially outward in accordance with the vis-viva equation (Chapter 2):

$v^2 / 2 = (G M / R) + $ constant

where v = $\Delta R/\Delta t$ is the radial speed.

(a) Show that this equation may be written in the form

$[(1 / R) (\Delta R/\Delta t)]^2 = (8 \pi G \rho / 3) + 2$ (constant)$/R^2$

Note that this is the equation that governs the expansion of our Universe and leads to the three cosmological models discussed in this chapter.

The vis-viva equation is given by Equation 1-34:

$v^2 = G(m_1 + m_2) [(2 / R) - (1 / a)]$

where for $M \equiv m_1 \gg m_2$, and a is a constant distance,

$v^2 = G M (2 / R) + (-G M / a)$

so

$v^2 / 2 = G M / R + $ constant .

Substituting v = $\Delta R/\Delta t$,

$(\Delta R/\Delta t)^2 = (2 G M / R) + $ (constant).

The mass can be written in terms of the density and volume,

$M = \rho (4/3) \pi R^3$

so

$(\Delta R/\Delta t)^2 = [(8/3) \pi G \rho R^3 / R] + $ (constant).

Dividing by R^2 gives

$[(1 / R) (\Delta R/\Delta t)]^2 = (8 \pi G \rho / 3) + [$(constant)$ / R^2]$

Note that this is Equation 25-4 which for the constant = 0 reduces to Equation 25-22, where

$$H = (1/R)(\Delta R/\Delta t)$$

and our constant term contains the geometry of the Universe (the factor "k" in Equation 25-4).

(b) From your knowledge of the vis-viva equation, show that the constant can be positive, zero, or negative; illustrate the evolution of R(t) in each case by drawing an approximate graph of R versus t. Comment on your results.

The constant in the vis-viva equation is related to the total energy (review Section 1-5D). If the total energy is less than zero, the orbit is bound; if it is greater than zero then the orbit is unbound (open). So, in considering the expansion of the Universe, if the constant term is negative the Universe is closed; if the constant is positive the Universe is open; and if the constant equals zero the Universe is flat. (Refer to Figure 25-2 for a graph of the scale of the Universe versus time for these three conditions.)

25-6. Demonstrate that the Universe is now matter-dominated. Argue that in the past it must at some time have been radiation-dominated. (*Hint:* Background radiation at about 3 K.)

The current values of the mass density and the mass equivalent radiation density are (Section 25-3):

$$\rho_{ro} = 4.5 \times 10^{-31} \text{ kg/m}^3 \quad \text{(2.7K cosmic background radiation)}$$

$$\rho_{mo} = 4 \times 10^{-28} \text{ kg/m}^3 \quad \text{(luminous matter)}$$

Because $\rho_{mo} \gg \rho_{ro}$, the Universe is matter-dominated.
However, in the past the Universe was radiation dominated.

To see how and when this occurred, we note that the radiation density decreases more rapidly than does the matter density with increasing radius. From Equation 25-27, the radiation density varies with the scale of the Universe by

$$\rho_r \propto R^{-4}$$

while the matter density decreases as the inverse of the volume

$$\rho_m \propto R^{-3}.$$

Since the scale of the Universe was less in the past, the ratio of the radiation density to the matter density increases as we consider times further in the past. Taking ratios, since

$$\rho_r = \rho_{ro}(R_o/R)^4$$
$$\rho_m = \rho_{mo}(R_o/R)^3$$

then

$$\rho_r/\rho_m = (\rho_{ro}/\rho_{mo})(R_o/R).$$

When $\rho_r > \rho_m$, the Universe was radiation-dominated (the "radiation era").

Matter and radiation decoupled when $\rho_m \approx \rho_r$, which occurred when the scale of the Universe was

$$\rho_r / \rho_m = (\rho_{ro} / \rho_{mo})(R_o / R) = 1$$

or $R = (\rho_{ro} / \rho_{mo}) R_o$.

Using the current values of the matter and radiation density,

$$R = [(4.5 \times 10^{-31} \text{ kg/m}^3) / (4 \times 10^{-28} \text{ kg/m}^3)] R_o$$
$$= 1.1 \times 10^{-3} R_o.$$

Since, for non-decelerating expansion, the scale of the Universe is proportional to the age, the radiation era occurred at an age of the Universe $\approx 10^{-3}$ the present value, or at a time 10 - 20 million yr (dependent on the value of H_o) after the Big Bang.

25-7. For a flat (k = 0) Universe, show that

$$t_o = (2/3) H_o^{-1}$$

where t_o is the age of the Universe. Evaluate t for the uncertainty in H_o.

[See also Equation 1-34 for derivation of the vis-viva equation.]

For a flat Universe where k = 0 (k is the constant in the second term of the vis-viva equation, Equation 25-4, which results in Equation 25-22), the vis-viva equation becomes (integrating Equation 25-2; see also Problem 25-5)

$$v^2 \equiv (dR/dt)^2 = 2GM/R$$

or

$$R^{1/2} dR = (2GM)^{1/2} dt.$$

Integrating from the Big Bang (time t =0) to the present, assuming that the mass of the Universe and the gravitational constant, G, both have remained constant:

$$\int_0^{R_o} R^{1/2} dR = (2GM)^{1/2} \int_0^{t_o} dt$$

$$[R^{3/2} / (3/2)]_0^{R_o} = (2GM)^{1/2} [t]_0^{t_o}$$

$$(2/3) R_o^{3/2} = (2GM)^{1/2} t_o$$

Now, $v = dR/dt = (2GM/R)^{1/2}$
and from the Hubble Law,
$v = HR$
so $(2GM/R_o)^{1/2} = H_o R_o$

$$H_o = (1/R_o)^{3/2} (2GM)^{1/2}$$

Substituting this relation into the derivation above gives

$$(2/3) t_o^{-1} = R_o^{-3/2} (2GM)^{1/2} = H_o$$

$$t_o = (2/3) H_o^{-1} \quad \text{for a flat Universe}$$

Adopting a value of the Hubble constant, $H_o = 50 \pm 5$ km / s · Mpc, the Hubble time, t_o, is given by (see also Problem 25-3):

$$H_o = 50 \text{ km/s} / [(3.09 \times 10^{13} \text{ km/pc}) \times (10^6 \text{ pc/Mpc})]$$
$$= 1.6 \times 10^{-18} \text{ s}^{-1} \qquad \text{(with a 10\% error)}$$
$$t_o \approx (2/3) H_o^{-1} = (2/3)\, 6.2 \times 10^{17} \text{ s} = 4.2 \times 10^{17} \text{ s}$$
$$= 1.3 \times 10^{10} \text{ yr} = 13 \text{ billion yr} \qquad \text{(with an error of 10\%) or}$$
$$t_o \approx 13 \pm 1.3 \text{ billion yr}$$

25-8. Follow the algebra in deriving Equation 25-17 from Equation 25-15.
Note: The text gives Equation 25-15 as

$$8\pi G P(t)/c^4 = -(k/R_o^2)(R_o/R) - (2/c^2)(d^2R/dt^2/R)$$
$$- (1/c^2)(dR/dt / R)^2 + \Lambda$$

This appears to be a typo, where the first term on the right hand side should have the quantity (R_o/R) squared, so Equation 25-15 should be:

$$8\pi G P(t)/c^4 = -(k/R_o^2)(R_o/R)^2 - (2/c^2)(d^2R/dt^2/R)$$
$$- (1/c^2)(dR/dt / R)^2 + \Lambda$$

Equation 25-16 gives the time-like equation
$$8\pi G U(t)/c^4 = (3k/R_o^2)(R_o/R)^2 + (3/c^2)(dR/dt / R)^2 - \Lambda$$
which can be rewritten as
$$(k/R_o^2)(R_o/R)^2 = 8\pi G U(t)/3c^4 - (1/c^2)(dR/dt / R)^2 + (1/3)\Lambda$$
Substituting this into the corrected version of Equation 25-15,
$$8\pi G P(t)/c^4 = -[8\pi G U(t)/3c^4 - (1/c^2)(dR/dt / R)^2 + (1/3)\Lambda]$$
$$- (2/c^2)(d^2R/dt^2/R) - (1/c^2)(dR\,(dR/dt / R)^2 + \Lambda$$
Rearranging, where the $(dR/dt / R)^2$ term cancels out,
$$8\pi G P(t)/c^4 + 8\pi G U(t)/3c^4 = -(2/c^2)(d^2R/dt^2/R) + [(-1/3) + 1]\Lambda$$
$$[8\pi G/c^4][P(t) + U(t)/3] = -(2/c^2)(d^2R/dt^2/R) + (2/3)\Lambda$$
$$(d^2R/dt^2/R) = [-4\pi G/c^2][P(t) + U(t)/3] - (c^2/3)\Lambda$$
which is Equation 25-17. QED

25-9. Follow the algebra in deriving Equation 25-18 from Equation 25-16.
Equation 25-16 is
$$8\pi G U(t)/c^4 = (3k/R_o^2)(R_o/R)^2 + (3/c^2)(dR/dt / R)^2 - \Lambda$$
adding this to the corrected Equation 25-15 *(see Note at beginning of solution to Problem 25-8 concerning typo in Equation 25-15 in text)*
$$8\pi G P(t)/c^4 = -(k/R_o^2)(R_o/R)^2 - (2/c^2)(d^2R/dt^2/R)$$

$$-(1/c^2)(dR/dt\,/\,R)^2 + \Lambda$$

gives

$$[8\pi G/c^4][P(t) + U(t)] = (2k/R_o^2)(R_o/R)^2 - (2/c^2)(d^2R/dt^2\,/\,R)$$
$$+ (2/c^2)(dR/dt\,/\,R)^2$$

Now, by the chain rule (Appendix 9-7A),

$$d[(dR/dt)/R]/dt = (dR/dt)[d(1/R)/dt] + (1/R)[d(dR/dt)/dt]$$
$$= -(1/R^2)(dR/dt)^2 + (1/R)(d^2R/dt^2)$$
$$= -(dR/dt\,/\,R)^2 + (d^2R/dt^2\,/\,R)$$

or, rearranging,

$$-(2/c^2)(d^2R/dt^2\,/\,R) + (2/c^2)(dR/dt\,/\,R)^2 = -(2/c^2)\,d[(dR/dt)/R]/dt$$

Substituting this equation in the previous equation containing P(t) and U(t) gives

$$[8\pi G/c^4][P(t)+U(t)] = (2k/R_o^2)(R_o/R)^2 - (2/c^2)\,d[(dR/dt)/R]/dt$$

(A) $\quad (1/c^2)\,d[(dR/dt)/R]/dt = (k/R_o^2)(R_o/R)^2 - [4\pi G/c^4][P(t)+U(t)]$

Taking Equation 25-16 and rearranging:

$$(8\pi G/c^4)U(t) = (3k)(1/R)^2 + (3/c^2)(dR/dt\,/\,R)^2 - \Lambda$$

Taking the time derivative of Equation 25-16:

(B) $\quad (8\pi G/c^4)\,dU(t)/dt = (-6k/R^3)(dR/dt) + (6/c^2)(dR/dt\,/\,R)\,d[(dR/dt)/R]/dt$

Inserting Equation (A) into (B):

$$(8\pi G/c^4)\,dU(t)/dt = (-6k/R^3)(dR/dt)$$
$$+ (6)(dR/dt\,/\,R)\{(k/R_o^2)(R_o/R)^2 - [4\pi G/c^4][P(t)+U(t)]\}$$

$$(8\pi G/c^4)\,dU(t)/dt = (-6k)(dR/dt)/R^3 + (6k)(dR/dt)/R^3$$
$$- [24\pi G/c^4](dR/dt\,/\,R)[P(t)+U(t)]$$

The first two terms on the right hand side cancel out:

$$(8\pi G/c^4)\,dU(t)/dt = -[24\pi G/c^4](dR/dt\,/\,R)[P(t)+U(t)]$$

Canceling out the term $8\pi G/c^4$

$$dU(t)/dt = -3(dR/dt\,/\,R)P(t) - 3(dR/dt\,/\,R)U(t)$$

Multiply by R^3 and rearranging,

$$R^3\,dU(t)/dt = -3P(t)R^2(dR/dt) - 3U(t)R^2(dR/dt)$$

$$(R^3)(dU(t)/dt) + U(t)(3R^2)(dR/dt) = -3P(t)R^2(dR/dt)$$

Noting that by the chain rule,

$$d[U(t)R^3]/dt = [dU(t)/dt](R^3) + U(t)(3R^2)(dR/dt)$$

And substituting into the previous equation:

$$d[U(t)R^3]/dt = -3P(t)R^2(dR/dt)$$

Since $d(R^3)/dt = 3R^2(dR/dt)$, this last equation becomes

$$d[U(t)R^3]/dt = -P(t)[d(R^3)/dt]$$

which is Equation 25-18. QED!

25-10. Evaluate Equations 25-22 and 25-23 using the value of the critical density given in the text.

Evaluating equation 25-22, for the Hubble constant in terms of the critical density, $\rho_c = 5 \times 10^{-27}$ kg/m^3:

$$H = [(8\pi G/3)\rho_c]^{1/2}$$
$$= [(8\pi/3)(6.67 \times 10^{-11} \text{ N m}^2/\text{kg}^2)(5 \times 10^{-27} \text{ kg/m}^3)]^{1/2}$$
$$= 1.67 \times 10^{-18} /\text{s}$$

To express H in more typical units of km/s·Mpc, where 1 Mpc = 3.086×10^{19} km,

$$H = (1.67 \times 10^{-18}/\text{s})(3.086 \times 10^{19} \text{ km/Mpc})$$
$$= 52 \text{ km/s} \cdot \text{Mpc}$$

Evaluating Equation 25-23, for the deceleration parameter in terms of the Hubble constant and critical density

$$q_0 = 4\pi G \rho / 3 H^2$$
$$= (4\pi/3)(6.67 \times 10^{-11} \text{ N m}^2/\text{kg}^2) \rho /(1.67 \times 10^{-18}/\text{s})^2$$
$$= (1.00 \times 10^{26} \text{ m}^3/\text{kg})\rho$$

If ρ is equal to the critical density ρ_c

$$q_0 = (1.00 \times 10^{26} \text{ m}^3/\text{kg})(5 \times 10^{-27} \text{ kg/m}^3) = 0.50$$

25-11. A friend of yours plans to give a lesson on cosmology at a local middle school. He wants to demonstrate the Big Bang expansion of the Universe by placing a handful of marbles, representing galaxies, in the center of a large bandana, representing the space between galaxies, and then disrupting the marbles. The outward motion of the marbles then illustrates the expansion of the Universe. Is it a good analogy? Is there a better way? Has your friend lost his marbles?
[See also Problem 22-6 for discussion of lecture demonstration models of the Big Bang.]
In this marble model the objects (marbles) in the Universe are moving, where in the Big Bang model it is the fabric of space-time which expands. The objects in the Universe (e.g. galaxies) are carried with this expansion, but are not themselves moving (on the cosmic scale). The model thus gives the mistaken impression that objects are moving away from each other, rather than the space between the galaxies increasing in scale. Models (see Problem 22-6) in which the fabric of space is expanding with the objects attached to the space are more realistic. Your friend has lost his marbles since they are not tied to the metric of space, and thus they will not return when the metric collapses in a closed universe model. The co-authors of this manual may have lost theirs!

25-12. At what times in the history of the Universe does the radiation temperature of the Universe correspond to the temperature of the Sun's core? Surface?
Since the temperature of the Sun's core and surface are greater than 3000 K, we are considering times during the radiation dominated era (see Section 25-4 for a discussion

of when the radiation-matter decoupling occurred). The temperature of the Universe during the radiation dominated era is given by Equation 25-32:

$$T(K) \approx 1.5 \times 10^{10} \, t^{-1/2}$$

$$t = (1.5 \times 10^{10})^2 \, T(K)^{-2} = 2.25 \times 10^{20} \, T(K)^{-2}$$

For temperatures corresponding to the core and surface temperatures of the Sun (see Figure 10-1 for core, Table 10-1 for surface), the corresponding times after the Big Bang are:

$$T_{core} \approx 1.6 \times 10^7 \, K$$

$$t = 2.25 \times 10^{20} \, (1.6 \times 10^7)^{-2}$$

$$= 8.79 \times 10^5 \, s = 2.8 \times 10^{-2} \, yr = 0.028 \, yr$$

and

$$T_{surface} \approx 5770 \, K$$

$$t = 2.25 \times 10^{20} \, (5.77 \times 10^3)^{-2}$$

$$= 6.76 \times 10^{12} \, s = 2.1 \times 10^5 \, yr = 210{,}000 \, yr$$

These values agree with the timeline in Figure 26-2.

Chapter 26: The New Cosmology

26-1. Assume that dust grains, whose characteristic size is 1 μm, are uniformly distributed throughout intergalactic space at a mean density of 10^{-27} kg/m^3 (the critical mass density).

(a) What is the number density (number per cubic meter) of this dust, and what is the average separation between grains?

Assuming that the grains are made of silicates with an average density of ρ = 2000 kg / m^3 (similar to such material on the Earth) and are spherical grains with diameter ≈1 μm. The mass of a grain is (Equation 2-1):

$$m_{dust} = <r> [(4\pi/3) R^3]$$
$$= (2 \times 10^3 \text{ kg/m}^3)(4\pi/3)(0.5 \times 10^{-6} \text{ m})^3 = 1.0 \times 10^{-15} \text{ kg}.$$

The number density is the mass density divided by the mass of a grain,

$$n = \rho / m_{dust}$$
$$= (1.0 \times 10^{-27} \text{ kg/m}^3) / (1.0 \times 10^{-15} \text{ kg/grain})$$
$$= 1.0 \times 10^{-12} \text{ grains/m}^3$$

The average distance between grains ($1/n^{1/3}$) in the intergalactic medium would be 10^4 m or 10 km!

(b) Compare your answers to part (a) with the number density and separation of the interstellar dust grains in our Galaxy.

The number density of H atoms in the interstellar medium is generally taken as (Section 15-2F):

$$n_H \approx 1/\text{cm}^3 = 10^6/\text{m}^3$$

corresponding to a mass density of the gas (assuming predominantly hydrogen)

$$\rho = n\, m_H \approx (10^6/\text{m}^3)(1.67 \times 10^{-27} \text{ kg}) \approx 2 \times 10^{-21} \text{ kg/m}^3.$$

The mass density of dust is typically 1% the mass density of gas (Section 15-1), so

$$\rho_{dust} \approx 2 \times 10^{-23} \text{ kg/m}^3.$$

Assuming the same characteristics as in part (a), the number density of dust in the interstellar medium is

$$n_{dust} = \rho / m_{dust}$$
$$= (2.0 \times 10^{-23} \text{ kg/m}^3) / (1.0 \times 10^{-15} \text{ kg/grain})$$
$$= 2.0 \times 10^{-8} \text{ grains/m}^3$$

[Section 15-1 gives the average number density of dust as $10^{-6}/\text{m}^3$.]

The average distance between grains ($1/n^{1/3}$) in the interstellar medium would be ≈ 400 m.

The number density of interstellar dust is 20,000 times greater than the intergalactic density.

(c) Show that this hypothetical intergalactic dust will drastically *redden* the stars observed in the

Andromeda Galaxy (such reddening is *not* observed in practice).

The interstellar dust extinction is about 1 magnitude per kpc. If the intergalactic dust density were 10^{-4} that of the interstellar medium, then the intergalactic extinction would be 1 magnitude per 10^4 kpc. If we were to observe the Andromeda Galaxy (distance ≈ 690 kpc -- Table 23-1), the intergalactic extinction would be ≈ 0.07 magnitudes. The interstellar reddening (color excess) is about 1/3 the visual extinction (Equation 15-4). The color excess due to the intergalactic dust would be ≈ 0.02 magnitudes. This reddening would be barely measurable for the nearby Andromeda Galaxy.

For more distant galaxies (e.g. d = 10 Mpc), the intergalactic extinction would be ≈ 1 magnitude and the reddening (color excess) an easily measurable ≈ 0.3 magnitude.

26-2. If intergalactic space is filled with **HII** at a temperature of 10^6 K (a plasma),

(a) What are the mean speed and mean kinetic energy (per particle) of these protons?

For a plasma of **HII** at 10^6 K, the mean speed and kinetic energy of the protons are given by Equation 8-28:

$$\langle KE \rangle = \langle (1/2) m v^2 \rangle = (3/2) k T$$

so $\langle KE \rangle = (3/2)(1.38 \times 10^{-23} \text{ J/K})(10^6 \text{ K}) = 2.1 \times 10^{-17}$ J

and $\langle v \rangle = [2(2.1 \times 10^{-17} \text{ J}) / (1.67 \times 10^{-27} \text{ kg})]^{1/2} = 1.6 \times 10^5$ m/s

(b) To what wavelength of electromagnetic radiation does this individual kinetic energy correspond? Could such radiation be detected from the surface of the Earth?

Using Equation 8-15:

$$E = h\nu = hc/\lambda$$

$\lambda = hc/E = (6.63 \times 10^{-34} \text{ J s})(3.0 \times 10^8 \text{ m/s}) / (2.1 \times 10^{-17} \text{ J})$

$= 9.5 \times 10^{-9}$ m $= 9.5$ nm $= 95$ Å

This corresponds to the ultraviolet portion of the electromagnetic spectrum. The Earth's atmosphere is opaque to his wavelength of radiation (see Table 8-1).

26-3. Electrons are the lightest stable particles made in the Big Bang. What is the *latest* time they could have been formed?

Equation 25-32 states that for the radiation dominated Universe (when electrons were formed) the temperature is related to time by:

$$T(K) = 1.5 \times 10^{10} \, t^{-1/2}$$

An electron-positron pair is created when a pair of γ-rays interact, having a total energy equal to $E = (m_{e^-} + m_{e^+}) c^2$. So a single γ-ray would require the mass/energy equivalent of an electron only,

$$E_\gamma = hc/\lambda = m_{e^-} c^2$$

so

$$h/\lambda = m_{e^-} c$$

Using the blackbody energy distribution (see Section 25-3) for the primordial fireball, the wavelength of peak emission corresponding to this energy of the Universe is, according to Wien's law (see also Equation 8-39):

$$\lambda_{max} = (2.90 \times 10^{-3} \text{ m K}) / T$$

so rearranging the two equations,

$$h T / (2.90 \times 10^{-3} \text{ m K}) = m_{e^-} c$$

$$T = (2.90 \times 10^{-3} \text{ m K}) m_{e^-} c / h$$

$$= (2.90 \times 10^{-3} \text{ m K})(9.11 \times 10^{-31} \text{ kg})(3.0 \times 10^8 \text{ m/s}) / (6.62 \times 10^{-34} \text{ J s})$$

$$= 1.2 \times 10^9 \text{ K} = 1.2 \text{ billion K}$$

The time after the Big Bang which corresponds to a temperature of 1.2 billion K is given by Equation 25-32:

$$T(K) \approx 1.5 \times 10^{10} \, t^{-1/2}$$

$$t \approx (1.5 \times 10^{10})^2 \, T(K)^{-2}$$

$$= (2.25 \times 10^{20})(1.2 \times 10^9)^{-2}$$

$$= 1.56 \times 10^2 \text{ s} \approx 160 \text{ s}$$

So electrons could have formed during the first ≈ 160 s (about 3 minutes) of the Big Bang, until the temperature dropped below 1.2 billion K. This time and temperature correspond to a critical time in the early Universe (Figure 26-1) when the abundances of the lighter elements were beginning to be set.

26-4. Use Equation 25-11 and Figure 25-2 to verify our statement in this chapter that Ω tends to evolve away from unity.

Equation 25-21 gives the deceleration parameter as

$$q_o = -(d^2R/dt^2) / (R_o H_o^2)$$

In the text, following Equation 25-23, it is stated that $\Omega = 2 q_o$ (with $\Omega = 1$ corresponding to a flat Universe), so

$$\Omega = -2 (d^2R_o/dt^2) / (R_o H_o^2) \propto (d^2R_o/dt^2) / R_o$$

Now, dR/dt is the slope of the model lines in Figure 25-2. From the graph, for the flat Universe, R(t) approaches 0 slope, or dR/dt approaches a constant. Therefore, d^2R/dt^2 approaches 0. Since $\Omega \propto (d^2R_o/dt^2) / R_o$, Ω would evolve away from 1 towards 0 in a flat Universe. For open or closed Universes, the same would occur.

26-5. If we use 10^{10} K as the critical temperature for photodissociation of deuterium, what is the binding energy of the deuteron?

The wavelength of maximum emission for a blackbody (e.g. the primordial fireball; see Section 25-3) is given by the Wien law (see also Equation 8-39):

$$\lambda_{max} = (2.90 \times 10^{-3} \text{ m K}) / T$$

$$= (2.90 \times 10^{-3} \text{ m K}) / (1 \times 10^{10} \text{ K})$$

$$= 2.90 \times 10^{-13} \text{ m}$$

Which corresponds to an energy (Equation 8-15)

$$E = hc/\lambda$$
$$= (6.62 \times 10^{-34} \text{ J s})(3 \times 10^8 \text{ m/s}) / (2.90 \times 10^{-13} \text{ m})$$
$$= 6.8 \times 10^{-13} \text{ J}$$
$$= (6.8 \times 10^{-13} \text{ J}) / (1.6 \times 10^{-19} \text{ J/eV}) = 4.3 \times 10^6 \text{ eV} = 4.3 \text{ MeV}$$

26-6. Section 26-7 outlines various stages in the history of the universe in time units. Estimate the corresponding density and radiation temperature for each of these times. Comment on the accuracy of your results.

The temperature at a given age in the radiation dominated Universe can be estimated from Equation 25-32

$$T(K) \approx 1.5 \times 10^{10} \, t^{-1/2}$$

and the density by Equation 25-30

$$\rho = (3/32 \pi G) \, t^{-2}$$
$$= [3/(32 \pi \times 6.67 \times 10^{-11})] \, t^{-2} \text{ kg/m}^3$$
$$= 4.5 \times 10^8 \, t^{-2} \text{ kg/m}^3$$

Evaluating for different times:

t (s)	T (K)	ρ (kg/m³)
10^{-45}	4.7×10^{32}	4.5×10^{98}
10^{-35}	4.7×10^{28}	4.5×10^{78}
10^{-32}	1.5×10^{26}	4.5×10^{72}
10^{-12}	1.5×10^{16}	4.5×10^{32}
10^2	1.5×10^9	4.5×10^4
$- 10^3$	$- 4.7 \times 10^8$	$- 4.5 \times 10^2$
10^{11}	4.7×10^4	4.5×10^{-14}
10^{16}	1.5×10^2	4.5×10^{-24}
10^{18}	1.5×10^1	4.5×10^{-28}
10^{40}	1.5×10^{-10}	4.5×10^{-72}

The calculated temperature and density, for the 8 stages discussed in Section 26-7, in the table above agree with the values read from Figure 26-2. (In comparing to Figure 26-2, note 1 kg/m³ = 10^{-3} g/cm³.)

26-7. The GUTS and inflation models help explain many of the problems of the standard Big Bang model. What problems still exist?

Inflation has eroded the value of our payment for this work, so we no longer have the GUTS to provide a more complete solution to this problem. We leave the reader with a few thoughts with which to pursue the ultimate answer to the ultimate question in this, our ultimate of instructor's manuals, for the penultimate of texts.

It is clear from reading Section 26-1 and Section 26-8 that five major problems confront cosmologists when the standard Big Bang model is used: isotropy and homogeneity of the Universe, how galaxy formation occurred in the early Universe, the flatness of the space-time metric, the net baryon number, and where are the Population III stars. The GUTS and inflation models, and the recent COBE results go a long way in solving these five earlier difficulties. The COBE results indicate that there was some inhomogeneity in the early Universe, which addresses the first two problems, though whether this observed inhomogeneity is large enough to explain the beginning of galaxy formation remains to be seen. A better understanding of the nature of the dark matter is of major importance.

bye!

Notes

Comments and Corrections to Problems

We summarize here our suggested changes to problems or equations which are used in solving problems. See our answers in this Manual for further information.
It is recommended that instructors copy these pages and distribute them to students.

Problem 6-4	The problem gives Pluto radius as 1120 km, while Figure 6-19 gives 1125 km and Appendix Table A3-3 gives the equatorial radius of Pluto as 1140 km. These slightly different values (determined by different observations and models) will produce slightly different results in calculations.
Problem 6-18	The students will need to know that the equation given on page 106, appears to be off by a factor of 10^6. The constant should be 4.8×10^2 for the units given for the variables IF the worked example for Jupiter on page 106 is to produce the answer given. The equation would thus be $$f_{max} \approx (4.8 \times 10^2) \, E_{[MeV]}^2 \, B_{[T]} \sin(\alpha) \text{ MHz}$$
Problem 7-8	The visual or infrared flux are needed -- assume flux as seen from Earth is 1.4×10^{-11} W/m^2
Problem 7-9	Need to know Herculina size -- assume a radius of 110 km
Problem 8-16	Because the text gives no specific information about the values of the statistical weights, g, the instructor should provide them (or the students will assume that they have value 1). For neutral hydrogen, $g_{n=1} = 2$, $g_{n=2} = 8$.
Problem 8-17	The constant A is not defined in the text, so the instructor must provide some information to the students. $$A = (2/h^3)(2 \pi m)^{3/2} (g_+ / g_0) \equiv A' \, (g_+ / g_0).$$ Taking $g_+ / g_0 = 1$, $A' \approx 9.4 \times 10^{55}$ kg$^{3/2}$ J^{-3} s^{-3}
Problem 8-18	Follow the same method as in Problem 8-17, making similar assumptions. For helium, take $g_+ / g_0 = 2$, so $$A \equiv A' \, (g_+ / g_0) \approx 1.88 \times 10^{56} \text{ kg}^{3/2} \text{ J}^{-3} \text{ s}^{-3}$$
Problem 9-14	To be consistent throughout the text, units should be "arcsec" instead of "arcseconds" and "arcmin" instead of "arcminutes"
Problem 9-16(b)	There is no numeric solution to this problem. Although this may be what the authors intended, it may be confusing to students.
Problem 9-18	"What is this *not*" ---> "Why is this *not*"
Problem 9-20	"Because at a site" ---> "At a site"
Problem 10-11	This problem may be difficult for students since it uses physical concepts not covered at this point in the text. Refer to Section 13-1.
Problem 10-16	κ is called the absorption coefficient in the problem, but is called opacity in the text (see page 203)
Problem 11-16	"that the star had a parallax" ---> "that the asteroid had a parallax"
Problem 12-16	The numbers are unrealistic
Problem 13-1(d)	"Ca II, H, and K" ---> "Ca II H and K"
Problem 13-7	"apparent magnitude-(B - V)" could be better written "apparent magnitude – (B-V)" [that is, different spacing would make it easier to read]
Problem 13-14(a)	"Figure 13-10" ---> "Figure 13-9,"
Problem 13-14(b)	"Figure 13-12 " ---> "Figure 13-10"
Problem 14-2	The Figure 14-4 that the problem refers to is not the Figure 14-4 in the text; the appropriate figure has been removed from this edition, and Figure 14-4 refers to the 3rd edition. The only figure which may be

	useful to the students is Figure A-18 in the Appendix A10-3
Problem 15-14(a)	Uses Equation 16-1 from the next chapter! The problem could instead refer to Equation 4-1
Problem 15-17	"Figure 15-22" ---> "Figure 15-21"
Problem 16-4	"Using Figures" ---> "Using Figure"
Problem 16-11(b)	"$r_o = 0.1$ R" ---> "$r_o = 0.1$ $R_\odot$"
Problem 17-16	Students will need to know that the text (page 340) gives the equation for time delay appropriate for cgs units. To use SI units, a factor of $1/4\pi\varepsilon_o$ must be included, where $\varepsilon_o = 8.85 \times 10^{-12}$ C^2/Nm^2. Also, the text has a factor of 2π in the numerator, which should be in the denominator. Use: $$t_2 - t_1 = (1/4\pi\varepsilon_o) [e^2/(2\pi m_e c)] [(1/f_2^2) - (1/f_1^2)] \, DM$$
Problem 17-17	Section 17-2D doesn't exist in this edition. From the 3rd edition: The binary system PSR +065564 has a period of revolution 24 hour 41 minute = 8.886×10^4 s, and an orbital semi-major axis 7.5×10^5 km.
Problem 17-20	"dialation" ---> "dilation"
Problem 18-11	The T Tauri stellar wind velocity is not given in the chapter, but it can be found in Problem 15-20 as ≈ 100 km/s.
Problem 18-14	Students will need to know that the equation for the mass function in Section 18-6A is incorrect. The equation should have P instead of P^2 on the right hand side.
Problem 18-15	The expansion velocity of SN 1987A is not given in this edition (Section 18-5E). The third edition gave it as 17000 km/s.
Problem 18-17	Students will need to know that Equation 18-2 in the text is incorrect. They can use the correct form of the equation given in the Key Equations at the end of the text.
Problem 19-15	In this problem and in Appendix Table A7-1, the symbol for the distance of the Sun from the center of the Galaxy should be R_0 instead of $R_\odot$.
Problem 20-4(b)	The problem uses $M_B = 4.7$, while Section 11-4C gives $M_{bol} = 4.75$
Problem 20-14	"Figure 20-4" ---> "Figure 20-8"
Problem 20-15	"Table 10-1" ---> "Table 10-2"
Problem 21-10	Students need information from Chapter 22 that they haven't seen yet!
Problem 20-11	Students should note that they will need Equation 15-10 (not Equation 16-19 as referenced in Section 20-3B)
Problem 21-12	The problem would be better is a redshift or velocity was given for the galaxy.
Problem 21-14	Students need information from Chapter 22 (Equation 22-3c in Section 22-2) that they haven't seen yet!
Problem 22-14	"maximum rate of decline" ---> "mean rate of decline"
Problem 23-1	Information on velocity dispersion needed for this problem has been deleted from this edition. Students will need to know the characteristic velocity dispersions in rich clusters is 400 - 1500 km/s
Problem 23-5	Students should note that they will need Equation 15-10 (not Equation 16-19 as referenced in Section 20-3B)
Problem 23-12	"Figure 23-3" ---> "Figure 22-3"
Problem 24-12	"marcsec" could be written better as "milli-arcsec"
Problem 25-8	Students should note that in Equation 25-15, the first term on the right hand side should be $-(k/R_0^2)(R_0/R)^2$ instead of $-(k/R_0^2)(R_0/R)$
Problem 25-9	Same note as Problem 25-8

Comments and Corrections to Text

We summarize here our suggested corrections to equations and tables in the body of the text. It is recommended that instructors copy these pages and distribute them to students.

pages 59 & 311 — The equations for P_c (Sections 4-3A and 16-1A) differ by a factor of 2 due to assumptions made in their derivation. Students should be careful of which approximation they use (and why they selected it!).

In the first instance, $P_c = (2/3) \pi G <\rho>^2 R^2$, while in the second instance, the approximation given is $P_c \approx G M <\rho> / R$ which can be converted to $P_c \approx G (4/3) \pi <\rho> R^3 <\rho> / R = (4/3) \pi G <\rho>^2 R^2$

page 86 — In the Concept Application, in the first equation the $<\rho>$ term should be squared and read $P_c = (1.4 \times 10^{-10}) <\rho>^2 R^2$ (see page 59). The rest of the solution thus becomes

$$P_c(\text{Venus}) / P_c(\text{Earth}) = <\rho(\text{Venus})>^2 R(\text{Venus})^2 / <\rho(\text{Earth})>^2 R(\text{Earth})^2$$

$$P_c(\text{Venus}) / P_c(\text{Earth}) = (0.95)^2 (0.95)^2 = 0.81$$

$$P_c(\text{Venus}) = (0.81)(1.7 \times 10^6 \text{ atm}) = 1.4 \times 10^6 \text{ atm}$$

The main point of the example remains unchanged -- the central pressures of Venus and Earth are about the same.

page 104 — Table 6-2: In bottom line of table, "(radii)" should be indented

page 106 — The example for Jupiter given at the bottom of the left column does not seem correct. For the parameters given and the equation for the peak frequency of the synchrotron emission given above the example, an energy of 140,000 MeV is calculated. Possibly the constant 4.8×10^{-4} is incorrect and should read 4.8×10^2 for the units specified. The equation would thus be

$$f_{max} \approx (4.8 \times 10^2) E_{[MeV]}^2 B_{[T]} \sin(\alpha) \text{ MHz}$$

(See Problem 6-18 solution in this Manual)

page 116 & A-8 — Figure 6-19 gives Pluto radius as 1125 km, Problem 6-4 gives 1120 km, and Appendix A3-3 gives 1140 km. These slightly different values (determined by different observations and models) are nearly the same, but may cause some confusion to students and will produce slightly different results from calculations in problems.

page 175 — Figure 8-16: The horizontal axis is log (v) not v. Or, the values should be labeled 10^8 instead of 8, etc.

pages 238 & 320 — Section 12-2B gives different values for the slope, α, of the mass - luminosity relation for stars than does Section 16-3B. In the later instance, a single slope is fit to all the data, while in the former instance the data is subdivided into sub-intervals. Students should be careful which approximation they use, and why.

pp 280, 447, A-17 — Table 14-1 gives the mass of the Milky Way Galaxy as $1 \times 10^{12} M_\odot$

Table 23-1 gives $4 \times 10^{11} M_\odot$

Appendix Table A7-1 gives $7 \times 10^{11} M_\odot$

Students should be careful which approximation they use, and understand what techniques or models were used to determine each value.

pages 280 & A-17	Table 14-1 gives the halo diameter of the Milky Way Galaxy as 100 kpc. Appendix Table A7-1 gives 120 kpc. Students should be careful which value they use, and understand what techniques or models were used to determine them.
page 291	Equation 15-7: The right hand side of this equation should be raised to the third power, $R_s = [...]^{1/3}$
pages 316 & A-13	Left column of page 316 and Appendix A5-1: Symbol for Argon is "Ar" not "A"
pages 340 & 349	The equation in the right column of page 340 that begins "$t_2 - t_1 = ...$" differs from the equation in the Key Equations and Concepts. Neither equation appears correct, though (see Problem 17-16). Use: $$t_2 - t_1 = (1/4\pi\varepsilon_0)[e^2/(2\pi m_e c)][(1/f_2^2) - (1/f_1^2)] \, DM$$
page 355	Left column, just before Equation 18-1, the density term should not be cubed: "$P^2 \propto R^3/<\rho>^3 R^3$" ---> "$P^2 \propto R^3/<\rho> R^3$"
page 356	Equation 18-2: should read "$M_V = -2.76 (\log P - 1.0) - 4.16$" [change = sign to - sign]
page 375	Equation 18-4: The right had side should read "$= P (V_c \sin i)^3 / (2\pi G)$, that is, the P term should not be squared
page 382	Left column, 4th line after Equation 19-2, "seconds per year" ---> "arcseconds per year"
page 398	Table 20-1: "T-Tauri" in first column should read "T Tauri" and "Rr Lyrae" in fifth column should read "RR Lyrae"
page 406	In the right column, "the free fall time (Equation 16-19)" should read "... (Equation 15-10)" [the section that contains this equation has been moved in this edition]
page 436	In the right column, after Equation 22-5, the text states that $v = cz$ is the redshift in ... units of "10^3 km/s." This should be in "km/s"
page 449	In right column, the distance to the Virgo cluster of galaxies is given as 15.7 Mpc, and the cluster's size is given as about 3 Mpc (based on a cluster angular extent of about 7°). Other estimates for the distance are more in the range of about 20 Mpc. (One of the projects of the *Hubble Space Telescope* is to determine more accurate distances to this and other clusters.)
page 466	Table 24-1: "C[III]" ---> "[C III]"
page 493	Equation 25-15: In the first term on the right hand side, the (R_0/R) ratio should be squared, so the first term should be: $-(k/R_0^2)(R_0/R)^2$ instead of $-(k/R_0^2)(R_0/R)$
page A-17	Appendix A7-1 and Problem 19-15: The symbol for the distance of the Sun from the center of the Galaxy should be R_0 instead of $R_\odot$.